Particle Technology
and Applications

GREEN CHEMISTRY AND CHEMICAL ENGINEERING

Series Editor: Sunggyu Lee
Ohio University, Athens, Ohio, USA

Materials in Biology and Medicine
Sunggyu Lee and David Henthorn

Particle Technology and Applications
Sunggyu Lee and Kimberly H. Henthorn

Lithium-Ion Batteries: Advanced Materials and Technologies
Xianxia Yuan, Hansan Liu, and Jiujun Zhang

Carbon-Neutral Fuels and Energy Carriers
Nazim Z. Muradov and T. Nejat Veziroğlu

Oxide Semiconductors for Solar Energy Conversion: Titanium Dioxide
Janusz Nowotny

Magneto Luminous Chemical Vapor Deposition
Hirotsugu Yasuda

Nuclear Hydrogen Production Handbook
Xing L. Yan and Ryutaro Hino

**Efficiency and Sustainability in the Energy and Chemical Industries:
Scientific Principles and Case Studies, Second Edition**
Krishnan Sankaranarayanan, Jakob de Swaan Arons, and Hedzer van der Kooi

Proton Exchange Membrane Fuel Cells: Contamination and Mitigation Strategies
Hui Li, Shanna Knights, Zheng Shi, John W. Van Zee, and Jiujun Zhang

Proton Exchange Membrane Fuel Cells: Materials Properties and Performance
David P. Wilkinson, Jiujun Zhang, Rob Hui, Jeffrey Fergus, and Xianguo Li

Solid Oxide Fuel Cells: Materials Properties and Performance
Jeffrey Fergus, Rob Hui, Xianguo Li, David P. Wilkinson, and Jiujun Zhang

Particle Technology and Applications

Edited by
Sunggyu Lee and
Kimberly H. Henthorn

CRC Press
Taylor & Francis Group
Boca Raton   London   New York

CRC Press is an imprint of the
Taylor & Francis Group, an **informa** business

CRC Press
Taylor & Francis Group
6000 Broken Sound Parkway NW, Suite 300
Boca Raton, FL 33487-2742

First issued in paperback 2017

© 2012 by Taylor & Francis Group, LLC
CRC Press is an imprint of Taylor & Francis Group, an Informa business

No claim to original U.S. Government works
Version Date: 20120202

ISBN 13: 978-1-138-07739-3 (pbk)
ISBN 13: 978-1-4398-8167-5 (hbk)

Visit the Taylor & Francis Web site at
http://www.taylorandfrancis.com

and the CRC Press Web site at
http://www.crcpress.com

Contents

Green Chemistry and Chemical Engineering Series Statement vii

Preface...ix

About the Editors xi

Contributors.. xiii

1 Particle Technology and Applications: Past, Present, and Future 1
Sunggyu Lee and Kimberly H. Henthorn

Section I   Characterization

2 Fractal Geometry Applications 9
Douglas K. Ludlow

3 Particle Cohesion 17
Kimberly H. Henthorn and Christine M. Hrenya

4 Particle–Particle Interaction 31
Anh V. Nguyen, Linh T. T. Tran and Jan D. Miller

Section II   Nanomaterials

5 Nanomaterials 53
David S. J. Arney, Jimmie R. Baran, Allen R. Siedle, and Matthew H. Frey

6 Nanostructured Materials 67
Vikrant N. Urade and Hugh W. Hillhouse

Section III   Solids Handling and Processing

7 Adsorption 87
Shivaji Sircar

8 Coal–Water Slurries 111
S. Komar Kawatra

9 Size Reduction 125
Sunil Kesavan

10 Pneumatic Conveying 145
Kimberly H. Henthorn

11 **Solid–Liquid Mixing: Numerical Simulation and Physical Experiments**............. 157
    Philippe A. Tanguy, Francis Thibault, Gabriel Ascanio, and Edmundo Brito-De La Fuente

12 **Fluidization**.. 181
    A.-H. Park and L.-S. Fan

13 **Fluidized Bed Reactor**.. 199
    John R. Grace, Jamal Chaouki, and Todd Pugsley

Section IV   Polymer and Supercritical Fluid Applications

14 **Powder Coating Application Processes**... 219
    Harry J. Lader

15 **Supercritical Carbon Dioxide Processing of Polymer–Clay Nanocomposites** 235
    Matthew J. Factor and Sunggyu Lee

16 **Infusion of Volatile Corrosion Inhibitor into Thermoplastic Resins and
    Films in a Supercritical Fluid Medium**.. 251
    Leah A. Taylor and Sunggyu Lee

Section V   Environment and Safety

17 **Electrostatic Precipitation**.. 281
    Kenneth R. Parker

18 **Dust Explosion Hazard Assessment and Control**... 301
    Vahid Ebadat

Index 319

Green Chemistry and Chemical Engineering

Series Statement

The subjects and disciplines of chemistry and chemical engineering have encountered a new landmark in the way of thinking about, developing, and designing chemical products and processes. This revolutionary philosophy, termed "green chemistry and chemical engineering," focuses on the designs of products and processes that are conducive to reducing or eliminating the use and generation of hazardous substances. In dealing with hazardous or potentially hazardous substances, there may be some overlaps and interrelationships between environmental chemistry and green chemistry. While environmental chemistry is the chemistry of the natural environment and the pollutant chemicals in nature, green chemistry proactively aims to reduce and prevent pollution at its very source. In essence, the philosophies of green chemistry and chemical engineering tend to focus more on industrial application and practice rather than academic principles and phenomenological science. However, as both chemistry and chemical engineering philosophy, green chemistry and chemical engineering derive from and build upon organic chemistry, inorganic chemistry, polymer chemistry, fuel chemistry, biochemistry, analytical chemistry, physical chemistry, environmental chemistry, thermodynamics, chemical reaction engineering, transport phenomena, chemical process design, separation technology, automatic process control, and more. In short, green chemistry and chemical engineering are the rigorous use of chemistry and chemical engineering for pollution prevention and environmental protection.

The Pollution Prevention Act of 1990 in the United States established a national policy to prevent or reduce pollution at its source whenever feasible. And adhering to the spirit of this policy, the Environmental Protection Agency (EPA) launched its Green Chemistry Program to promote innovative chemical technologies that reduce or eliminate the use or generation of hazardous substances in the design, manufacture, and use of chemical products. The global efforts in green chemistry and chemical engineering have recently gained a substantial amount of support from the international community of science, engineering, academia, industry, and governments in all phases and aspects. Some of the successful examples and key technological developments include the use of supercritical carbon dioxide as green solvent in separation technologies, application of supercritical water oxidation for destruction of harmful substances, process integration with carbon dioxide sequestration steps, solvent-free synthesis of chemicals and polymeric materials, exploitation of biologically degradable materials, use of aqueous hydrogen peroxide for efficient oxidation, development of hydrogen proton exchange membrane (PEM) fuel cells for a variety of power generation needs, advanced biofuel productions, devulcanization of spent tire rubber, avoidance of the use of chemicals and processes causing generation of volatile organic compounds (VOCs), replacement of traditional petrochemical processes by microorganism-based bioengineering processes, replacement of chlorofluorocarbons (CFCs) with nonhazardous alternatives, advances in design of energy efficient processes, use of clean alternative and renewable energy sources in manufacturing, and much more. This list, even though it is only a partial compilation, is undoubtedly growing exponentially.

This book series on Green Chemistry and Chemical Engineering by CRC Press/Taylor & Francis is designed to meet the new challenges of the twenty-first century in the chemistry and chemical engineering disciplines by publishing books and monographs based on cutting-edge research and development to effect reducing adverse impacts on the environment by chemical enterprise. To achieve this, the series will detail the development of alternative sustainable technologies that will minimize the hazard and maximize the efficiency of any chemical choice. The series aims at delivering readers in academia and industry with an authoritative information source in the field of green chemistry and chemical engineering. The publisher and its series editor are fully aware of the rapidly evolving nature of the subject and its long-lasting impact on the quality of human life in both the present and future. As such, the team is committed to making this series the most comprehensive and accurate literary source in the field of green chemistry and chemical engineering.

Sunggyu Lee

Preface

Particle technology has been practiced in a variety of forms and applications throughout the world over the course of human history. Changes and breakthroughs, both recorded and unrecorded, have been achieved in regard to the use of particulate systems in industrial processes and in the daily lives of humans. Consequently, particle science and technology have become well-studied and intensely researched subjects in modern technological society. Many books, journals, conference papers, and patents related to the body of knowledge of particle science and technology can be found in the literature. While these articles are widely available and varied in terms of the specific fields of application, levels of scientific background required of readers, scope and nature of the topical treatments, and presentation of relevant theories and principles, there is a dearth of comprehensive books that deal with up-to-date practical applications of particle technologies developed and practiced in the chemical, petrochemical, energy, mechanical, and materials industries. This book is intended to address such a need in the field by providing readers with guided examples of particle technology and its advanced industrial applications. Necessary scientific background of particle technology and relevant technological details of its application areas are given in order to help readers grasp the specific details of the applied technology, upon the outcomes of which the advanced particle technology can have an impact.

The chapters of this book were written by a select group of field experts. Although the level of the book is intended for readers who have a background in college-level chemistry and physics, the value of this book may be more appreciated by graduate students working on diverse scientific and engineering problems, researchers and inventors in related fields, and practicing engineers and scientists in the chosen application areas. It is the editors' wish that this book contributes to scientists, engineers, researchers, and industrialists in their technological thinking, tackling challenges in novel process and product development, and devising and inventing new methodologies in clean technology development.

This book is published as a spin-off volume of the *Encyclopedia of Chemical Processing* and is based on encyclopedia articles recently published in the particle technology area. This book is also published as a book of the *Green Chemistry and Chemical Engineering* book series.

Sunggyu "KB" Lee
Athens, Ohio

Kimberly H. Henthorn
Terre Haute, Indiana

About the Editors

Dr. Sunggyu Lee earned both his Bachelor's and Master's degrees in chemical engineering from Seoul National University, Seoul, Korea, in 1974 and in 1976, respectively. He received his Ph.D. in chemical engineering from Case Western Reserve University, Cleveland, Ohio in 1980. He started his professorial career with The University of Akron in 1980 and was promoted to full professor in 1988. From 1988–1997, he served as Robert Iredell Professor and Chairman of chemical engineering as well as the Founding Director of the Process Research Center. From 1997–2005, he held positions of chairman and C. W. LaPierre Professor of chemical engineering at the University of Missouri. From 2006–10, he was with Missouri University of Science and Technology, where he established the Laboratory for Transportation Fuels and Polymer Processing. Since 2010, he has held positions as the Russ Ohio Research Scholar and professor of chemical and biomolecular engineering, Ohio University, Athens, Ohio. He has established the Sustainable Energy and Advanced Materials (SEAM) Laboratory as a stand-alone off-campus laboratory of excellence.

Dr. Lee has authored 7 books, 9 monographs, 9 book chapters, 144 refereed journal articles, and over 360 proceedings and conference papers. Titles of his published books include *Methanol Synthesis Technology*, *Oil Shale Technology*, *Alternative Fuels*, *Methane and Its Derivatives*, *Handbook of Environmental Technology*, and *Handbook of Alternative Fuel Technologies*. He is the editor of the *Encyclopedia of Chemical Processing* (5 volumes with 350 chapters), published in 2006. He has received 30 U.S. patents based on his inventions, mainly in the areas of clean alternative energy, functional polymers, and supercritical fluid technology. Most of his inventions are being commercially utilized. He has received and directed over 100 research grants/contracts from both industrial and governmental agencies totaling over $19M as principal investigator and co-principal investigator. His specialties are in the areas of alternative fuels, supercritical fluid technology, chemical process engineering and reactor design, polymer synthesis and processing. He has guided over 90 advanced degree students and 24 postdoctoral fellows as major advisor.

Dr. Kimberly Henthorn earned her Bachelor's degree in chemical engineering from Rose-Hulman Institute of Technology, Terre Haute, Indiana in 1999 and her Ph.D. in chemical engineering from Purdue University, West Lafayette, Indiana in 2004. She joined the Chemical and Biological Engineering Department at Missouri University of Science of Technology (formerly the University of Missouri-Rolla) in 2004, where she was promoted to associate professor in 2010. Since 2010, she has held the position of associate professor of chemical engineering at Rose-Hulman Institute of Technology. Dr. Henthorn specializes in the areas of particle characterization, solids entrainment and transport, and two-phase microfluidic flows.

Contributors

David S. J. Arney
3M Company, St. Paul, Minnesota, U.S.A.

Gabriel Ascanio
URPEI, Department of Chemical
Engineering, Ecole Polytechnique,
Montreal, Quebec, Canada

Jimmie R. Baran
3M Company, St. Paul, Minnesota, U.S.A.

Edmundo Brito-De La Fuente
Departamento de Alimentos y
Biotecnología UNAM, México, México

Jamal Chaouki
Ecole Polytechnique, Montreal, Quebec,
Canada

Vahid Ebadat
Chilworth Technology, Inc., Princeton,
New Jersey, U.S.A.

Matthew J. Factor
Department of Chemical and Biological
Engineering, Missouri University of
Science and Technology, Rolla, Missouri,
U.S.A.

L.-S. Fan
Department of Chemical and Biomolecular
Engineering, The Ohio State University,
Columbus, Ohio, U.S.A.

Matthew H. Frey
3M Company, St. Paul, Minnesota, U.S.A.

John R. Grace
University of British Columbia, Vancouver,
British Columbia, Canada

Kimberly H. Henthorn
Department of Chemical and Biological
Engineering, Missouri University
of Science and Technology, Rolla,
Missouri, and Rose-Hulman Institute of
Technology, Terre Haute, Indiana, U.S.A.

Hugh W. Hillhouse
Purdue University, West Lafayette,
Indiana, U.S.A.

Christine M. Hrenya
Department of Chemical and Biological
Engineering, University of Colorado,
Boulder, Colorado, U.S.A.

S. Komar Kawatra
Department of Chemical Engineering,
Michigan Technological University,
Houghton, Michigan, U.S.A.

Sunil Kesavan
Akebono Corporation, Farmington Hills,
Michigan, U.S.A.

Harry J. Lader
Harry Lader and Associates, Inc.,
Cleveland, Ohio, U.S.A.

Sunggyu Lee
Department of Chemical and Biomolecular
Engineering, Ohio University, Athens,
Ohio, U.S.A.

Douglas K. Ludlow
Chemical and Biological Engineering,
University of Missouri-Rolla, Rolla,
Missouri, U.S.A.

Jan D. Miller
Department of Metallurgical Engineering,
University of Utah, Salt Lake City, Utah,
U.S.A.

Anh V. Nguyen
Discipline of Chemical Engineering, The
   University of Newcastle, Callaghan,
   New South Wales, Australia

A.-H. Park
Department of Chemical and Biomolecular
   Engineering, The Ohio State University,
   Columbus, Ohio, U.S.A.

Kenneth R. Parker
Ken Parker Consultant APC, West
   Midlands, U.K.

Todd Pugsley
University of Saskatchewan, Saskatoon,
   Saskatchewan, Canada

Allen R. Siedle
3M Company, St. Paul, Minnesota, U.S.A.

Shivaji Sircar
Chemical Engineering Department, Lehigh
   University, Bethlehem, Pennsylvania,
   U.S.A.

Philippe A. Tanguy
URPEI, Department of Chemical
   Engineering, Ecole Polytechnique,
   Montreal, Quebec, Canada

Leah A. Taylor
Department of Chemical Engineering,
   University of Missouri, Columbia,
   Missouri, and NIC Industries,
   White City, Oregon, U.S.A.

Francis Thibault
URPEI, Department of Chemical
   Engineering, Ecole Polytechnique,
   Montreal, Quebec, Canada

Linh T. T. Tran
Discipline of Chemical Engineering, The
   University of Newcastle, Callaghan,
   New South Wales, Australia

Vikrant N. Urade
Purdue University, West Lafayette,
   Indiana, U.S.A.

1

Particle Technology and Applications: Past, Present, and Future

Sunggyu Lee and Kimberly H. Henthorn

CONTENTS

Definition and Scope 1
Historical Perspectives 2
Current Industrial Practices... 2
Significance and Future Trends.. 5

Definition and Scope

Particle science and technology is, simply speaking, the study of particles. Particle science and technology may be defined as the science and engineering dealing with and related to materials dispersed in a continuous fluid or medium. In this context, a particle is defined as a small, discretely identifiable entity that has an interface with the surrounding environment or has a separate domain with respect to the continuous medium. Particles may be in the form of solids or liquid droplets and include any material dispersed in a continuous medium or fluid. The most typical continuous media are air, gas, water, liquid solvent, gel, and polymeric resins. Examples of particles and their applications are found everywhere in daily life as well as in chemical, petrochemical, environmental, pharmaceutical, and metallurgical processes. They include ground coal particles, rubblized oil shale rocks, crushed torrefied biomass, multigrain flour for baking, baking powder, ground pepper, table salt crystals, glass beads, graphite lubricant, ground or pelletized polymer resin, ground natural rubber latex, ground spent tire rubber, deformable polymer droplets in a solvent, oil droplets in water, particulate matter in air, aerosol droplets in air, porous metal oxide catalysts, crystallites precipitated onto the internal surface of catalyst carriers, crystals formed from a supersaturated solution, polymer flakes formed from a supercritical fluid solution at its snow point, carbonaceous residues from coking processes, nanostructured particles dispersed in polymeric matrix, and much more.

Particles can be grown from embryos to desired dimensions and shapes of crystals, obtained or extracted from naturally occurring materials, generated by size reduction from much larger materials and aggregates, created by management and/or manipulation of phases of materials, and produced as products or by-products of chemical and biological transformations and atomic interactions.

Historical Perspectives

Particle technology in its most general sense has long been practiced in every aspect of human life, even though the terminology as a scientific subject started to appear only recently. Humans have empirically found and sometimes perfected various applications of particle processing for enhancement of quality of life and have developed and accumulated knowledge for the processes that work best for specifically desired outcomes. Some classical examples of particle processing in human history include the following:

1. Utilization of flours and grains to bake breads, brew liquors, and prepare other food products.
2. Discovery of a variety of natural and synthetic seasonings, and finding that the powder form is more effective in generating flavors and creating desirable tastes.
3. Discovery of natural dyes and pigments and finding ways of applying them to their garments, artifacts, and buildings. Humans have used both organic and inorganic materials for these purposes.
4. Extraction of vegetable oils from oil seeds. Humans use these oils for cooking and lighting as well as for food. In most occasions, the seeds are first crushed before the oil is squeezed out. Humans have further enhanced the processes by introducing heat, pressure, and selective solvents.
5. Use of herbal medicines, both raw and processed. Humans have gradually found that ground dry leaves are more effective, have learned how to efficiently mix different ingredients together, and have developed effective methods for preparing pellets, pills, and pastes.
6. Use of earthenware and ceramic products and the development of very fine forms of pottery techniques. Humans have learned the crucial importance of the particle size and mixing of ingredients in the formulation.
7. Use of sands, clays, adobes, and mud for desired needs and applications.
8. Utilization of mining, ore processing, and purification technologies, all of which involve particle processing to certain degrees.

Current Industrial Practices

More recent examples of particle technology are found in advanced processing within chemical, petrochemical, pharmaceutical, and environmental industries, which have resulted in higher-quality products and novel materials. Some examples are listed below.

1. Finely pulverized coal particles, finer than 200 U.S. standard mesh, are used effectively for coal gasification in entrained flow reactors. Finer particle sizes permit rapid gasification reactions without a significant mass transfer barrier and facilitate efficient handling of gas–solid reaction systems in continuous flow reactors.
2. Coal–water and coal–oil mixtures have advantageous properties that include petroleum-like combustion characteristics and pumpability. The slurry preparation

method is based on the dispersion of finely pulverized, demineralized, and desulfurized coal particles in liquid suspensions. Stability of such coal slurry fuels is one of the most essential parameters of their performance in internal combustion engines as well as wide acceptance as an alternative fuel.

3. Hydraulic fracturing, hydrofracking, or fracking of oil shale bed for shale gas production has proven to be technoeconomically feasible with advances in long horizontal drilling and chemical-aided fracturing technologies. Effective liberation of gaseous light hydrocarbon ingredients entrapped in shale rock matrices has become a principal issue in shale gas commercialization. The Marcellus Formation of the eastern United States, specifically the states of New York, Pennsylvania, and Ohio, is attracting a great deal of commercial interest as a new domestic source of natural gas.

4. Vegetable oil can be extracted from oil seeds by several different methods, including solvent extraction, expeller method, and supercritical extraction. To maximize extraction efficiency, oil seeds are usually "crushed," thus decreasing particle size and lowering the mass transfer resistance of oil removal. While the expeller method has long been used in diverse forms, solvent extraction technology including supercritical fluid extraction is more recent.

5. Algae biodiesel technology involves algae harvesting as an initial step. The term *algae harvesting* technically refers to the concentration of a fairly diluted (ca. 0.02–0.06 wt.% total suspended solids, TSS) algae suspension until a slurry or paste containing 5–25% TSS or higher is obtained. Specific harvesting methods depend primarily on the type of algae and its growth media. The most common harvesting processes include the following: 1) microscreening; 2) flocculation; and 3) centrifugation. The three methods represent different unit operations of filtration, flotation, and centrifugation, respectively. All processing steps require energy-efficient particle processing technology involving a very dilute suspension. Commercial success will largely depend on the energy-efficient algae harvesting process.

6. Methanol is commercially synthesized from synthesis gas over a copper-based catalyst in a process route called low pressure methanol synthesis technology. The average copper crystallite size of freshly reduced commercial methanol catalyst is around 35–37 Å, and any growth from this size typically represents a sign of thermal deactivation, often referred to as aging. The preparation of coprecipitated $CuO/ZnO/Al_2O_3$ catalyst, copper crystallite size growth via thermal aging, and in situ regeneration of spent catalyst via repeated oxidation/reduction cycles are direct applications of particle science, and there still is room for improvement.

7. Methanol synthesis using synthesis gas has been conventionally carried out in the vapor phase, where synthesis gas is reacted over a copper-based catalyst at 230–250°C and 50–80 atm. The liquid-phase methanol synthesis process is a more recent technology, where the methanol synthesis reaction is carried out in a catalyst–inert oil slurry, thus enhancing the reactor heat transfer characteristics while sacrificing some of the mass transfer conditions. To alleviate the mass transfer barriers due to the presence of an inert oil phase in the reactor, finer catalyst particles are utilized in a highly concentrated catalyst slurry. The added benefits of such a process configuration include the ability to convert CO-rich synthesis gas into methanol at a higher single-pass conversion, which is more suitable for coal- and biomass-based synthesis gas.

8. Processing of natural rubber latex to manufacture surgical and examination rubber gloves requires a latex dispersion of narrowly distributed ingredient particle size (~2 micrometer) for the highest product quality as measured by uniform thickness, high tensile strength, antiwebbing, absence of pin holes, etc.

9. Conventional powder coatings, including acrylics, epoxy, polyester, and hybrids, are preferentially prepared in the ingredient particle size range of 15–20 micrometers for most effective coating applications. Manufacturing operations involving compounding, extrusion, size reduction, and particle size classification are built upon advanced particle technology.

10. Powdery graft copolymers are effective in blend compatiblization of two different polymeric resins that would otherwise be incompatible. Incompatible polymer blends will result in phase separation, which leads to premature product failure. The resultant compatibilization due to the incorporation of synergistically effective graft copolymers comes from the reduction in interfacial tension and enhanced steric stabilization of blend components, besides the enhanced material affinity imparted by the reactive functional groups of the graft copolymer. Practical applications invariably use fine particles of graft copolymers to maximize the interfacial properties of the material.

11. Breathable films find specialty applications in personal care, health, and automotive industries. In general, breathable polymer films provide a barrier to liquid transmission through the film, while air or gaseous flow across the film is still allowed. Such films can be prepared using particle technology, in which fine particles are first distributed in the polymer film and the filled film is evenly stretched, thus creating microcavities around the filled particles and allowing gas transmission across the film. Such films are often referred to as microporous films.

12. Volatile corrosion inhibitor films are good examples of industrial products resulting from a synergistic combination of particle technology, chemistry involving materials with controlled volatility, and polymer film technology. Inorganic crystals such as sodium nitrite, when dispersed in a thermoplastic polymeric film matrix, have electrochemical properties that can be exploited to create a corrosion-inhibiting environment for metallic parts inside the film packaging. In such applications, the inorganic crystals as fine particles have a role as specialty functional fillers.

13. Clay is abundantly available and environmentally benign. Natural clays typically exist as agglomerations of stacked platelets that exhibit thixotropic characteristics, thus manifesting a reduction in viscosity under applied stress. The addition and incorporation of clay in an appropriate polymeric system can impart a significant effect on the mechanical, thermal, and gas-barrier properties of the base polymer. Due to the substantial enhancement of polymer properties by adding such a common material, polymer–clay nanocomposites have become of great interest to researchers.

14. Fluoropolymers can be produced via free radical polymerization of their respective monomers in a supercritical fluid medium. The technology is similar to that of precipitation polymerization, wherein grown polymeric molecules are precipitated out of the supercritical fluid mixture containing monomers, initiators, growing oligomers, and polymers. The advantages include the controllability of molecular weight distribution of produced polymers and avoidance of emulsion

polymerization. A similar process technology can also be applied to the polymerization of polymethylmethacrylate. The morphology of product polymer particles can also be manipulated using advanced particle processing techniques.

15. A supercritical fluid is a substance that exists at a temperature and pressure above its critical point. Supercritical fluids exhibit extraordinary physicochemical properties that are highly desirable for advanced particle processing. These properties include high organic solubility, controllable solubility toward inorganic salts, swelling ability, altered polarity, improved lubricity, enhanced material diffusivity, ability to lower the glass transition temperature for polymeric substances, and more. Successful examples include infusion of inorganic crystals into a polymeric matrix, preparation of polymer foams, precipitation polymerization in supercritical fluids, and pretreatment and processing of nanostructured materials.

16. Advanced particle technologies are actively being developed for coating, treating, and modifying micron-sized, submicron-sized, and nanoscale particles for use in drug delivery systems, personal care products, and pharmaceutical applications.

Significance and Future Trends

As illustrated above, particle technology has found applications in many areas of chemical, agricultural, petrochemical, pharmaceutical, materials, energy, and environmental industries. The benefits include property enhancement in conventional and novel materials and products, new capabilities in material processing, pinpoint selectivity in chemical reactions and advances in catalysis, multiple functionality embedded in composite materials, utilization and reutilization of natural and man-made resources, advanced separation technologies including gas cleanup and purification, development of new devices and analytical equipment, and much more.

Particle technology R&D uses a variety of processing and analytical equipment. Particle-related equipment may be classified into several groups including the following: 1) size reduction; 2) particle and slurry transportation; 3) mixing and compounding; 4) separation; 5) size classification; 6) particle characterization; 7) surface science and analysis; and 8) microscopic instruments. Advances in these machineries, devices, and instruments have made the practical applicability range of particle science and technology from as fine as 1 nanometer to as large as 10 cm in dimension. Furthermore, industrial operations involving particle size management and particulate separation often are the most energy-intensive steps of the entire process technology. Future process development efforts in diverse practical applications will be spent on the particle technology aspect and its optimization for their ultimate commercial success.

Section I

Characterization

2

Fractal Geometry Applications

Douglas K. Ludlow

CONTENTS

Introduction 9
Fractal Geometry 9
    Fractal Surfaces 11
    Aggregates 12
    Dissolutions and Etchings 13
    Diffusion and Reactions 13
    Nature Inspired Chemical Engineering 14
Conclusions 14
References 15

Introduction

The concept of the non-Euclidian geometry, known as fractal geometry was introduced in the late 1970s. During the 1980s and 1990s there was an explosion of scientific and engineering research and journal publications involving fractals as the concept of nature inspired geometry was tried in numerous applications. The beauty of the underlying mathematics and the ability to describe complex spatial and temporal structures utilizing the concepts of fractal geometry inspired a generation of researchers seeking applications and uses of fractal geometry. Originally, it was thought to be able to fill the need and provide the ability to model complex systems and morphologies and was nearly considered a panacea. Time and maturity of the topic has tempered some of the initial exuberance; however, numerous applications have been found in areas of science, engineering, economics, etc. Because many of the systems important to chemical processing occur at the molecular scale where morphologies are fractal over the length scale range of interest, there have been several important applications of fractal geometry in chemical systems.

Fractal Geometry

Benoit Mandelbrot's *The Fractal Geometry of Nature* made the world aware of the underlying universal order that can be found in many spatial and temporal phenomena.[1] Prior to that, such phenomena at best could be referred to as "complex" and described in terms such as "rough," "ramified," or "fragmented." The presence of scaling laws had been recognized for a number of individual objects and systems, but the significance of the concepts of fractal geometry was to unify these and to recognize the importance of the underlying symmetry. Many objects and phenomena scale with noninteger exponents. The concept of fractal geometry implies the invariance of details when magnified. This symmetry is essential in understanding the object or phenomena, and helps to describe, characterize, and to measure. Many objects in nature exhibit self-similar features and are of fractal shape. A mathematical fractal object has details at all scales, whatever the magnification, and is invariant under its generating transformation. These objects show both self-similarity and self-affinity. The mathematics of fractal geometry is yet to be fully developed. Until now the use of fractals in applied sciences and engineering has been mostly limited to phenomenological descriptions related to the empirical discovery of power laws. Nearly a decade after Mandelbrot's seminal work, scientists began reporting applications for these concepts. *The Fractal Approach to Heterogeneous Chemistry* described the application of several fractal concepts to chemistry.[2] A few years later, *Fractals in Chemistry, Geochemistry and Biophysics* made further inroads in the application of fractal geometry concepts to chemistry and chemical processes.[3] Half a decade later *Fractals in Chemistry* followed and had the advantage in that the field had matured somewhat, so that critical reviews and opinions could be expressed concerning at least some of the aspects of the use of fractals in chemistry in the chemical processing industry.[4]

During the later part of the 1980s, as Mandelbrot's concepts reached a wider audience, sessions in the American Physical Society, the American Chemical Society, and the American Institute of Chemical Engineers gave many optimistic hopes for the application of the relationship:

$$\text{Property} \propto \text{Scale}^{\beta}$$

to numerous different phenomena. It almost appeared that many thought the formalism of fractals and the exponential scale β (slope of power-law relationship) was a panacea to reduce many complex situations to one simple explanation, a fractal dimension. However, deeper thought and time have shown that in many cases the exponential scaling factor β is not truly a fractal dimension, and that fractals do not answer well all of the questions initially thought to be accessible through their formalism. *Nature Inspired Chemical Engineering* by Coppens points out not only some of the limitations but also many of the exciting vistas that can occur through the application of the concepts of fractal geometry to applied physical science and engineering.[5] Indeed, fractals are everywhere, but not everything is fractal: A noninteger exponent is not necessarily a fractal dimension. The distinction is not critical if the noninteger scaling exponent is to be used only for phenomenological descriptions, but is crucial in more advanced, deductive modeling and in building new applications.

The key to applying fractal geometry to any field, including chemical engineering, is to understand the underlying assumptions. Some property X vs. some scale δ, may be plotted as $\log(X)$ vs. $\log(\delta)$ and can be fitted to a straight line, so that $X \sim \delta^{\beta}$. Typically, this scaling relationship is only determined over a fairly narrow range of scales, because is it difficult to get experimental data over a wide range of scales. No object or phenomenon in nature is scaling over an infinite range of scales as is the case in a mathematical fractal object. The limit of $\delta \to 0$, necessary in mathematical theorems involving fractal measures, is impossible to take in practice. In the nonmathematical world, scaling can only be measured within a finite range, and if the range of the measurements is wide enough, the curve of log(X) vs. log(δ) will deviate from a straight line outside of the interval of $[\delta_{min}, \delta_{max}]$, the fractal scaling range. When this concept is ignored in applications of fractal geometry the results are misunderstood and even misused. If the experimentally available interval is too narrow, say less than a factor of 10, then there may be a noninteger exponent that might be wrongly considered a fractal dimension when, owing to the limited experimental range, it is unclear whether there is an underlying self-similarity or self-affinity to the phenomena.

Fractal Surfaces[6]

One place where the concepts of fractal geometry have found use is in the quantitative description of surfaces and surface ruggedness. In nature, most surfaces are not smooth except over some limited ranges. In the classical, "what is the length of the coastline of England?" problem, it is evident, that the total length of the coastline determined will be dependent on the length of the ruler that you use to measure it. With the ruggedness of the actual coastline, if your ruler length is 10 km, many features that are smaller will be disregarded. If a smaller ruler is used, more of the detail will be followed. It is found that the length of the coastline measured scales to the length of the ruler in a way that can be related to a fractal dimension. This same concept can be applied to surfaces. If a rugged surface area is to be determined by measuring the number of equal sized particles that just cover it (monolayer coverage), then the surface area determined will depend on the size of the particles used as the yardstick. Thus, the fractal dimension of a surface can be estimated by determination of the number of objects (molecules) of a given size that would completely cover the surface for a series of different sized objects. This concept has been used in several different techniques to determine the fractal dimension of a surface, such as gas adsorption using a homologous series of adsorbates, liquid adsorption of a series of different sized polymers or latex beads, etc.

The surface property of a solid is characterized by the nature of the surface boundary. The surface boundary is expected to be related to the underlying geometric nature of the surface, hence its fractal dimension. Many properties of the solid depend on the scaling behavior of the entire solid and of the pore space. The distribution of mass in the porous solid and the distribution of pore space may also reflect the fractal nature of the surface. If the mass and the surface scale are alike, that is, have the same power-law relationship between the radius of a particle and its mass, then the system is referred to as a mass fractal. In a similar manner if the pore volume of porous material has the same power–law relationship between the pore volume and radius as that of the surface, then it is described as a pore fractal.

Powdered and porous materials can also be characterized from additional techniques that do not depend on probing the surface with a molecule. Image analysis techniques (but still using the concept of varying the length of the ruler) have been developed. Techniques have also been developed to determine a surface fractal dimension from scattering experiments using neutrons, visible light, and x-rays. Although no solid surface is fractal over all scales in a mathematical sense, many surfaces in the range where molecules interact (single-surface molecules up through macropore networks in the particles) can be described and characterized with a fractal dimension. This has been one of the useful applications of fractal theory, being able to quantitatively describe surface/interface ruggedness by the fractal dimension. Prior to fractal analysis, surfaces could be described as being more rugged, or more ramified, or more pitted than other surfaces. Using fractal analysis, the fractal dimension of the surface can be measured and quantified.

Aggregates

Another area where the application of fractal geometry concepts has provided insight into physical phenomena is in the characterization and understanding of aggregation and growth. Aggregates are formed from smaller, nearly identical particles. Because mechanisms for formation contain random elements, we can expect three main structural phenomena arising in aggregated objects. First, mass fractality, because the process of formation often leads to more or less diaphanous aggregates. Second, fractal pore domains consisting of interstitial void elements of a wider range of linear size. Third, surface fractality from the interfaces between pore and mass regions.

Fractal geometry has been found to be very useful in describing aggregates.[2–4,7] For several decades it has been known that diffusion and randomness play an important part of aggregate growth. In the simplest form diffusion-limited aggregation (DLA), a randomly drifting/diffusing particle is in a system with a preexisting nucleus/aggregate. When the diffusing particle hits the nucleus/aggregate, then there is an instantaneous and irreversible attachment. As more and more particles stick to the growing nucleus a branching, tree-like structure is formed. For a mass-distance scaling relationship, the number of particles, N, within a given radius, R, scales by:

$$N \sim R^D$$

Image analysis of two-dimensional representations of aggregates, such as planar projections can be used to determine a fractal dimension using the technique above. Numerous computational and theoretical studies of DLA clusters generated on computers have found that in DLA structures in which there is a high sticking probability, the fractal dimension, D, of a DLA cluster formed in a two-dimensional lattice is close to 1.7 and for a three-dimensional lattice, it is close to 2.5. As the sticking probability is decreased and as relaxation of the structures is allowed, the numerical models indicate for the two-dimensional lattice systems that the scaling relationship will approach 2.

Many experimental aggregate structures show this branch-like nature. Besides being formed in cases of diffusion limited growth, there will also be varying kinetics of aggregation, which can be classified into two kinds, slow and fast aggregation, each with a different rate-limiting physics. Systems that demonstrate either diffusion limited and/or reaction limited aggregation form fractal particles. Aggregation of solid particles or monomers in the liquid or in the gas phase far from saturation forms mass fractals. Of course if there is some form of relaxation or equilibrium of the structures formed the fractal nature of the particles will change. Aggregation conditions closer to saturation lead to denser structures, with a smooth or fractally rough surface. In the latter case these are surface fractals. It has been demonstrated that when the planar projections of numerous different aggregate particles are analyzed using the scaling relationship, the dimension determined is in the range of $D \sim 1.77$. If the same types of particles are formed under conditions where the kinetics of aggregation are slow, then more compact particles are formed and the planar projections give a scaling close to 2. Hence, the mass scaling relationship with the size of aggregates gives one indication of the kinetics of formation. There are several books dedicated to the study and aspects of fractal growth of aggregates and this has been one area where the concepts of fractal geometry have found many applications.

Techniques have been developed to determine the fractal dimension of experimental aggregate particles in solution using small-angle scattering techniques, from x-ray, light, or even neutron sources. In these techniques the scattering intensity, $I(q)$, is proportional to the scattering vector, q, raised to the mass fractal dimension by:

$$I(q) \sim q^{-D_m}$$

This technique has been successfully applied in numerous studies to determine the fractal dimension of aggregate particles.

Closely related to the aggregates formed in colloidal aggregation in gels and sols is the deposition of aggregate particles during electrodeposition. The complex structures formed from electrodeposition have been analyzed using fractal methods. There have been several studies in which dentritic structures are formed that are statistically simple and self-similar, i.e., fractal in nature. These are typically formed under conditions of diffusion-limited growth. The growth of deposits on the cathodes of batteries is often the main factor limiting the lifetime of the batteries and other electronic parts. There have been studies that show that the fractal nature of the electrodes and/or deposits on the electrodes affects the efficiency of the battery.

Dissolutions and Etchings

The chemical reaction between a solid and a reactive fluid is of interest in many areas of chemical engineering.[8] The kinetics of the phenomenon is dependent on two factors, namely, the diffusion rate of the reactants toward the solid/fluid interface and the heterogenous reaction rate at the interface. Reactions can also take place within particles, which have accessible porosity. The behavior will depend on the relative importance of the reaction outside and inside the particle. Fractal analysis has been applied to several cases of dissolution and etching in such natural occurring caves, petroleum reservoirs, corrosion, and fractures. In these cases fractal theory has found usefulness for quantifying the shape (line or surface) with only a few parameters: the fractal dimension and the cutoffs. There have been some attempts to use a fractal dimension for reactivity as a global parameter. Finally, fractal concepts have been used to aid in the interpretation of experimental results, if patterns quantitatively similar to DLA are obtained.

Diffusion and Reactions

One area that has seen much interest by chemists and chemical engineers is the study of diffusion and reaction on fractals. Many of the heterogenous reaction systems contain structures that have a fractal-like nature. Several numerical studies of diffusion and transport on fractals have been completed with the aim of being able to describe the effect of fractal structure on flow and diffusion of species. Utilizing concepts of fractal lattices as models for more realistic porous media, several numerical studies have been carried out to characterize the flow and diffusion through porous media. Numerical studies of the diffusion and reaction of reactant systems occurring in a fractal space vs. a traditional Euclidean space has led to predictions of anomalous diffusion and reaction kinetics. The application of fractal concepts has been useful in modeling permeation and percolation through porous structures.

Nature Inspired Chemical Engineering

One of the positive by-products of the application of fractal theory to various different physical phenomena has been the realization that many structures found in nature demonstrate a fractal nature over a limited range and more importantly that these self-similar and self-affine structures have a natural optimization. For instance a tree has to control the essential processes at the scale of the elementary microunits such as the cells in the leaves and this is realized in the specific structure of the veins within the leaves. Yet, the tree can only function well if the nutrients can access the leaves easily, thus there is a different fractal structure of the trunk and branches. We find similar behavior in lungs and kidneys. Through the fractal structure of the tree branches, there is a fast and uniform transport to the microunits or cells, covering a high surface area or volume, from one or a few points (the stem). These hierarchical structures are found throughout nature because they lead to easy transport of molecules and energy and also give mechanical strength and flexibility while making efficient use of available materials. Researchers have successfully applied such concepts to the design of fluid injectors for fluidized bed with the injector having a fractal shape. It was found that the fractal, nature inspired, injectors gave more uniform fluid flow in the solids with greater mixing and contact.

Conclusions

In the mid-1980s, there was an explosion of research seeking applications for fractal theory to various aspects of physical phenomenon in physics, chemistry, and engineering. The sessions at the professional conference that dealt with fractals would leave one almost breathless with the vast variety of topics and experimental results that were being analyzed in terms of fractal concepts. It seemed that everywhere we looked we found fractals and that many phenomena were scaled by a power-law relationship. It has been said that indeed, exuberance sort of ran away from sober reflection, and that perhaps the scaling exponent was taken too frequently for a fractal dimension without deeper investigation.[4] Perhaps, the concept of fractals was just filling the human need to be able to explain complex and seemingly intractable systems in a simple manner. Time has found that fractals are neither simple nor have they answered all of the questions initially thought to be accessible through their formalism. Nevertheless, a fractal approach to the modeling of many phenomena has been successful and helpful in our understanding of these phenomena. At the same time, and despite many theoretical highlights, real technological applications of fractals are still strikingly scarce for such a fundamental, universal concept. Many natural objects and phenomena, such as trees and turbulence, are scaling within a finite scaling range. Fractals become useful when recognizing this finite range, and measuring the scaling range by correctly interpreting the measurement results, remembering that too narrow of a range may lead to a bias on the value of the dimension determined. Fractal geometry, like Euclidean geometry, is useful within the range it applies to. Many important physical, chemical, biological, geological, economical, and other phenomena occur exactly within the range where the underlying structure is geometrically scaling and hereby fractal.

References

1. Mandelbrot, B.B. *The Fractal Geometry of Nature*. W.H. Freeman & Company: New York, 1977.
2. Avnir, D. Ed. *The Fractal Approach to Heterogeneous Chemistry, Surfaces, Colloids, Polymers*. John Wiley & Sons: New York, 1989.
3. Birdi, K.S. *Fractals in Chemistry, Geochemistry, and Biophysics: An Introduction*. Plenum Press: New York, 1993.
4. Rothschild, W.G. *Fractals in Chemistry*. John Wiley & Sons: New York, 1998.
5. Coppens, M.-O. *Nature Inspired Chemical Engineering*. Delft University Press: Delft, Netherlands, 2003.
6. Russ, J.C. *Fractal Surfaces*. Plenum Press: New York, 1994.
7. Vicsek, T. *Fractal Growth Phenomena*, 2nd Ed.; World Scientific: Singapore, 1992.
8. Giona, M.; Biardi, G. *Fractals and Chaos in Chemical Engineering*. World Scientific: Singapore, 1997.

3

Particle Cohesion

Kimberly H. Henthorn and Christine M. Hrenya

CONTENTS

Introduction 17
Sources of Cohesion 18
    Van der Waals Forces 18
    Liquid Bridges 19
    Electrostatics 19
Experiments 20
    Individual Particles 20
        Force Measurements 20
        Collision-Based Experiments 21
    Many-Particle Systems 21
        Pickup Velocity 21
        Pneumatic Conveying 24
Mathematical Models 25
Dem Simulations 25
Continuum Theory 26
Conclusions 27
References 27

Introduction

Flows composed of "massive" solid particles are ubiquitous in numerous natural settings (landslides, avalanches, ice floes, planetary rings, etc.) and industries (energy, pharmaceuticals, chemicals, foodstuffs, mining, etc.). For the purposes of this work, "massive" refers to particles in which the Stokes number (St) is relatively high, where the Stokes number is defined as the ratio of particle inertia to fluid viscosity effects. Namely, $St = \rho_p d_p V_{rel}/\mu$, where ρ_p and d_p are the material density and diameter of the solid particle, V_{rel} is the relative approach velocity between two particles, and μ is the viscosity of the surrounding fluid. For large St, particle–particle collisions dominate the interactions, whereas lubrication forces dominate at small St. Note that the large St flows considered here include both *granular* systems, in which the role of the interstitial fluid is essentially negligible, and *gas–solid* systems, in which the drag force between the two phases plays a role in addition to particle–particle contacts. Systems with lower St which are not considered here include aerosols, colloidal systems, and liquid–solid suspensions.

In such inertia-dominated systems, a wide range of particle–particle (inelastic collisions, frictional contacts, van der Waals forces, electrostatics, etc.) and particle–fluid interactions (drag, shear lift, virtual mass, etc.) may be present. The complex and coupled nature of these interactions gives rise to a host of surprising phenomena such as clustering, agglomeration, species segregation, or demixing.[1–12] Accordingly, the detailed prediction of such systems is non-trivial at best.

In the current effort, focus is placed on cohesive interactions experienced between solid particles. Generally speaking, cohesion refers to an attractive force between particles. As a result of this attraction, particles will flow differently and may form agglomerates, which can be dynamic or fairly stable in nature. Furthermore, depending on the magnitude of the cohesive force and the energy input into the system (e.g., the speed of a rotating blade in a mixer), the agglomerate may be composed of a few particles or many particles. For example, a vibrated, gas-fluidized bed of nanometer-to-micron-sized particles can be designed to achieve the fluidization of agglomerates, where the agglomerates are composed of several particles and are transient in nature. In this system, the cohesive force arises from van der Waals forces between particles. At the other end of the spectrum, a sandcastle serves as an example of a fairly stable agglomerate composed of a large number of particles. In this system, the liquid bridges between particles give rise to the cohesion.

This chapter is intended to serve as an overview of cohesive, inertial flows of solid particles. The outline is as follows. In the second section, the major sources of cohesion are introduced: van der Waals forces, liquid bridges, and electrostatics. The third section contains a review of experimental techniques used to quantify cohesion, ranging from particle-level measurements to bulk measurements of many-particle systems. An overview of the incorporation of cohesive forces into discrete-particle simulations is given in the next section. Finally, efforts to include cohesive forces into continuum models of particulate flows are described in the last section.

Sources of Cohesion

The cohesive forces covered here include van der Waals forces, liquid bridges, and electrostatic forces.[13,14] Other types such as magnetic forces, sintering, and steric forces may be present in some systems, but are not considered here owing to space limitations.

van der Waals Forces

In its typical usage, the van der Waals force refers to the collection of dipole–dipole, dipole–non-polar, and dispersion forces experienced between two *molecules*. For two *solid particles*, van der Waals forces refer to the interparticle force arising from the combined effect of all molecular interactions between the two particles. Although several approaches are available for approximating the resulting particle–particle force,[15] a straightforward and fairly accurate method is obtained via Hamaker theory, in which the pair potential (potential energy) associated with each pair of molecules from different solid bodies is assumed to be additive and non-interacting. For two solid spheres of diameter d_p and separation distance a, the resulting van der Waals force is

$$F_{vdw} = \frac{A d_p}{12 a^2}$$

(1)

where A is the Hamaker constant. The Hamaker constant depends on the material making up the solid phase and interstitial fluid phase, and is typically of the order of 10^{-20} J for solids acting across air. Analogous expressions for other particle geometries and values of the Hamaker constant for specific materials are available from Israelachvili.[15]

Liquid Bridges

When particles are "wetted"—completely or partially covered with a thin layer of liquid, as opposed to being fully immersed in a liquid—a liquid bridge forms between the two particles. The presence of such a bridge generally leads to an attractive force between the two particles, owing to static and/or dynamic effects, each of which are detailed below.

Static forces include the force due to the pressure deficiency in the bridge itself, the surface tension force at the gas–liquid–solid interface, and buoyancy due to partial immersion of the particles. Although the latter two are fairly straightforward to prescribe, the force arising from the pressure deficiency requires solution of the Laplace–Young equation, which has been evaluated numerically using an array of approximations.[16] As an example of an expression for the surface tension and pressure deficiency forces between two spherical particles, referred to collectively as the capillary force, Pitois, Moucheront, and Chateau[17]

$$F_{cap} \approx 2\pi R\sigma\cos\theta\left[1 - \frac{1}{\sqrt{1+(2V/\pi Ra^2)}}\right] \tag{2}$$

where R is the particle radius, σ is the liquid surface tension, θ is the solid–liquid contact angle (angle between the lines tangent to the liquid bridge and particle surface at the point of contact), V is the bridge volume, and a is the separation distance.

Unlike static forces, dynamic forces are present only when the particles are in motion relative to one another, and arise from the effects of viscosity. For the case of low Reynolds number (creeping) flow, lubrication theory can be used to approximate the viscous force between two wetted particles as

$$F_{vis} = \frac{-3\pi\mu R^2 V_{rel}}{2a} \tag{3}$$

where V_{rel} is the normal, relative approach velocity between the two particles. A more complete description of the interaction between wetted particles in relative motion is given by elastohydrodynamics,[18,19] which couples the viscous fluid forces ("hydrodynamics") with particle deformation ("elasto").

Electrostatics

Electrostatic attraction and repulsion arise from surface charging resulting from the preferential transfer of electrons from one body to another. This imbalance of charges typically occurs in solids handling through triboelectrification, which is the transfer of electrons because of the contact or rubbing of two surfaces. The magnitude of the electrostatic force between two bodies is proportional to the product of their two charges and is inversely proportional to the square of the distance between them. However, predicting the degree of charging for a given solids population is extremely difficult, and in most cases impossible, even in well-controlled environments. Unlike most adhesion forces, electrostatic interactions can act over relatively long distances and do not require direct contact between the bodies.

Electrostatic interactions can become especially problematic when handling insulating materials such as glass. Conducting materials, such as metals, are able to dissipate charge efficiently, so triboelectrification is typically less of a problem when handling these materials. Insulating materials, on the other hand, impede the flow of electrons across their surfaces, so transferred charges tend to remain at the point of contact.

Individual particles with different properties, such as varying sizes or differing materials, are particularly prone to electrostatically driven cohesion. However, the most common reason for particle charging during handling is contact between the particles and another solid surface, such as the walls of a conveying pipe or the surface of a bin or hopper, particularly if either material is insulative. One way to decrease the effect of electrostatic charging in solids handling is to increase the humidity of the system. Since water is a conductor, moisture gradually removes excess charge from the surface of the particles. However, a large increase in humidity may promote the existence of liquid bridges, thereby giving rise to another source of cohesion.

Experiments

Experimental characterization of particle cohesion typically leads to measurements of combined interparticle forces. Careful selection of system parameters, however, can minimize or eliminate the effects of one or more forces, leading to more refined estimates of individual sources of particle–particle interactions. These experiments can focus on interactions between two or three particles or may rely on measurements from bulk particle samples.

Individual Particles

Force Measurements

Single-particle cohesion or adhesion measurements are critical to the understanding of particle contact mechanisms. It should be noted that although single-particle measurements are not easily related to the bulk behavior of the material, recent work has attempted to link the two.[20] One of the simplest, single-particle, cohesion-related properties to measure is the normal interparticle force, or pull-off force, between two like particles or between a particle and a surface.

Atomic force microscopy (AFM) is an experimental technique commonly used to characterize surface topography, but can also be used to quantify pull-off force. In this method, a single particle is supported on an AFM stage, and the force required to separate the particle from the AFM cantilever or a like particle mounted on the cantilever is measured. A force–distance curve is typically generated in such experiments, which relates the force exerted on the cantilever (deflection) to its distance from the sample surface. These curves exhibit hysteretic behavior because of the "snapping on" of the cantilever as it approaches the sample and the sudden decrease in force as the two surfaces separate. The pull-off force can be estimated once the effect of cantilever compliance is taken into consideration.

Material characteristics such as yield stress and surface energy can be determined from the force–distance curve and from the absolute value of the pull-off force. However, it should be noted that measurement of particle cohesion and adhesion using AFM is still not fully understood. This technique is not well-suited for certain highly cohesive materials because of the nature of the measurement, and other factors, such as cantilever properties, must be taken into consideration when extracting meaning from the acquired data.[21–23]

In addition to AFM, another experimental technique capable of force vs. distance measurements involves the use of a high-precision scale. This method has been applied to a system of two particles joined by a liquid bridge in which a stationary particle is attached to the scale, while the motion of the other particle is precisely controlled via a motor-driven differential micrometer screw.[17]

Finally, force vs. time measurements between two cohesive particles are an alternative to the force vs. distance measurements described above. Methods for obtaining such data include impact load cells[24] as well as piezosensors embedded in the particles themselves.[25]

Collision-Based Experiments

As an alternative to the detailed force profile measurements described above, the role of cohesive forces can also be assessed via measurements of pre- and postcollisional velocities of particles involved in a collision. The ratio of rebound velocity to the impact velocity, or the restitution coefficient e, follows, where $e = 0$ is indicative of two particles forming an agglomerate and $e > 0$ is indicative of particles that separate after the collision. For example, Davis and coworkers[18,26,27] have used strobed imaging in both constrained (pendulum-based) and unconstrained experiments to obtain such measurements for the case of wetted particles in which viscous effects dominate.

Many-Particle Systems

Pickup Velocity

Pickup velocity, defined as the fluid velocity required to entrain a particle from rest, has applications in a number of areas, including environmental studies, pulmonary drug delivery, and pharmaceutical processing.[28] In two-phase conveying systems, fluid velocities below the particle pickup velocity can result in clogged pipelines or channels, and velocities much larger than what is required can lead to particle attrition, excessive pipeline abrasion, and unnecessary energy consumption. Therefore, knowledge of a material's pickup velocity is of great interest to those designing particle conveying processes. However, pickup velocity is a complex function of many variables including particle properties, such as size, shape, density, and electrostatic behavior, and fluid properties, such as density and viscosity. Measurements of pickup velocity have revealed qualitative information about the relative importance of some interparticle forces, which has led to more refined models and a better understanding of how to control cohesive materials with a particular set of properties.

Experiments have shown the existence of a minimum pickup velocity as a function of particle size for a range of particle shapes and materials entrained into a gas stream.[28–30]

Above 40 µm, the entrainment of large, non-cohesive particles is dominated by inertial forces, so pickup velocity increases with increasing particle size. However, interparticle forces such as van der Waals and electrostatic forces become more significant for particles less than 40 µm, so pickup velocity increases as particle size decreases. Hayden, Park, and Sinclair[28] found that pickup velocity stays relatively constant for particles less than 20 µm because of dominant van der Waals forces and the interaction between particle size and degree of particle packing. It was also found that irregularly shaped particles generally had a higher pickup velocity and larger deviations in repeated trials because of particle interlocking and increased interparticle contact area.

Particle entrainment has been modeled by many investigators over the past several decades, and requires knowledge of fluid behavior as well as system and particle material properties. Although most real relevant systems are composed of many particles, a simplified model of a single particle resting on a flat surface can be developed to explain the phenomena seen in experimental observations. Cabrejos and Klinzing[31,32] proved that the horizontal forces acting on a single particle become balanced before the vertical forces, resulting in rolling or sliding (rather than lifting) at the point of incipient motion. However, some define particle entrainment to be the onset of vertical movement into a flowing fluid, which can be modeled through a simple particle force balance:

$$F_g + F_a = F_l + F_b \tag{4}$$

where F_g, F_a, F_l, and F_b represent the gravitational, adhesion, lift, and buoyancy forces, respectively.[28,29] Figure 3.1 shows the principal forces acting on a single particle resting on a flat surface. For particles small enough to be completely embedded in the laminar boundary layer, the velocity profile appears linear, as seen in the figure.[29] The adhesion

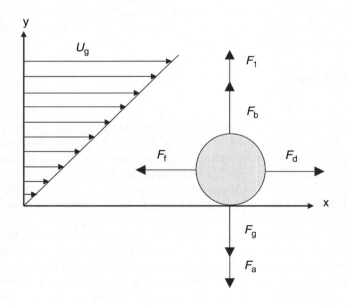

FIGURE 3.1
Forces acting on a single particle in shear flow: F_f = frictional force; F_d = drag force; F_l = lift force; F_b = buoyancy force; F_g = gravitational force; F_a = adhesive forces; U_g = fluid velocity. (From Incipient Motion of Solid Particles in Pneumatic Conveying.[29])

term in this model describes the attractive and repulsive forces between a particle and a flat surface; for many-particle systems, this term is replaced by the cohesive force between like particles. Hayden, Park, and Sinclair[28] and Cabrejos[29] showed that this preliminary model underpredicts the pickup velocity for many-particle systems, mainly because of the simplification of the adhesion term.

In some instances, it is necessary to predict the onset of horizontal particle motion. For example, particle detachment in a microchannel is typically characterized by the axial movement of a particle by fluid flow. Sliding occurs when the drag force imparted by the flowing fluid is large enough to overcome the frictional force between the particle and the surface on which it rests. A simple force balance can be applied here as well, where the frictional component is a function of the vertical forces and includes buoyancy, lift, adhesion, and gravitational forces.

If particle motion is initiated by rolling, a balance of moments rather than a balance of forces is used to characterize the system.[33] In order for a particle to begin rolling, the moment induced by the drag force from the fluid must be large enough to overcome the moment from the frictional force between the particle and the surface on which it rests.[34] The rolling moment (RM) is the ratio of the hydrodynamic rolling moment to the adhesion resting moment, and must be greater than or equal to 1 for particle motion to be initiated by rolling:

$$\text{RM} = \frac{F_d(1.399d_{\text{p}}/2 - \alpha)}{(F_{\text{a}} + F_{\text{g}} - F_{\text{b}} - F_{\text{l}})(a_{\text{eq}}/2)} \tag{5}$$

where d_{p} is the diameter of the particle, a_{eq} is the contact diameter, and α is the distance of approach. The distance of approach is given by

$$\alpha = \frac{d_{\text{p}}}{2} - \left[\left(\frac{d_{\text{p}}}{2} \right)^2 - \left(\frac{a_{\text{eq}}}{2} \right)^2 \right]^{1/2} \tag{6}$$

The contact diameter, which is the effective length of contact between the particle and the surface on which it rests, is given by

$$a_{\text{eq}} = \left(\frac{6\pi W_A d_{\text{p}}^2}{4K} \right)^{1/3} \tag{7}$$

where W_A is the work of adhesion and K is the composite Young's modulus.

As previously mentioned, these models give a simplified view of most real systems, which are typically composed of many particles. In many-particle systems, the geometry of the particle–particle contacts needs to be considered. Typically, these are simplified to take the form of a single particle resting on a tripod of three other particles. The cohesive forces are then summed over all the particle contacts. Several models have been developed for particle entrainment from a bed of particles, including the work of Wicks[35] and Stevenson, Cabrejos, and Thorpe[36] reported that although particle shape affects the pickup velocity for a single particle resting on a flat surface, there is no apparent shape effect for the pickup of a particle resting on a bed of similar particles. This conclusion contradicts those found by Cabrejos and Hayden et al., who showed that pickup velocity in a multiple particle system is dependent on particle shape for particles having mean diameters less than 100 μm.

Pneumatic Conveying

Although pneumatic conveying is very common in industrial particulate handling, downtime due to blockages and particle adhesion to pipelines is a major problem. In addition, the success rate for the start-up of pneumatic conveying processes is only 60%, as compared to a 90% success rate for other types of operations.[37] As discussed in section on "Electrostatics," particle aggregation and particle adhesion to pipe surfaces can be problematic in pneumatic conveying systems because of enhanced electrostatic charging resulting from particle–particle and particle–wall contacts. Some work has been done to characterize the nature of charging in conveying lines, since the current knowledge is limited and prediction is crude at best.

The addition of a powdered antistatic additive such as Larostat® (BASF, Florham Park, New Jersey, U.S.A.) can minimize electrostatic attractions, but has been found to be relatively ineffective for particles smaller than 20 μm. This is because fine particles tend to coat the inner surface of the pipe, reducing the particle–wall contact for the bulk population and leaving contact only between like materials.[38] However, the adhesion of fine particles onto larger particles can be used to minimize cohesion during conveying. Zhu, Yao, and Wang[39] found that Larostat particles, which are commonly used to reduce electrostatic charging, can adhere to the pipe wall and aggregate with larger conveyed material, so that particle–particle and particle–wall interactions are effectively collisions between like Larostat particles (which are not insulating and thus not prone to charging).

This concept is also used to improve flowability of fine drug particles for use in dry powder inhalers.[40] Active drug particles are typically less than 5 μm in size and are fairly cohesive. Larger biodegradable excipient particles (>30 μm) are mixed with the active drug to act as carrier particles. The drug particles adhere to the surface of the free-flowing excipient, greatly improving their flow properties during mixing and delivery. During pulmonary drug delivery, the smaller particles on the surface of the excipient are easily entrained into the air stream and are separated from the excipient particle in the turbulent inhalation air flow before being carried into the deep lung.

In addition to electrostatic charging, other sources of particle cohesion can affect particle mobility in pneumatic conveying systems. Liang et al.[41] found that increased moisture content during the dense-phase pneumatic transport of pulverized coal particles promotes the formation of liquid bridges, leading to decreased flowability and dispersibility for a given superficial velocity. As the cohesive forces increase, fine particles aggregate, leading to a larger effective particle size. This results in an increase in the solids friction coefficient and viscosity, and the particle mass flow rate decreases. In addition, pressure fluctuations are dampened and pressure drop per unit length decreases. Liang et al. found that in their system, a moisture content above 6% typically resulted in pipe blockage.

van der Waals forces dominates over electrostatic forces for particles less than 20 μm in size.[28] In general, the effect of van der Waals forces is enhanced in non-spherical particles as result of an increase in available contact area over which they can act. However, the structure of a non-spherical particle or aggregate plays a large role in the cohesive properties of the material. Keuter et al.[42] studied van der Waals-induced aggregation of carbon black particles during pneumatic conveying. They characterized the shape of several types of aggregates, which was quantified by the degree of branching, and investigated the effect of aggregate structure on caking and aggregate enlargement after conveying. It was concluded that highly branched aggregates were not as likely to adhere to the pipe wall or cohere into larger aggregates as their less branched counterparts. This is because the more linear structure of a lightly branched aggregate offers more available particle–particle and particle–wall contact points, increasing the likelihood of cohesion and adhesion due to van der Waals forces.

Mathematical Models

Generally speaking, cohesion has been incorporated into two types of mathematical models for flows of solid particles: discrete-particle simulations and continuum theory. The former, also known as the discrete element method (DEM) or molecular dynamics (MD), treats particles as individual entities and tracks their motion via the solution of force balances for each particle present in the system.[43] The latter uses various forms of averaging to treat the particles as a continuum phase, resulting in mass, momentum, and energy balances for the particle phase,[44–46] which mimic those developed for single-phase fluids (e.g., Navier–Stokes equations for Newtonian fluids). A disadvantage of DEM models is the high computational requirement needed to solve a large number of force balances, making them intractable for typical unit operations found in industry, which may contain billions and billions of particles. The computational requirement for continuum models is more modest since they involve a single balance equation for each variable, though a disadvantage is the need for constitutive relations for those terms arising from the averaging process (e.g., stress tensor associated with the particle phase). For this reason, both modeling approaches remain active areas of research.

Dem Simulations

Because of the straightforward nature of implementing cohesive forces into DEM simulations of both granular and gas–solid flows, examples of such are prevalent in the literature and are too numerous for a thorough review here. Generally speaking, two different approaches have been taken for the incorporation of cohesion into DEM simulations: the incorporation of a generic cohesive force or the incorporation of a specific type of cohesive force (van der Waals, liquid bridging, etc.).

An example of a generic cohesive force is the square-well model,[47,48] in which the cohesive force is represented as an impulsive force that occurs at a preset distance from the particle surface. This impulsive force is depicted in Figure 3.2, which shows the potential energy ϕ as a function of the separation distance R. Note that the corresponding force F is defined as $F = -d\phi/dR$, so the force is only non-zero at $2r_{outer}$ (cohesive contact at preset, non-zero separation distance) and when the particles are in physical contact at $R = 2r_{inner}$ (repulsive force upon contact). An advantage of this representation is that it can be incorporated into hard-sphere (or event-driven) simulations (which are more efficient than their soft-sphere counterparts) because of the binary and instantaneous nature of the cohesive force between particles, and is thus also a candidate for incorporation into a

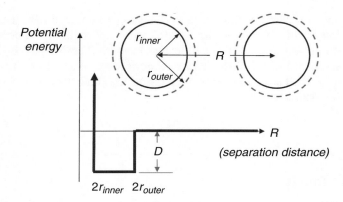

FIGURE 3.2
Potential energy as a function of particle separation distance for the square-well model.

kinetic-theory-based continuum description of cohesive particles. A disadvantage, however, is the need for the specification of model parameters (r_{outer} and well depth D in Figure 3.2), which do not map directly to physical, measurable quantities. Nonetheless, a mapping strategy has been developed, which is shown to be successful for systems with dynamic agglomeration,[48] though differences arise when applied to systems with sustained particle contacts.[49]

Unlike the generic forces described above, DEM models specific to a type of cohesion (van der Waals, liquid bridging, etc.) have also been widely utilized. The more rigorous treatment of cohesive forces can lead to some drawbacks, however. One drawback that can be easily overcome is the limiting behavior of some models in which the cohesive force approaches infinity as the separation distance goes to zero (see Eqs. 1 and 3 above). This situation is typically handled by assigning a cutoff distance, usually an estimate for the interatomic distance for solids; the force for separation distances below this cutoff is then set to remain constant at the value obtained at the cutoff. Another drawback, as alluded to in the previous paragraph, is the dependency of the cohesive force on separation distance. This continuous behavior precludes the use of hard-sphere simulations, so the more computationally demanding soft-sphere treatment is required. Finally, and perhaps most importantly, simplifications to the rigorous cohesive description are often required for incorporation into a DEM code. For example, in the description of lubrication effects, it is often assumed that lubrication forces and particle deformation are not coupled,[17,50,51] since a coupled approach requires simplifications and/or a numerical solution to the governing differential equations.[18,19,52] The validity of many of these simplified approaches remains to be tested.

Continuum Theory

Unlike their slower counterparts that are characterized by multiparticle enduring contacts, the continuum description of rapid (collision-dominated) granular and gas–solid flows is more advanced. In particular, an analogy between the solid particles and molecules in a gas allows for a treatment using the rigors of kinetic theory.[1,8,44,46] Nonetheless, extension to flows of cohesive particles is in its infancy, because of assumptions utilized in kinetic-theory-based models. In particular, these models are based on an assumption of binary, instantaneous collisions between particles, so the incorporation of an additional force that varies continuously with distance is non-trivial at best. To date, the only formal incorporation of cohesion into a kinetic-theory framework was performed by Kim and Arastoopour.[53] Their model is based on a contact bonding energy, which is activated at particle contact. Several simplifications were made in their derivation, including the treatment of a many-particle agglomerate as a single sphere with the same total mass, and the lack of a deagglomeration mechanism.

Conclusions

Cohesion plays a critical role in solids handling and processing. Electrostatic interactions, van der Waals forces, and liquid bridging are typically the major contributors to the overall cohesion force. This chapter has given a brief overview of experimental characterization and modeling of inertia-dominated, cohesive-particle systems. Generally speaking, more experimental effort is needed at the particle-level to characterize the cohesive interaction in a detailed nature (force vs. separation distance, prediction of agglomeration and deagglomeration, etc.) so that existing DEM models can be tested and improved as necessary. A more challenging prospect involves linking the micro (particle-level) behavior to the macro (bulk) behavior. DEM models are capable of providing valuable information in this regard, though their high computational overhead precludes the simulations of many industrial-size units, and the incorporation of cohesive interactions in continuum theory is in its infancy. Nonetheless, such models are needed to predict the cohesion-induced behaviors observed experimentally in pneumatic transport systems, fluidized beds (not covered here owing to space constraints), and other common particle processes, including low Stokes number flows such as aerosols, colloidal systems, and liquid–solids suspensions.

References

1. Campbell, C.S. Rapid granular flows. Annu. Rev. Fluid Mech. **1990**, *22*, 57–92.
2. Bridgwater, J. Mixing and segregation mechanisms in particle flow. In *Granular Matter: An Interdisciplinary Approach*. Mehta, A., Ed., Springer-Verlag: New York, **1994**, 161–193.
3. Jaeger, H., Nagel, S., Behringer, R. The Physics of granular materials. Phys. Today, April **1996**, 32–38.
4. Ottino, J., Khakhar, D. Mixing and segregation of granular materials. Annu. Rev. Fluid Mech. **2000**, *32*, 55–91.
5. Sundaresan, S. Perspective: Modeling the hydrodynamics of multiphase flow reactors: Current status and challenges. AIChE J. **2000**, *46*, 1102–1105.
6. Sundaresan, S. Some outstanding questions in handling of cohesionless particles. Powder Tech. **2001**, *115*, 2–7.
7. Muzzio, F., Shinbrot, T., Glasser, B. Powder technology in the pharmaceutical industry: The need to catch up fast. Powder Tech. **2002**, *124* (1–2), 1–7.
8. Goldhirsch, I. Rapid granular flows. Ann. Rev. Fluid Mech. **2003**, *35*, 267–293.

9. Sundaresan, S., Eaton, J., Koch, D., Ottino, J. Appendix 2: Report of study group on disperse flow. Int. J. Multiphas. Flow **2003**, *29*, 1069–1087.

10. Curtis, J., van Wachem, B. Modeling particle-laden flows: A research outlook. AIChE J. **2004**, *50* (11), 2638–2645.

11. Goldhirsch, I., Noskowicz, S., Bar-Lev, O. Theory of granular gases: Some recent results and some open problems. J. Phys.: Condens. Matter **2004**, *17*, 2591–2608.

12. Wassgren, C., Curtis, J. The application of computational modeling to pharmaceutical materials science. MRS Bull. **2006**, *31*, 900–904.

13. Visser, J. An invited review: van der Waals and other cohesive forces affecting powder fluidization. Powder Tech. **1989**, *58*, 1–10.

14. Seville, J., Willett, C., Knight, P. Interparticle forces in fluidisation: A review. Powder Tech. **2000**, *113*, 261–268.

15. Israelachvili, J. *Intermolecular and Surface Forces, 2nd Ed.*, Academic Press: New York, **1992**.

16. Lian, G., Thornton, C., Adams, M. A theoretical study of the liquid bridge forces between two rigid spherical bodies. J. Colloid Interface Sci. **1993**, *161*, 138–147.

17. Pitois, O., Moucheront, P., Chateau, X. Liquid bridge between two moving spheres: An experimental study of viscosity effects. J. Colloid Interface Science **2000**, *231*, 26–31.

18. Davis, R., Rager, D., Good, B. Elastohydrodynamic rebound of spheres from coated surfaces. J. Fluid Mech. **2002**, *468*, 107–119.

19. Kantak, A., Davis, R. Elastohydrodynamic theory for wet oblique collisions. Powder Tech. **2006**, *168* (1), 42–54.

20. Jones, R. From single particle AFM studies of adhesion and friction to bulk flow: Forging the links. Granular Matter **2003**, *4*, 191–204.

21. Ouyang, Q., Ishida, K., Okada, K. Investigation of micro-adhesion by atomic force microscopy. Appl. Surf. Sci. **2001**, *169–170*, 644–648.

22. Sader, J., Chon, J., Mulvaney, P. Calibration of rectangular atomic force microscope cantilevers. Rev. Sci. Instr. **1999**, *70* (10), 3967–3969.

23. Marti, O. Measurement of adhesion and pull-off forces with the AFM. In *Modern Tribology Handbook*. Bhushan, B., Ed. CRC Press: Boca Raton, FL, **2001**, 617–635.

24. Tavares, L. Breakage of single particles: Quasi-static. In *Handbook of Particle Technology: Particle Breakage*. Salman, A., Hounslow, M., Ghadiri, M., Ed. Elsevier: Boston, 2007.

25. Daraio, C., Nesterenko, V., Herbold, E., Jin, S. Strongly nonlinear waves in a chain of Teflon beads. Phys. Rev. E, **2005**, *72*, art. no. 016603.

26. Kantak, A. Davis, R. Oblique collisions and rebound of spheres from a wetted surface. J. Fluid Mech. **2004**, *509*, 63–81.

27. Kantak, A., Galvin, J., Wildemuth, D., Davis, R. Low-velocity collisions of particles with a dry or wet wall. Microgravity Sci. Tech. **2005**, *17* (1), 18–25.

28. Hayden, K., Park, K., Sinclair, J. Effect of particle characteristics on particle pickup velocity. Powder Tech. **2003**, *131* (1), 7–14.

29. Cabrejos, F. *Incipient Motion of Solid Particles in Pneumatic Conveying*. Masters thesis, University of Pittsburgh: Pittsburgh, 1991.

30. Kalman, H., Satran, A., Meir, D., Rabinovich, E. Pickup (critical) velocity of particles. Powder Tech. **2005**, *160*, 103–113.

31. Cabrejos, F., Klinzing, G. Incipient motion of solid particles in horizontal pneumatic conveying. Powder Tech. **1992**, *72*, 51–61.

32. Cabrejos, F., Klinzing, G. Pickup and saltation mechanisms of solid particles in horizontal pneumatic transport. Powder Tech. **1994**, *79*, 173–186.

33. Abd-Elhady, M., Rindt, C., Wijers, J., van Steenhoven, A. Removal of Particles from Powdery Fouled Surfaces. Proceedings Twelfth International Heat Transfer Conference Grenoble, France 2002.

34. Essaway, A. *Microparticle Detachment from Surfaces by Fluid Flow*. PhD Thesis, University of Notre Dame, 2004.

35. Wicks, M. Transport of solids at low concentration in horizontal pipes. In *Advances in Solid-Liquid Flow in Pipes and its Application*. Zandi, I., Ed. Pergamon Press: Oxford, 1971, 101–124.

36. Stevenson, P., Cabrejos, F., Thorpe, R. Incipient motion of particles on a bed of like particles in hydraulic and pneumatic conveying. 4th World Congress of Particle Technology Sydney, Australia, 2002.
37. Merrow, E. Linking R-and-D to problems experienced in solids processing. Chem. Eng. Prog. **1985**, *81* (5), 14–22.
38. Wang, F.-J., Zhu, J.-X., Beeckmans, J. Pressure gradient and particle adhesion in the pneumatic transport of cohesive fine powders. Int. J. Multiphas. Flow **2000**, *26*, 245–265.
39. Zhu, K., Yao, J., Wang, C. On the electrostatics of pneumatic conveying of granular materials. D Space (MIT) URI, http://hdl.handle.net/1721.1/3939, 2004 (accessed September 2008).
40. Bennett, F., Carter, P., Rowley, G., Dandiker, Y. Modification of electrostatic charge on inhaled carrier lactose particles by addition of fine particles. Drug Dev. Ind. Pharm. **1999**, *25* (1), 99–103.
41. Liang, C., Zhao, C., Chen, X., Pu, W., Lu, P., Fan, C. Flow characteristics and Shannon entropy analysis of dense-phase pneumatic conveying of pulverized coal with variable moisture content at high pressure. Chem. Eng. Technol. **2007**, *30* (7), 926–931.
42. Keuter, H., Limper, A., Wehmeier, A., Riedemann, T., Freitag, K. *Increase in fines content and adhesion behavior in the pneumatic conveying of CB.* Rubber World. August **2001**, 38–43.
43. Pöschel, T., Schwager, T. *Computational Granular Dynamics*. Springer-Verlag: New York, 2005.
44. Gidaspow, D. *Multiphase Flow and Fluidization*. Academic Press: San Diego, 1994.
45. Jackson, R. *The Dynamics of Fluidized Particles*. Cambridge University Press: New York, 2000.
46. Brilliantov, N., Pöschel, T. *Kinetic Theory of Granular Gases*. Oxford University Press: New York, 2004.
47. Weber, M., Hoffman, D., Hrenya, C. Discrete-particle simulations of cohesive granular flow using a square-well potential. Granular Matter **2004**, *6*, 239–254.
48. Weber, M., Hrenya, C. Square-well model for cohesion in fluidized beds. Chem. Eng. Sci. **2006**, *61*, 4511–4527.
49. Weber, M., Hrenya, C. Computational study of pressure-drop hysteresis in fluidized beds. Powder Tech. **2007**, *177* (3), 170–184.
50. Ennis, B., Tardos, G., Pfeffer, R. A microlevel-based characterization of granulation phenomena. Powder Tech. **1991**, *65*, 257–272.
51. Liu, L., Litster, J., Iveson, S., Ennis, B. Coalescence of deformable granules in wet granulation processes. AIChE J. **2000**, *46* (3), 529–539.
52. Lian, G., Adams, M., Thorton, C. Elastohydrodynamic collisions of solid spheres. J. Fluid Mech. **1996**, *311*, 141–152.
53. Kim, H., Arastoopour, H. Extension of kinetic theory to cohesive particle flow. Powder Tech. **2002**, *122* (1), 83–94.

4

Particle–Particle Interaction

Anh V. Nguyen, Linh T. T. Tran, and Jan D. Miller

CONTENTS

Introduction 31
Van der Waals Interactions Between Colloidal Particles 32
    Hamaker Microscopic Approach 32
    Lifshitz Macroscopic Approach 33
    Approximate and Simplified Equations for Van der Waals Interaction Between
    Spheres 36
Electrostatic Double-Layer Interaction 39
    Approximate Models for EDL Interaction Between Two Spheres 40
    Exact Numerical Solutions for EDL Interaction Between Two Spheres 42
Limits of the Derjaguin Approximation Probed with AFM Tips 46
Conclusions 47
Acknowledgments 48
References 48

Introduction

Particle–particle interaction is central to a wide range of engineering applications and processing industries. Examples include coagulation, flocculation, dispersion, emulsification, and froth flotation. In these applications, the particle size is small, and the overall particulate behavior is determined by forces associated with the surface properties rather than those related to mass or volume. The surface properties of a particle in a liquid medium are the result of a complex interaction between molecules, atoms, and ions at the particle surface and in the surrounding liquid. If a number of particles are present, interactions also take place between particles at short separation distances, and it is this interaction that is of most interest as it can determine the overall stability or instability of dispersions and/or suspensions.

It is customarily assumed that the overall particle–particle interaction can be quantified by a net surface force, which is the sum of a number of independent forces. The most often considered force components are those due to the electrodynamic or van der Waals interactions, the electrostatic double-layer interaction, and other non-DLVO interactions. The first two interactions form the basis of the celebrated Derjaguin–Landau–Verwey–Overbeek (DLVO) theory on colloid stability and coagulation. The non-DLVO forces are usually determined by subtracting the DLVO forces from the experimental data. Therefore, precise prediction of DLVO forces is also critical to the determination of the non-DLVO forces. The surface force apparatus and atomic force microscopy (AFM) have been used to successfully quantify these interaction forces and have revealed important information about the surface force components. This chapter focuses on improved predictions for DLVO forces between colloid and nano-sized particles. The force data obtained with AFM tips are used to illustrate limits of the renowned Derjaguin approximation when applied to surfaces with nano-sized radii of curvature.

This chapter is organized in the following paragraphs: First, the advanced determination of van der Waals interaction between spherical particles is described. Second, the relevant approximate expressions and direct numerical solutions for the double-layer interaction between spherical surfaces are reviewed. Third, the experimental data obtained for AFM tips having nano-sized radii of curvature and the DLVO forces predicted by the Derjaguin approximation and improved predictions are compared. Finally, a summary of the review and recommended equations for determining the DLVO interaction force and energy between colloid and nano-sized particles is included.

Van der Waals Interactions Between Colloidal Particles

There are essentially two approaches for the determination of the van der Waals interaction energy and forces. In the first, due largely to Hamaker,[1] the interaction between macroscopic bodies is calculated by a pairwise summation of all the relevant microscopic interactions, which are assumed to be nonretarded and additive. In reality, the microscopic interactions are retarded by the nearby atoms and molecules, which are included in the classical theories on retardation via the finite speed of light propagating in dispersed media.[2] The second, more rigorous, approach developed by Lifshitz[3] is based on quantum-mechanics theory and depends entirely on the macroscopic electrodynamic properties of the interacting particles and the media such as dielectric constants and refractive indices. The electric fields established by the fluctuating dipoles are considered to interact both constructively and destructively. The result of these fluctuating and many-body interactions is the formation of a standing wave between the bodies whereby only certain modes, or frequencies of electromagnetic radiation may pass through. The summation of all fluctuation modes in the electromagnetic field gives the van der Waals interaction energy.[4] The full theory is complicated, and this is probably the reason that the additive approach of Hamaker is still used by many scientists. In some cases, the results of the Hamaker theory are accurate when an experimentally determined Hamaker constant is used. For many practical systems of colloidal particles, the Hamaker constant is not known but the Hamaker prediction can be useful if the (effective) Hamaker constant (function) is determined from the Lifshitz continuum approach.

Hamaker Microscopic Approach

In the Hamaker approach, the energies of all the atoms in one body with all the atoms in the other one are summed, leading to an integral expression for the interaction energy, E, between the two macroscopic bodies: $E_{vdW} = -\oint\oint (C\rho_1\rho_2/r^6)dv_1 dv_2$, where C is the coefficient of the appropriate interaction between elementary particles separated by distance r and dv_i ($i = 1, 2$) the volume elements of bodies 1 and 2 at distance r, and with densities of atoms or molecules of ρ_i. The problem appears simple since it only requires the calculation of a closed integral. However, the results in closed analytical forms are available only for some simple systems.[4] The Hamaker prediction for interaction energy splits into two terms: the Hamaker constant described by $A = \pi^2 C \rho_1 \rho_2$ and the second term that represents the geometrical dependence of the van der Waals energy.

For the van der Waals energy per unit area, E_{vdW}^*, between two half-spaces separated by a distance h, the Hamaker theory gives

$$E_{vdW}^* = -\frac{A}{12\pi h^2} \tag{1}$$

The interaction force, F, is then determined by the first derivative of the energy with respect to the distance by $F(h) = -dE/dh$.

The Hamaker interaction energy for two spheres separated by the inter-surface shortest distance, h, is described by

$$
\begin{aligned}
E_{vdW} = -\frac{A}{6}\Bigg\{ &\frac{2R_1 R_2}{r^2 - (R_1 + R_2)^2} + \frac{2R_1 R_2}{r^2 - (R_1 - R_2)^2} \\
&+ \ln \frac{r^2 - (R_1 + R_2)^2}{r^2 - (R_1 - R_2)^2} \Bigg\}
\end{aligned}
\tag{2}
$$

where R_1 and R_2 are the radii of two spheres and r the inter-center distance, $r = R_1 + R_2 + h$. We will see later that Eqs. (1) and (2) are useful for combining the Hamaker microscopic and Lifshitz macroscopic theories.

Lifshitz Macroscopic Approach

The description of the van der Waals interaction based on the Lifshitz[3,5] approach is now sufficiently advanced to provide accurate predictions for the complete interaction energy. For the geometry of two half-spaces, the exact theory is available in a formulation suited for computational purposes.[4] In parallel with work on planar systems, there has been a focus on the interaction between spheres.[4,6–8] These developed theories have been used as the exact solutions in the validation of the approximate predictions using the Hamaker approach. The significant contribution of the continuum approach to our understanding of the van der Waals interaction lies in the reliable prediction of the Hamaker constant. The interaction energy for two half-spaces and two spheres is summarized below.

The van der Waals interaction energy, E_{132}^*, per unit area between two half-spaces 1, 2 immersed in a medium 3 as a function of the separation, h, is described by the Lifshitz macroscopic theory as follows:[4]

$$
\begin{aligned}
E_{132}^*(h) = \frac{k_B T}{8\pi h^2} \sum_{N=0}^{\infty} \int_{x_N}^{\infty} x \ln\Big[&\big(1 - y_{13} y_{23} e^{-x}\big) \\
&\times \big(1 - z_{13} z_{23} e^{-x}\big)\Big] dx
\end{aligned}
\tag{3}
$$

where $y_{a3} = \dfrac{x\varepsilon_a - s_a\varepsilon_3}{x\varepsilon_a + s_a\varepsilon_3}$, $z_{a3} = \dfrac{x - s_a}{x + s_a}$, $\xi_N = 2N\pi k_B T/\hbar$, $x_N = 2h\xi_N \sqrt{\varepsilon_3}/c$, $\varepsilon_a = \varepsilon_a(i\xi_N)$, and $s_a^2 = x^2 + x_N^2\{\varepsilon_a/\varepsilon_3 - 1\}$. In these equations, the subscript $a = 1$ to 3, $i = \sqrt{-1}$, c is the speed of light, k_B Boltzmann's constant, T the absolute temperature, $\hbar$ the Planck constant divided by 2π, and $i\xi_N$ are the discrete equally spaced imaginary frequencies. The prime on the summation symbol indicates that the zero-frequency term, which accounts for the contributions due to the orientation and induction interactions, is divided by 2. Eq. (3) shows that the interaction energy depends on electromagnetic fluctuations, via the relative permittivity, $\varepsilon(i\xi_N)$, of all three materials.

For two spheres separated by the (shortest) distance, h, the interaction energy is given by the Lifshitz macroscopic theory:[6,9]

$$E_{132}(h) = k_B T \sum_{N=0}^{\infty}{}' \sum_{m,n=1}^{\infty} (2m+1)(2n+1)\Delta_m\Delta_n$$

$$\times \sum_{\mu=-\mu_m}^{\mu_m} V_{mn}^{\mu}\left(K_3 r\right) V_{nm}^{-\mu}\left(K_3 r\right) \tag{4}$$

where

$$\Delta_m = \frac{\left(x_3^2 - x_1^2\right)m i_m(x_3)i_m(x_1) + x_3 x_1\left[x_1 i_m(x_1)i_{m-1}(x_3) - x_3 i_m(x_3)i_{m-1}(x_1)\right]}{\left(x_3^2 - x_1^2\right)m k_m(x_3)i_m(x_1) - x_3 x_1\left[x_1 i_m(x_1)k_{m-1}(x_3) + x_3 k_m(x_3)i_{m-1}(x_1)\right]}$$

$$V_{mn}^{\mu}(x) = U_{mn}^{\mu}(x) + \frac{n-\mu+1}{(n+1)(2n+1)} x U_{mn+1}^{\mu}(x)$$

$$-\frac{n+\mu}{n(2n+1)} x U_{mn-1}^{\mu}(x),$$

$$U_{mn}^{\mu}(x) = \left(\frac{2}{x}\right)^{\mu} \sum_{\upsilon=0}^{\upsilon_m} S_{mn}^{\mu\upsilon} k_{m+n-\mu-2\upsilon}(x),$$

$$V_{mn}^{-\mu}(x) = \frac{(n-\mu)!(m-\mu)!}{(n+\mu)!(m+\mu)!} V_{mn+1}^{\mu}(x),$$

and

$$S_{mn}^{\mu\upsilon} = \frac{\Gamma(m-\upsilon+1/2)\Gamma(n-\upsilon+1/2)\Gamma(m+\upsilon+1/2)\times(m+n-\upsilon)!(m+n-\mu-2\upsilon+1/2)}{\Gamma(m+n-\mu-\upsilon+3/2)\Gamma(\mu+1/2)\Gamma(1/2)\times(m-\mu-\upsilon)!(n-\mu-\upsilon)!\upsilon!}$$

In these equations, $K_a = \left(\xi_N/c\right)\sqrt{\varepsilon_a\left(i\xi_N\right)}$ $(a = 1, 2, 3)$, $x_a = K_a R$, ε_a is the permittivity, $\mu_m = \min(m, n)$, $\upsilon_m = \min(m - \mu, n - \mu)$, r the inter-center distance, $r = R_1 + R_2 + h$, Γ the gamma function, and i_m and k_m are the modified spherical Bessel functions of the order m and of the first and second kind, respectively. The susceptibility function Δ_m is purposely used for particle 1 with radius $R = R_1$. The susceptibility function Δ_n is used for particle 2 and can be described similarly with x_1 replaced by x_2 and $R = R_2$. There is a typographic error in Δ_m given in:[9] the term in the brackets must be multiplied by the particle radius. The modified spherical Bessel functions are not uniquely defined in the literature. In this chapter, the functions are defined by the respective modified Bessel functions: $i_m(x) = \sqrt{\pi/(2x)}\,I_{m+1/2}(x)$ and $k_m(x) = \sqrt{2/(\pi x)}\,K_{m+1/2}(x)$. Because the definition of modified spherical Bessel functions of the third kind is not unique in the literature, the Bessel functions of the third kind are not used in the above equations.

The convergence of Langbein's solution described by Eq. (4) is slow. Pailthorpe and Russel[9] evaluated Eq. (4) by computing the difference between the retarded and nonretarded interactions, which converges much faster. They then evaluated the nonretarded interaction by using Love's prediction,[7] which is faster than evaluating Langbein's nonretarded expression, and added this to the difference. Figure 4.1 shows the results of the calculation and comparison with different models.

The full spectra of permittivities are required in the calculation of the van der Waals interaction energy using Eqs. (3) and (4). For water and a few materials, the dependence of the permittivity, ε_a, on the sampling frequency, ξ_N, is available (see Figure 4.2).[10–12] For highly polar liquids such as water, the relaxation in the microwave and infrared appears significant and the oscillator model for the spectrum of water permittivity has a number of terms as described by Eq. (5), and the model parameters are given in Table 4.1.

$$\varepsilon(i\xi) = 1 + \frac{d_m}{1 + \xi\tau_m} + \sum_j \frac{f_j}{\omega_j^2 + g_j\xi + \xi^2} \tag{5}$$

For other materials, the permittivity spectra can be approximately determined using the refractive index, $\sqrt{B}$, and the static dielectric constant, $\varepsilon(0)$, as follows:

$$\varepsilon(i\xi_N) = \begin{cases} \dfrac{B + \left(\xi_N/\omega\right)^2}{1 + \left(\xi_N/\omega\right)^2} & \text{for } N \geq 1 \\[2ex] \varepsilon(0) & \text{for } N = 0 \end{cases} \tag{6}$$

where ω is the characteristic relaxation frequency of the UV region, which is $\sim 2 \times 10^{16}$ rad/s.[14] The nonzero sampling frequency ξ_N ($N \geq 1$) begins at $\xi_1 = 2.4 \times 10^{14}$ rad/s and is closely spaced in the UV region; the relaxation in the UV dominates the portion of the dielectric spectra most important for the van der Waals interaction. For most nonmetallic materials with simple spectra, a single UV relaxation may be sufficient for the Hamaker constant calculation.[12] Thus, the simple oscillator model described by Eq. (6) is sufficient for determining the van der Waals energy. In this simplification, water presents an important exception: the permanent dipole contribution of water is only taken into account in the static dielectric constant but is ignored in calculating $\varepsilon(i\xi)$ at nonzero frequencies. The best fit with the available permittivity for water gives $B = 1.887$, which is slightly higher than the square of the refractive index of water. Parameters of Eq. (6) for some materials are given in Table 4.2. The simplified model given by Eq. (6) also allows simple equations for van der Waals energy to be developed.

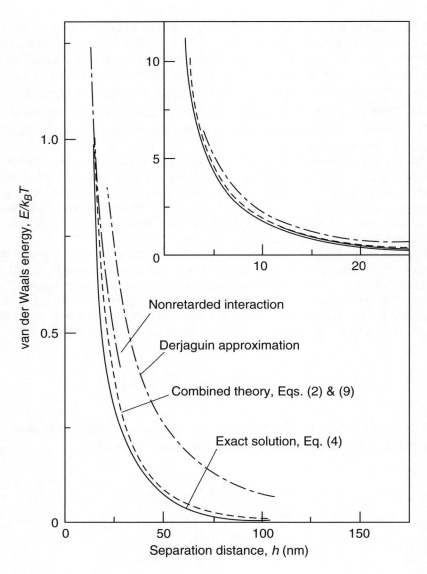

FIGURE 4.1
Comparison between approximate models and the exact solution for the van der Waals energy for two polystyrene spheres of radius $R = 250$ nm in water. (From Nguyen.[10])

Approximate and Simplified Equations for Van der Waals Interaction Between Spheres

A number of approximate models have been developed for the calculation of the van der Waals interaction energy between two spherical particles. The most well-known approximation was provided by Derjaguin[15] who made the approximate estimation of the interaction energy or the force between two spherical particles or curved surfaces from the corresponding interaction energy per unit area between two infinite parallel flat plates. In terms of force, $F(h)$, between two spherical particles with radii R_1 and R_2, and energy per unit area between parallel flat plates, $E^*(h)$, the Derjaguin approximation gives

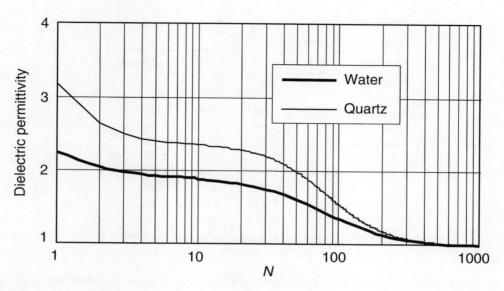

FIGURE 4.2

Dielectric permittivity $\varepsilon(i\xi_N)$ vs. the index N of the equally spaced real frequency ξ_N for water and crystalline quartz. (From Nguyen.[10])

$$F(h) = 2\pi \frac{R_1 R_2}{R_1 + R_2} E^*(h) \tag{7}$$

The interaction energy between the particles is then obtained by integrating Eq. (7) from infinity to the separation distance h. The Derjaguin approximation can be applied to van der Waals, double-layer, and many other interactions. For van der Waals interaction, either Eq. (1) or (3) for $E^*(h)$ can be substituted into Eq. (7) to determine the force between two spheres.

TABLE 4.1

Parameters for the dielectric function $\varepsilon(i\xi)$ [Eq. (5)] for water at 20°C

Region	d_m	τ_m (s/rad)	f_j (rad²/s²)	ω_j (rad/s)	g_j (rad/s)
Microwave	75.5	0.94×10^{11}			
Infrared			9.57×10^{11}	3.19×10^{13}	2.28×10^{13}
			5.31×10^{12}	1.05×10^{14}	5.77×10^{13}
			1.97×10^{12}	1.40×10^{14}	4.25×10^{13}
			8.66×10^{12}	3.04×10^{14}	3.80×10^{13}
			2.13×10^{13}	6.38×10^{14}	8.50×10^{13}
Ultraviolet			4.95×10^{15}	1.25×10^{16}	9.57×10^{14}
			5.88×10^{15}	1.52×10^{16}	1.28×10^{15}
			1.82×10^{16}	1.73×10^{16}	3.11×10^{15}
			9.66×10^{16}	2.07×10^{16}	5.92×10^{15}
			1.73×10^{17}	2.70×10^{16}	1.11×10^{16}
			3.69×10^{16}	3.83×10^{16}	8.11×10^{15}

(From Nguyen[10] and Parsegian and Weiss.[13])

TABLE 4.2

Parameters $\varepsilon(0)$, B, and ω in Eq. (6)

Materials	B	$\varepsilon(0)$	$\omega(10^{16}$ rad/s)
Water	1.887	80	2.068
Air	1	1	—
Crystalline quartz	2.359	4.29	2.032
Fused quartz	2.098	3.80	2.024
Fused silica	2.098	3.81	2.033
Calcite	2.516	8.20	1.897
Calcium fluoride	2.036	7.36	2.368
Sapphire	3.071	11.6	2.017
Polystyrene	2.447	2.56	1.393
Tetradecane	2.041	2.04	1.661

(From Nguyen,[10] Hough and White,[12] and Israelachvili.[14])

As long as the range of the interaction and the separation h is much less than the radius of curvature of the system, it is a valid approximation for interaction between surfaces quadratically curved in the vicinity of the point of closest approach. The condition makes curvature effects, higher than second order (via truncation of a Taylor series expansion), on the approximated energy significantly small.[16] However, the Derjaguin approximation should not be adequate for submicron-sized particles. Shown in Figure 4.1 is the deviation of the Derjaguin approximation from Langbein's full macroscopic prediction described by Eq. (4) for van der Waals interaction between 250 nm radii particles. There are two important improvements of the Derjaguin approximation for submicron-sized particles.

The first improvement is based on the discovery that Eq. (2) of the Hamaker microscopic theory for spheres agrees with Eq. (4) of the continuum macroscopic theory when the Hamaker constant, A, in Eq. (2) is determined from Eq. (3) of the Lifshitz theory for parallel flat plates (Figure 4.1). The combined Hamaker–Lifshitz function, $A(h)$, can be obtained by comparing the right-hand sides of Eqs. (1) and (3), giving

$$A_{132}(h) = -\frac{3k_B T}{2} \sum_{N=0}^{\infty}{}' \int_{x_N}^{\infty} x \ln\left[\left(1 - y_{13}y_{23}e^{-x}\right) \times \left(1 - z_{13}z_{23}e^{-x}\right)\right] dx \qquad (8)$$

Eq. (8) can be simplified to provide an approximate prediction suitable for engineering calculations. Using Eq. (6), the simplification of Eq. (8) gives

$$A_{132}(h) = A_{132}^0 + A_{132}^{\xi}(h) \qquad (9)$$

The first and second terms on the right-hand side of Eq. (9) represent the zero-frequency (separation-independent) and the nonzero-frequency (dispersion) parts described by:[17]

$$A_{132}^0 = \frac{3k_B T}{4} \sum_{m=1}^{\infty} \left\{ \frac{\varepsilon_1(0) - \varepsilon_3(0)}{\varepsilon_1(0) + \varepsilon_3(0)} \frac{\varepsilon_2(0) - \varepsilon_3(0)}{\varepsilon_2(0) + \varepsilon_3(0)} \right\}^m \bigg/ m^3 \qquad (10)$$

$$A_{132}^{\xi}(h) = \frac{3\hbar\omega}{8\sqrt{2}} \frac{(B_1 - B_3)(B_2 - B_3)}{(B_1 - B_2)} \times \left\{ \frac{I_2(h)}{\sqrt{B_2 + B_3}} - \frac{I_1(h)}{\sqrt{B_1 + B_3}} \right\} \tag{11}$$

Functions $I_1(h)$ and $I_2(h)$ in Eq. (11) describe the retardation effects and are defined by $I_j(h) = [1 + (h/\lambda_j)^q]^{-1/q}$, where $q = 1.185$. The characteristic wavelengths, λ_j, are measured in units of length by: $\lambda_j = \frac{c}{\pi^2\omega} \sqrt{\frac{2}{B_3(B_j + B_3)}}$, where c is the speed of light in vacuum. Unlike the London wavelength in the classical theories on retardation,[2] which is ~100 nm, λ_j is on the order of 5 nm, depending on the refractive index of the interacting materials. For two identical particles, Eq. (11) reduces to $A_{131}^{\xi}(h) = \frac{3\hbar\omega}{16\sqrt{2}} \frac{(B_1 - B_3)^2}{(B_1 + B_3)^{3/2}} \left\{ 1 + (h/\lambda)^q \right\}^{-1/q}$.

The van der Waals interaction energy in salt solutions is reduced.[4] In this case, Eq. (9) for the Hamaker–Lifshitz function is recast to give

$$A_{132}(\kappa, h) = A_{132}^0 (1 + 2\kappa h)e^{-2\kappa h} + A_{132}^{\xi}(h) \tag{12}$$

where κ is the Debye constant. The zero-frequency term of the Hamaker–Lifshitz function is strongly influenced by salts whereas the dispersion (nonzero frequency) term remains unaffected, as the electrolyte ions do not respond to frequencies higher than 10^{14} rad/s. Therefore, the van der Waals forces across an electrolyte solution will effectively be determined solely by the London dispersion interaction.

Electrostatic Double-Layer Interaction

Since particles in a polar solvent, like water, become electrically charged, electrostatic double-layer interaction (EDL) is important in determining particle–particle interactions. The Poisson–Boltzmann (PB) equation is used to describe the double-layer interaction. For $z : z$ valence salt solutions, the PB equation yields

$$\nabla^2 \left(\frac{ez\psi}{k_B T} \right) = \kappa^2 \sinh \left(\frac{ez\psi}{k_B T} \right) \tag{13}$$

where ψ is the electrostatic potential in the solution and e is the charge of an electron. The Debye constant, κ, is defined by: $\kappa = \left\{ \frac{2000cN_A e^2 z^2}{\varepsilon\varepsilon_0 k_B T} \right\}^{1/2}$, where c is the salt concentration in mole/L, N_A the Avogadro number, ε_0 the permittivity of the vacuum, and ε is the dielectric constant (the relative permittivity) of the salt solution. The EDL force acting on a particle is determined by integrating the stress tensor, $\mathbf{T}$, over the particle surface:[18]

$$\mathbf{F}_{edl} = \oint \mathbf{T} \cdot \mathrm{d}s \tag{14}$$

where ds is the normal vector of the surface element. **T** is determined by the tensor of the hydrostatic (osmotic) pressure and the electric stress tensor by:

$$\mathbf{T} = \left[\varepsilon\varepsilon_0 \kappa^2 \left(\frac{k_B T}{ez} \right)^2 \sinh^2 \left(\frac{ez_i \psi}{2k_B T} \right) + \frac{\varepsilon\varepsilon_0}{2} E^2 \right] \mathbf{I} - \varepsilon\varepsilon_0 \mathbf{EE} \tag{15}$$

where $\mathbf{E} = -\nabla\psi$ is the electric field, E the magnitude of $\mathbf{E}$, $\mathbf{I}$ the unit tensor, and $\mathbf{EE}$ describes the dyadic product of the two vectors $\mathbf{E}$ and $\mathbf{E}$. The integration of Eq. (14) gives the net force acting along the particle center-to-center line as by symmetry all components perpendicular to the centerline integrate to zero.

In Eq. (15), the electrostatic potential, ψ, is for the overlapping electric double layer of the interacting particles. Numerous models have been created to predict the overlapping field electrostatic potential between parallel plates. However, calculation of the EDL interaction for the common geometry of two spheres has not been satisfactorily resolved, due mainly to the nonlinear partial differential terms in Eq. (13) arising because of the three-dimensional geometry of the system. As a consequence, a number of approximate and numerical models have been developed for the calculation of the EDL interaction between two spheres. These models are briefly described below.

Approximate Models for EDL Interaction Between Two Spheres

The first approximate models are obtained by solving the linearized Eq. (13) for planar parallel surfaces and determining the EDL force and energy between spherical particles using the Derjaguin approximation. For low potential, the right-hand side of Eq. (13) can be linearized by $\sinh z = z + O(z^3)$. Eq. (13) reduces to a one-dimensional differential equation, which can be solved analytically using appropriate boundary conditions at the surfaces. The solution for ψ is then substituted into Eq. (15) to determine the pressure, which can be integrated to obtain the force and energy between spheres, using the Derjaguin approximation described by Eq. (7). The obtained models are summarized in Table 4.3.

The second group of EDL models are obtained by solving the nonlinear PB Eq. (13) for planar parallel surfaces using elliptic functions and integrals,[20–22] and determining the EDL force and energy between spherical particles using the Derjaguin approximation.[15,20,22] While the models given in Table 4.3 are only applied to low surface potentials, the models with elliptic functions and integrals in this second group are not limited by the condition of low surface potentials. The interaction force and energy obtained for spherical particles are only restricted by the applicability of the Derjaguin approximation, i.e., the particle size must be significantly larger than the separation distance between the surfaces. This condition is not satisfied by submicron- and nano-sized particles because the range of double-layer interaction in water often exceeds a few 100 nm.

The third group of the approximate models includes various improvements of the Derjaguin approximation, linearization, and approximate solutions of PB Eq. (13) for spherical particles. The first improvement on the Derjaguin approximation for the interaction energy between identical spheres was probably obtained by the Debye–Hückel linearization and the superposition approximation,[23] given by:

$$E(h) = 4\pi\varepsilon\varepsilon_0 R\psi_s^2 \frac{R+h}{2R+h} \times \ln\left[1 + \frac{R}{R+h} \exp(-\kappa h) \right] \tag{16}$$

TABLE 4.3

Results of Debye–Hückel linearization of Eq. (13) and Derjaguin approximation

Boundary Conditions	Force and Energy Between Spheres with Radii R_1 and R_2
Constant surface potentials: $\psi(0) = \psi_1$ and $\psi(h) = \psi_2$ Hogg–Healy–Fuerstenau (HHF) model[19]	$F^{\psi} = \varepsilon\varepsilon_0 \kappa \dfrac{2\pi R_1 R_2}{R_1 + R_2} \dfrac{2\psi_1\psi_2 \exp(\kappa h) - \psi_1^2 - \psi_2^2}{\exp(2\kappa h) - 1}$
	$E^{\psi} = \varepsilon\varepsilon_0 \dfrac{\pi R_1 R_2}{R_1 + R_2} \left[4\psi_1\psi_2 \, a\tanh\left(e^{-\kappa h}\right) + \left(\psi_1^2 + \psi_2^2\right)\ln\left(1 - e^{-2\kappa h}\right) \right]$
	ψ_1 and ψ_2 are the particle surface potentials
Constant surface charges: $\left(\dfrac{d\psi}{dx}\right)_{x=0} = -\kappa\psi_{1\infty}$ $\left(\dfrac{d\psi}{dx}\right)_{x=h} = \kappa\psi_{2\infty}$	$F^{\sigma} = \varepsilon\varepsilon_0 \kappa \dfrac{2\pi R_1 R_2}{R_1 + R_2} \dfrac{2\psi_{1\infty}\psi_{2\infty} \exp(\kappa h) + \psi_{1\infty}^2 \psi_{2\infty}^2}{\exp(2\kappa h) - 1}$
	$E^{\sigma} = \varepsilon\varepsilon_0 \dfrac{\pi R_1 R_2}{R_1 + R_2} \left[4\psi_{1\infty}\psi_{2\infty} \, a\tanh\left(e^{-\kappa h}\right) - \left(\psi_{1\infty}^2 + \psi_{2\infty}^2\right)\ln\left(1 - e^{-2\kappa h}\right) \right]$
	$\psi_{1\infty}$ and $\psi_{2\infty}$ are the surface potential at infinite separation distance (of isolated particles)
Constant surface potential and charge: $\psi(0) = \psi_1$ $\left(\dfrac{d\psi}{dx}\right)_{x=h} = \kappa\psi_{2\infty}$	$F^{\sigma\psi} = \varepsilon\varepsilon_0 \kappa \dfrac{2\pi R_1 R_2}{R_1 + R_2} \dfrac{2\psi_1\psi_{2\infty} \exp(\kappa h) + \psi_1^2 - \psi_2^2}{\exp(2\kappa h) + 1}$
	$E^{\sigma\psi} = \varepsilon\varepsilon_0 \dfrac{\pi R_1 R_2}{R_1 + R_2} \left[4\psi_1\psi_{2\infty} \, a\tanh\left(e^{-\kappa h}\right) + \left(\psi_1^2 - \psi_{2\infty}^2\right)\ln\left(1 + e^{-2\kappa h}\right) \right]$

For two spheres of different radii, R_1 and R_2, and at a constant surface charge interaction, the improved Derjaguin approximation in the framework of the Debye–Hückel linearization can be expressed as:[24]

$$E^{\psi}(h) = \varepsilon\varepsilon_0 \frac{\pi R_1 R_2}{R_1 + R_2 + h} \left[4\psi_1\psi_2 \, a\tanh\left(e^{-\kappa h}\right) + \left(\psi_1^2 + \psi_2^2\right)\ln\left(1 - e^{-2\kappa h}\right) \right] \tag{17}$$

Eq. (17) includes the separation distance in the sum of the particle radii. In the limit of large particles, namely, $R_1 + R_2 \gg h$, Eq. (17) reduces to the Hogg–Healy–Fuerstenau expression for $E^{\psi}(h)$, given in Table 4.3.

For the interaction between identical spheres at moderate surface potentials ($y_\infty \le 4$), the improved Derjaguin approximation can be described by:

$$E^{\psi}(h) = \varepsilon\varepsilon_0 \frac{64\pi R^2}{2R + h} \left(\frac{k_B T}{ez}\right)^2 \times a\tanh^2\left[e^{-\kappa h/2} \tanh\frac{y_\infty}{4} \right] e^{\kappa h} \ln\left(1 + e^{-\kappa h}\right) \tag{18}$$

Eq. (18) is claimed to be valid for all κh.[24] Note that in the limit of small surface potentials, Eq. (18) does not reduce exactly to Eq. (16), which was derived based on the superposition approximation typically used in the limit of large κh. However, if the particle radius is significantly larger than the range of the EDL interaction, Eqs. (16) and (18) in the limit of small potentials reduce to the expression:

$$E(h) = 2\pi\varepsilon\varepsilon_0 R\psi_s^2 \ln[1 + \exp(-\kappa h)], \tag{19}$$

which was first derived for two identical spheres at the constant surface potential interaction using the Derjaguin approximation and the Debye–Hückel linearization by Derjaguin in 1939.[20]

There are two other developments in predicting electrostatic double-layer interactions between spheres. Firstly, an approximate expression for the correction to the Derjaguin approximation has been using a series expansion and the Debye–Hückel linearization.[25] Comparison with the exact numerical results has found that the derived expressions for the EDL interaction at constant surface potential are in good agreement with the exact numerical results. However, for constant surface charge a poorer approximation is obtained at small separation distances. This is most probably due to the inaccuracy of the Debye–Hückel linearization under the condition of the constant surface charge interaction. The study highlights the ongoing difficulty in improving the Derjaguin approximation for the EDL interaction at constant surface charge for particles in the colloidal size range. Secondly, the Derjaguin approximation is actually the zero-order approximation in terms of curvatures of the particle surfaces. Improvements to this model include the effect of curvature. For instance, in the case of interaction at constant surface potential, the additional corrections to the known HHF expression are the second and higher order terms of $1/(\kappa R)$.[26] However, the corrections often contain many terms and series expansions, approaching the complication of direct numerical computations.

Exact Numerical Solutions for EDL Interaction Between Two Spheres

The direct numerical solutions of PB equation for spheres have been reported by a number of researchers, including.[27] In the numerical computation, PB Eq. (13) is conveniently expressed and solved in the bispherical coordinates. Due to the rotational symmetry of the interaction along the centerline, Eq. (13) simplifies into

$$\frac{\partial}{\partial \xi}\left(\frac{\sin \eta}{\cosh \xi - \cos \eta}\frac{\partial \phi}{\partial \xi}\right) + \frac{\partial}{\partial \eta}\left(\frac{\sin \eta}{\cosh \xi - \cos \eta}\frac{\partial \phi}{\partial \eta}\right) = \frac{c^2 \sin \eta}{(\cosh \xi - \cos \eta)^3}\sinh \phi \qquad (20)$$

where the reduced potential, ϕ, is defined by $\phi = ez\psi/k_B T$. The bipolar coordinates (η, ξ) are directly related with the reduced cylindrical coordinates (r, z), where the z-axis is the particle inter-center line. One obtains: $r = \dfrac{c\sin \eta}{\cosh \xi - \cos \eta}$ and $z = \dfrac{c\sinh \xi}{\cosh \xi - \cos \eta}$. The origin of the cylindrical coordinates on the centerline is determined by the inter-center distance, $d = h + R_1 + R_2$, between the spheres and the distances, d_1 and d_2, of the center of each sphere to the origin. We have $d_1 = \dfrac{d^2 + R_1^2 - R_2^2}{2d}$ and $d_2 = \dfrac{d^2 + R_2^2 - R_1^2}{2d}$, where R_1 and R_2 are the sphere radii, and $d_1 > 0$ and $d_2 < 0$. In the bipolar coordinates, the sphere surfaces are described by the constancy of the ξ-coordinate, i.e., by $\xi = \alpha_1$ for sphere 1 and $\xi = \alpha_2$ for sphere 2, where $\alpha_1 = a\cosh\dfrac{d^2 + R_1^2 - R_2^2}{2dR_1}$ and $\alpha_2 = -a\cosh\dfrac{d^2 + R_2^2 - R_1^2}{2dR_2}$. The length constant, c, in Eq. (20) is defined by $c = R_1 \sinh \alpha_1 = -R_2 \sinh \alpha_2$.

Eq. (20) can be discretized in the computational domain bounded by $\xi = \langle\alpha_2, \alpha_1\rangle$ and $\eta = \langle 0, \pi\rangle$. The discretization can be carried out using either the finite volume or finite difference scheme, producing a system of nonlinear equations for the reduced potential in the computational domain, which can be solved employing the Newton–Ralphson method or other relaxation techniques. The boundary conditions of constant surface potentials are described by $\phi = \phi_1$ at the surface of sphere 1 with $\xi = \alpha_1$ and $\eta = \langle 0, \pi \rangle$, and $\phi = \phi_2$ at the surface of sphere 2 with $\xi = \alpha_2$ and $\eta = \langle 0, \pi \rangle$. The boundary conditions of constant surface charge are described by $\left(\dfrac{\cosh \xi - \cos \eta}{c} \dfrac{\partial \phi}{\partial \xi} \right)_{\xi = \alpha_1, \alpha_2} = \pm \dfrac{\sigma_{1,2} z e}{\varepsilon \varepsilon_0 \kappa k_B T}$ at the surface of spheres 1 and 2 with the surface charges, σ_1 and σ_2, respectively.

The general behavior of the numerical results of Eq. (20) for the potential distribution between the particle surfaces is shown in Figure 4.3 for the constant surface potential interaction between two spherical particles with radii of 20 nm and surface potentials of 50 mV. The effect of the interaction is to distort the potential profiles around the particles. At a separation distance of about 10-particle radii (~200 nm), the potential profiles are almost identical to those around the isolated particles. However, the distortion of the potential profiles becomes stronger with decreasing the separation distance between the surfaces. The strong distortion of the potential profiles increases the repulsive interaction between the particles as shown below.

Once the potential field in the computational domain is available, the EDL force can be determined by integrating the stress over the sphere surface using Eqs. (14) and (15). In the bipolar coordinates, the normal vector of the surface element in

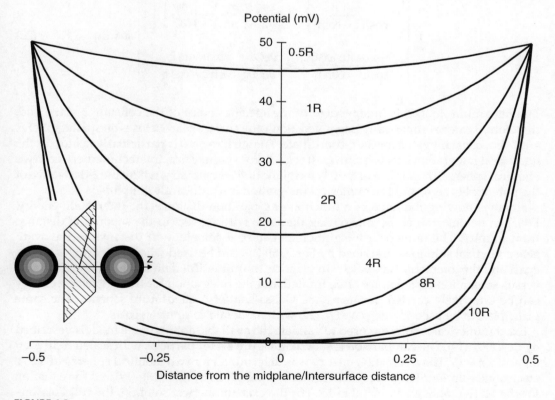

FIGURE 4.3
Numerical results of Eq. (13) for the potential distribution on the inter-center line of two identical spherical particles with radii $R = 20$ nm and surface potentials of 50 mV, showing the distribution at the different inter-surface distances: 0.5, 1, 2, 4, 8, and $10R$. The distance from the ($z = 0$) midplane is normalized by dividing by the inter-surface distance.

Eq. (14) yields $ds = -i_\xi 2\pi \left(\dfrac{c}{\cosh \xi - \cos \eta} \right)^2 \sin \eta \, d\eta$, where the unit vector, i_ξ, is pointed towards the surface of one of the spheres. The potential gradient is described by $E = -\dfrac{\cosh \xi - \cos \eta}{c} \left(\dfrac{\partial \psi}{\partial \eta} i_\eta + \dfrac{\partial \psi}{\partial \xi} i_\xi \right)$, where i_η is another unit vector of the η-coordinate. Combining Eqs. (14) and (15) gives

$$F = 2\pi \int_0^\pi \left[\left(\left\{ \frac{c}{\cosh \xi - \cos \eta} \right\}^2 \Pi + \frac{1}{2} \varepsilon \varepsilon_0 \left\{ \left(\frac{\partial \psi}{\partial \xi} \right)^2 - \left(\frac{\partial \psi}{\partial \eta} \right)^2 \right\} \right) i_\xi + \varepsilon \varepsilon_0 \frac{\partial \psi}{\partial \eta} \frac{\partial \psi}{\partial \xi} i_\eta \right] \sin \eta \, d\eta \quad (21)$$

where Π describes the first term (the osmotic pressure) in the brackets in Eq. (15). The net force acting on the inter-center line is obtained by multiplying Eq. (21) with the unit vector, i_z, of the z-axis. Since $i_\eta \cdot i_z = \dfrac{\sin \eta \sinh \xi}{\cosh \xi - \cos \eta}$ and $i_\xi \cdot i_z = \dfrac{\cos \eta \cosh \xi - 1}{\cosh \xi - \cos \eta}$, Eq. (21) gives

$$F = 2\pi \int_0^\pi \left[\begin{array}{l} \left(\left\{ \dfrac{c}{\cosh \xi - \cos \eta} \right\}^2 \Pi + \dfrac{\varepsilon \varepsilon_0}{2} \left\{ \left(\dfrac{\partial \psi}{\partial \eta} \right)^2 - \left(\dfrac{\partial \psi}{\partial \xi} \right)^2 \right\} \right) \\ \times \dfrac{1 - \cos \eta \cosh \xi}{\cos \xi - \cosh \eta} + \varepsilon \varepsilon_0 \dfrac{\partial \psi}{\partial \eta} \dfrac{\partial \psi}{\partial \xi} \dfrac{\sin \eta \sinh \xi}{\cosh \xi - \cos \eta} \end{array} \right] \sin \eta \, d\eta \quad (22)$$

The integral in Eq. (22) is independent of the specific choice of the constant ξ-plane and, therefore, it can be numerically integrated along any surface that encloses one of the spheres and has a constant value for the ξ-coordinate. The surface $\xi = 0$ is particularly useful for the numerical integration because $\sinh \xi = 0$ at $\xi = 0$. For symmetrical interactions between two identical spheres, the surface of $\xi = 0$ is the plane of the symmetry, where the ξ-derivative of the potential is zero, and the numerical integration is significantly simplified.

Finally, knowing the force as a function of separation distance, the interaction energy, $E(h)$, can be determined by integrating the force with respect to the separation distance from infinity. The numerical integration can start at some cut-off distance as the available analytical solutions described by Eqs. (16)–(19) can be used when the particles are far apart and the interaction is weak.[20] In practice, the numerical data for the interaction force versus separation distance are often limited and the integration for the interaction energy can be effectively carried out using the Gauss-Laguerre quadrature scheme. The same principle can be used to integrate Eq. (22) with moderate computing costs.

Exact numerical results are used to validate the available approximate models described by Eqs. (16)–(19). The comparison is shown in Figure 4.4 for particles with scaled radii $R\kappa = 0.1$ and $R\kappa = 15$. The interaction energy was determined for two identical spheres in a $z : z$ electrolyte solution. The approximate solutions are given by Eqs. (16)–(19) and the equation for the HHF model given in Table 4.3. For the exact numerical solution, the full Poisson–Boltzmann equation was discretized and solved by the finite volume method. The results have been plotted for two particle sizes $\kappa R = 0.1$ (Figure 4.4A) and $\kappa R = 15$ (Figure 4.4B).

It can be seen that for both scaled particle sizes the predictions for $G(h)$ from Eqs. (16)–(18) match closely the exact numerical solution. For the result of the Derjaguin Eq. (19) and

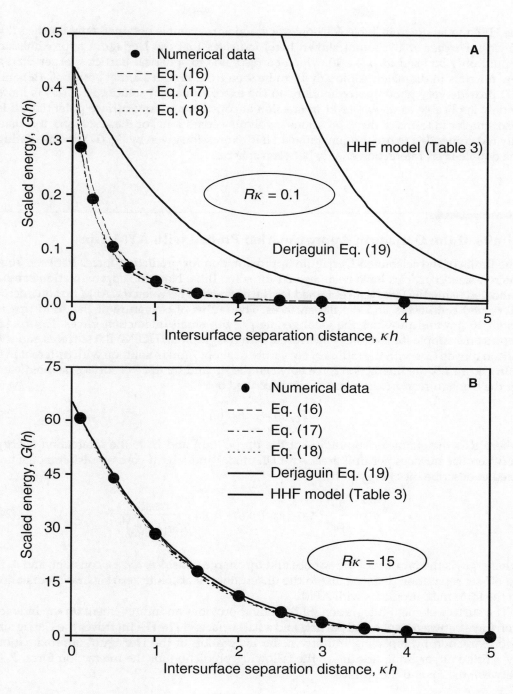

FIGURE 4.4

Comparison of the model improvements on the Derjaguin approximation to the exact numerical computational results of the full Poisson–Boltzmann equation for two spheres with the scaled radius $R\kappa = 0.1$ and $R\kappa = 15$ and constant surface potential $\psi_s ez/(k_B T) = 1$. The scaled energy, $G(h)$, on the vertical axis is defined by

$$G(h) = E(h) \Bigg/ \left\{ \frac{\varepsilon\varepsilon_0}{\kappa} \left(\frac{k_B T}{ez} \right)^2 \right\}.$$

the HFF model given in Table 4.3, however, good agreement is obtained only for the $\kappa R = 15$ case. Further analysis (not shown here) indicates that the Derjaguin approximation should only be used for $\kappa R \geq 10$, which corresponds to spherical particles larger than a few microns in diameter. Although it can be seen from Figure 4.4 that both Eqs. (16) and (17) provide very good approximations to the exact numerical solution, Eq. (17) is more widely applicable to double-layer interaction at constant and low surface potentials. It is also simpler in terms of the expressions involving separation. For these reasons, it is usually recommended to use the generalized HHF expression given by Eq. (17) for calculating the double-layer interaction energy between spheres.

Limits of the Derjaguin Approximation Probed with AFM Tips

The limits of the celebrated Derjaguin approximation for predicting forces between submicron-sized particles have been argued for some time. Now the approximation can be validated using the force data obtained for the interaction between the AFM tips on microfabricated cantilevers and the flat surfaces. The radius of curvature of the AFM tips is about 10 nm and provides the ideal geometry with small interaction forces. Figure 4.5 shows an example for the forces measured with the graphite (HOPG) flat surfaces and the silicon nitride tips with the radius of curvature of about 7 nm in solution with different pH.

In Figure 4.5, the interaction force between the tip and the HOPG surface, as predicted by the Derjaguin approximation, F_{DA}, is described by:[28]

$$F_{DA}(h) = 2\pi R E^*(h) \tag{23}$$

where R is the characteristic radius of the tip (~7 nm) and E^* is the interaction energy between flat surfaces per unit area. An EDL model and a hard-core repulsion model are used to describe the energy E^* by:[28]

$$E^*(h) = \frac{2\pi}{\varepsilon\varepsilon_0\kappa}(\sigma_1\sigma_2)\exp[-\kappa(h-h_0)] + \frac{const}{72\pi(h-h_0)^8} \tag{24}$$

where $\sigma_1\sigma_2$ is the product of the surface and tip charge densities, *const* a constant, and h_0 is an off-set separation distance due to the uncertainty of absolute zero separation distance in the force measurements with AFM.

The surface element integration (SEI) method provides an improvement on the interaction force between a spherical particle and a flat surface.[29] The SEI improves the Derjaguin approximation by replacing infinity in the integration of the Derjaguin approximation by a finite upper limit, leading to the following prediction for the interaction force, F_{SEI}, between the tip and the surface:

$$F_{SEI}(h) = 2\pi\int_0^R \frac{\partial}{\partial h}\left\{E^*(D_+) - E^*(D_-)\right\}r\,dr \tag{25}$$

where the energy E^* is given by Eq. (24) and $D_\pm = h + R \pm \sqrt{R^2 - r^2}$. The integration can be calculated numerically.

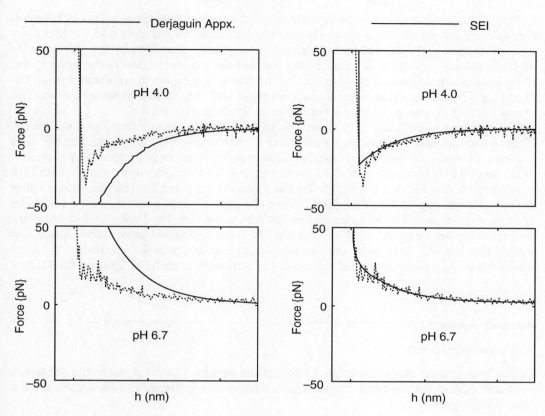

FIGURE 4.5
Experimental data (dotted lines), and the Derjaguin approximation (solid lines on the left diagrams) and its SEI
improved prediction (solid lines on the right diagrams) for the interaction force vs. separation distance between
a silicon nitride AFM tip and an HOPG surface in 1 mM NaCl solutions ($1/\kappa$ = 9.6 nm). (From Todd and Eppell.[28])

The solid lines in Figure 4.5 show Eqs. (23) and (25) with the model parameters obtained
by regression analysis. Despite the best-fit procedure employed, the Derjaguin approximation shows significant deviation from the experimental data at short separation distances.
The Derjaguin approximation cannot model the interaction under the given conditions
when the radius of the surface curvature is similar to the range of the double-layer forces
($1/\kappa$ = 9.6 nm). In this case, the upper limit of the integration of the Derjaguin approximation cannot approach infinity and the SEI approximation, which considers the finite limit
of the integration, agrees with the experimental data.

Conclusions

Improvements in the DLVO force predictions for the van der Waals and electrostatic double-
layer interactions between small spherical particles have been critically reviewed. The
van der Waals interaction energy between two spheres can be determined from the exact

Langbein theory. The approximate equations for van der Waals interaction energy between two spheres can be obtained by combining the Hamaker microscopic and Lifshitz macroscopic theories. The Hamaker–Lifshitz function, $A_{132}(h)$, described by Eq. (12) depends on the separation distance (accounting for the retardation effects due to the finite speed of light) and salt concentration (accounting for the screening effects of electrolytes). The Hamaker–Lifshitz function is used in conjunction with Eq. (2) to determine the van der Waals interaction energy and force between small spherical particles.

It was shown in the analysis that the Poisson–Boltzmann Eq. (13) developed within the Gouy–Chapman theory provides a useful tool for predicting the EDL interaction between particles. Comparison with the numerical solution of the Poisson–Boltzmann equations and experimental data obtained with AFM tips shows that the Derjaguin approximation will fail to predict the double-layer interaction between nano-sized particles. The predictions using the approximate expressions described by Eqs. (16)–(18) for small particles were found to be an improvement over the Derjaguin approximation, when compared with the exact numerical data obtained by numerically solving the full Poisson–Boltzmann equation. It was also shown that the SEI technique provides significant improvement for predicting the force between a surface with nano-sized radius of curvature such as an AFM tip and a flat surface.

Acknowledgments

The authors gratefully acknowledge the Australian Research Council, the National Science Foundation, and the Research Project Committee of the University of Newcastle for the financial support.

References

1. Hamaker, H.C. The London-van der Waals attraction between spherical particles. Physica. **1937**, *4*, 1058–1072.
2. Casimir, H.B.G., Polder, D. The influence of retardation on the London-van der Waals forces. Physical Review **1948**, *73*, 360–372; Overbeek, J.T.G. The interaction between colloidal particles. In *Colloid Science*. Kruyt, H.R., Ed. Elsevier: Amsterdam, 1952, 266–270; Schenkel, J.H., Kitchener, J.A. A test of the Derjaguin-Verwey-Overbeek theory with a colloidal suspension. Trans. Faraday Soc. **1960**, *56*, 161–173.
3. Lifshitz, E.M. The theory of molecular attractive forces between solid bodies. J. Exp. Theor. Phys. USSR **1955**, *29*, 83–94.
4. Mahanty, J.H., Ninham, B.W. *Colloid Science: Dispersion Forces*. Academic Press: London, 1977.
5. Dzaloshinskii, I.E., Lifshitz, E.M., Pitaerskii, L.P. The general theory of van der Waals forces. Adv. Phys. **1961**, *10*, 165–209.
6. Langbein, D. *van der Waals Attraction*. Springer Verlag: Berlin, 1974.
7. Love, J.D. On the van der Waals force between two spheres or a sphere and a wall. J. Chem. Soc. Faraday Trans. 2 **1977**, *73* (5), 669–688.
8. Kiefer, J.E., Parsegian, V.A., Weiss, G.H. Some convenient bounds and approximations for the many body van der Waals attraction between two spheres. J. Colloid Interface Sci. **1978**, *67* (1), 140–153.

9. Pailthorpe, B.A., Russel, W.B. The retarded van der Waals interaction between spheres. J. Colloid Interface Sci. **1982**, *89* (2), 563–566.

10. Nguyen, A.V. Improved approximation of water dielectric permittivity for calculation of Hamaker constants. J. Colloid Interface Sci. **2000**, *229* (2), 648–651.

11. Nguyen, A.V., Schulze, H.J. *Colloidal Science of Flotation*. Marcel Dekker: New York, 2004.

12. Hough, D.B., White, L.R. The calculation of Hamaker constants from Lifshitz theory with applications to wetting phenomena. Adv. Colloid Interface Sci. **1980**, *14* (1), 3–41.

13. Parsegian, V.A., Weiss, G.H. Spectroscopic parameters for computation of van der Waals forces. J. Colloid Interface Sci. **1981**, *81*, 285–289; Roth, C.M., Lenhoff, A.M. Improved parametric representation of water dielectric data for Lifshitz theory calculations. J. Colloid Interface Sci. **1996**, *179* (27), 637–639.

14. Israelachvili, J.N. *Intermolecular and Surface Forces*. Academic Press: London, 1992.

15. Derjaguin, B. Untersuchungen ueber die Reibung und Adhaesion IV. Theorie des Anhaften kleiner Teilchen. Kolloid-Z. **1934**, *69*, 155–166.

16. White, L.R. On the Deryaguin approximation for the interaction of macrobodies. J. Colloid Interface Sci. **1983**, *95* (1), 286–288.

17. Nguyen, A.V., Evans, G.M., Schulze, H.J. Prediction of van der Waals interaction in bubble-particle attachment in flotation. Int. J. Miner. Process. **2001**, *61* (3), 155–169.

18. Derjaguin, B.V. On the repulsive forces between charged colloid particles and on the theory of slow coagulation and stability of lyophobic sols. Trans. Faraday Soc. **1940**, *36*, 203–215; Bell, G.M., Levine, S., McCartney, L.N. Approximate methods of determining the double-layer free energy of interaction between two charged colloidal spheres. J. Colloid Interface Sci. **1970**, *33* (3), 333–360; Russel, W.B., Saville, D.A., Schowalter, W.R. *Colloidal Dispersions*. Cambridge University Press: Cambridge, 1989.

19. Hogg, R., Healy, T.W., Fuerstenau, D.W. Mutual coagulation of colloidal dispersions. Trans. Faraday Soc. **1966**, *62*, 1638–1651.

20. Derjaguin, B.V. *Theory of Stability of Colloids and Thin Films*. Plenum: New York, 1989.

21. Chan, D.Y.C. A simple algorithm for calculating electrical double layer interactions in asymmetric electrolytes. Poisson–Boltzmann theory. J. Colloid Interface Sci. **2002**, *245* (2), 307–310; Devereeux, O.F., deBruyn, P.L. *Interaction of Plane Parallel Double Layers*. MIT Press: Cambridge, 1963.

22. Verwey, E.J.W., Overbeek, J.T.G. *Theory of the Stability of Lyophobic Colloids*. Elsevier: Amsterdam, 1948.

23. McCartney, L.N., Levine, S. An improvement on Derjaguin's expression at small potentials for the double layer interaction energy of two spherical colloidal particles. J. Colloid Interface Sci. **1969**, *30*, 345–362.

24. Sader, J.E., Carnie, S.L., Chan, D.Y.C. Accurate analytic formulas for the double-layer interaction between spheres. J. Colloid Interface Sci. **1995**, *171* (1), 46–54.

25. Glendinning, A.B., Russel, W.B. The electrostatic repulsion between charged spheres from exact solutions to the linearized Poisson–Boltzmann equation. J. Colloid Interface Sci. **1983**, *93*, 95–111; Carnie, S.L., Chan, D.Y.C. Interaction free energy between identical spherical colloidal particles: the linearized Poisson–Boltzmann theory. J. Colloid Interface Sci. **1993**, *155* (2), 297–312.

26. Chan, D.Y.C., White, L.R. The electrostatic interaction between spherical colloidal particles–a comment on the paper by Barouch et al. J. Colloid Interface Sci. **1980**, *74* (1), 303–305; Ohshima, H., Chan, D.Y.C., Healy, T.W., White, L.R. Improvement on the Hogg–Healy–Fuerstenau formulas for the interaction of dissimilar double layers. II. Curvature correction to the formula for the interaction of spheres. J. Colloid Interface Sci. **1983**, *92* (1), 232–242.

27. Chan, B.K.C., Chan, D.Y.C. Electrical double-layer interaction between spherical colloidal particles: an exact solution. J. Colloid Interface Sci. **1983**, *92* (1), 281–283; Palkar, S.A., Lenhoff, A.M. Energetic and entropic contributions to the interaction of unequal spherical double layers. J. Colloid Interface Sci. **1994**, *165* (1), 177–194; Qian, Y., Bowen, W.R. Accuracy assessment of numerical solutions of the nonlinear Poisson–Boltzmann equation for charged colloidal particles. J. Colloid Interface Sci. **1998**, *201* (1), 7–12; Carnie, S.L, Chan, D.Y.C., Stankovich, J.

Computation of forces between spherical colloidal particles: Nonlinear Poisson–Boltzmann theory. J. Colloid Interface Sci. **1994**, *165* (1), 116–128; Stankovich, J., Carnie, S.L. Electrical double layer interaction between dissimilar spherical colloidal particles and between a sphere and a plate: nonlinear Poisson–Boltzmann theory. Langmuir **1996**, *12* (6), 1453–61.

28. Todd, B.A., Eppell, S.J. Probing the limits of the Derjaguin approximation with scanning force microscopy. Langmuir **2004**, *20*, 4892–4897.

29. Bhattacharjee, S., Elimelech, M. Surface element integration: a novel technique for evaluation of DLVO interaction between a particle and a flat plate. J. Colloid Interface Sci. **1997**, *193* (2), 273–285.

Section II

Nanomaterials

5

Nanomaterials

David S. J. Arney, Jimmie R. Baran, Allen R. Siedle, and Matthew H. Frey

CONTENTS

Introduction 53
Nanoparticles in Surface and Interfacial Phenomena 53
Gold Nanoparticle Catalysts 57
    Catalyst Syntheses and Characterization 58
        Chemical Methods 58
        Sputtering Methods 59
        Basic Studies 59
        Hydrogenation 60
        Silane Chemistry 60
        Other Applications 61
Nanophase Glass-Ceramics 61
Conclusions 63
References 63

Introduction

Industrial interest in nanomaterials derives from the novel properties they exhibit. These are defined for this chapter as materials having engineered discrete particulate domains with diameters in the range of 1 nm to a few hundred nanometers. These domains may appear in many forms, such as dispersions of nanoparticles in a liquid, on surfaces, or embedded in a continuous matrix. The unique properties of nanomaterials are a consequence of the small size and extremely large interfacial areas. In this regime, dramatic variations in the chemical and physical properties of a material may be effected. Representative examples of size-critical properties, enabling new industrial applications, reviewed in this chapter include surface and interfacial, catalytic, optical, and mechanical.

Nanoparticles in Surface and Interfacial Phenomena

Small particles have been known for years to stabilize dispersions, foams, and emulsions.[1] Two prominent theories have been put forth to explain this phenomenon. One is based on the "surface activity" of the particles, while the other is based on viscosity control of the continuous phase.

The first theory claims that the particles are "surface active but, unlike surfactants, are not amphiphilic."[2] Partially modified particles preferentially partition the interface between the two phases, with the key physical characteristic for stabilization being the contact angle of the individual particle with the interface. Particles having a 90° contact angle (approximately equal amounts of hydrophobic and hydrophilic character) with the interface will create the most stable emulsions. The particles used in these studies are nanometer sized (~10–30 nm), but this is the primary particle size. The particles are actually agglomerated and are typically several hundred nanometers in size. This distinction is important for Eq. (1), which is used to support the proposed mechanism:

$$E = \pi r^2 \gamma_{\alpha,\beta} (1 \pm \cos\theta)^2 \tag{1}$$

where E is the energy required to remove a particle from the interface between the two liquid phases, r the radius (in meters) of the particle, $\gamma_{\alpha,\beta}$ the interfacial tension between the two phases, and θ the contact angle the particle makes with the interface.

This implies that the larger the particle (up to where gravity becomes a factor), the more stable the emulsion. It is also implicit that a particle with a too high (or too low) contact angle or of too small size should not be held at the interface and would be located in the bulk. This theory concludes that particles bound to the interface will stabilize systems, while those that are not bound to the interface cannot be used for stabilization.

Proponents readily admit that there is no significant reduction in surface tension when the particles are incorporated into the fluids. No arguments are proposed as to how the particles are bound to the surface. Their presence cannot be detected by surface tension measurements.

There is little applicability of this mechanism to stabilization by small particles. For instance, using the values exemplified earlier, the energy required to remove a particle with a diameter of 200 nm (approximate actual size of the particles in the above study) and a contact angle of 150° from a water/toluene interface (interfacial tension = 0.036 N/m) is 4927 kT, while a 5 nm particle in the same system has a binding energy of 3 kT. Therefore, a 200 nm particle will be "irreversibly" bound to the interface, while a 5 nm particle should not be held at the interface and if stabilization occurs, it must take place by a different mechanism.

An alternative theory based on results using unmodified, nanometer-sized, individual silica particles in an aqueous system was proposed almost 15 years ago.[3] Using reflected light microinterferometry, it was determined that the nanoparticles formed ordered structures inside the thin liquid films located between the dispersed phase regimes. This mechanism relies on purely physical effects for stabilization, not on thermodynamic/surface tension phenomena. To briefly summarize this work, the diagram in Figure 5.1 will be useful.

The large circles represent a dispersed phase (bubbles, droplets, or solids), while the small ones represent surface modified nanoparticles (SMNs) dispersed in the continuous phase that can be found between dispersed phase domains. The diagram on the left illustrates that the SMNs must be dispersed in the continuous phase. If they are not well dispersed or are agglomerated, they will not work as efficiently.

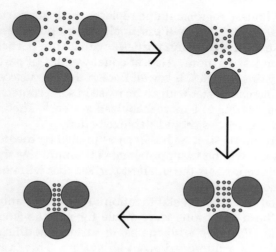

FIGURE 5.1
Proposed stabilization mechanism. (*View this art in color at www.dekker.com.*)

In the next diagram, the thin liquid film located between the dispersed phase regimes begins to drain and the discontinuous phase regimes begin to approach each other. When this happens, the SMNs are forced into the restricted volume of the thinning film, resulting in an ordered structure. The SMNs begin acting as steric barriers to coalescence or flocculation. As this ordering takes place, a localized concentration of the SMNs occurs—the liquid drains faster than the SMNs. This results in a very concentrated dispersion in the localized environment between domains. As with any concentrated dispersion, the viscosity in this region (only) increases. With the viscosity increasing in the region between domains, drainage is slowed and the system is further stabilized.

In the third diagram, the local concentration increases and the SMNs actually create well-ordered layers between the dispersed phase regimes while the liquid continues to drain. The SMNs cannot get out of the way fast enough, which is more commonly known as "depletion flocculation."[4]

Depletion flocculation occurs when two particles approach each other to within a distance that is smaller than the particle size so that no other particle can fit into the space between them. The osmotic pressures between the particles and in the rest of the dispersion are not balanced, and this pressure difference pushes the two particles toward each other. This leads to a further increase in viscosity and stability.

Finally, in the last diagram, drainage cannot be completely halted and layers of SMNs will eventually drain away. This stepwise draining is attributed to the existence of an osmotic pressure gradient—SMNs leave the film and "vacancies" appear in their place. The higher chemical potential of the SMNs causes the SMNs to slowly diffuse to the meniscus, leaving behind vacancies. The vacancies reach an equilibrium concentration that can only be increased by the further removal of SMNs from the film. This is known as the diffusive osmotic model.[5]

These diagrams demonstrate that the SMNs should be well dispersed in the continuous phase, monodisperse, and small in size. This can be visualized if one assumes that the SMNs in the diagrams are 5 nm particles. Thus, there would be four layers of particles between the dispersed phase regions. If one replaces the 5 nm particles with 20 nm particles, then a single 20 nm particle will essentially take up the same space (linearly) as four 5 nm particles. Replacing the 5 nm particles with 20 nm particles makes the path to coalesce or flocculation less tortuous. Also, at equal weights of particles, there are fewer 20 nm particles than 5 nm particles. It is well known that the viscosity of a dispersion of smaller particles will increase faster with an increase in solid content than that of a dispersion of larger particles, leading to a more stabilized system.[6] Therefore, with everything else being equal, smaller particles should stabilize better.

There are limitations on the size. As has been estimated by theoretical calculations, for spherical particles, ~4 nm will be the approximate minimum size that should stabilize an emulsion or foam. This represents the smallest particle that will not be enveloped by the amplitude of surface corrugations.[3]

Work was conducted in emulsion stabilization using only nanoparticles. It was surprising to find that these emulsions were stable for months (sometimes years) without high shear mixing. Particles with the same surface modification would stabilize a water-in-oil system and a fluorocarbon-in-oil system. This is nearly unheard of in emulsion science. Typically for a surfactant to be useful, it must be located at the interface between two phases, meaning that it must have a limited amount of solubility in both phases. Because water and fluorochemicals are vastly different in their chemical potentials, it is unlikely that a single surfactant would stabilize both types of systems. Figure 5.2 shows optical micrographs of these multiple emulsions. On the left is one of the more interesting "micelle" shapes. On the right is the actual breaking of one of the emulsion droplets.

This led to the inclusion of two types of particles into the emulsions and the formation of multiple emulsions. For example, hydrophobically modified particles were dispersed in toluene, while hydrophilically modified particles were dispersed in water. Emulsification of this system produced water-in-toluene-in-water or toluene-in-water-in-toluene multiple emulsions. Formulations of one type over the other were achieved by changing the water/toluene ratio or hydrophobic/hydrophilic particle ratio. Fluorocarbon-in-toluene-in-water multiple emulsions were made by adding a fluorochemical fluid. Figure 5.3 is an optical micrograph of a multiple emulsion containing Fluorinert® FC75 in toluene in water.

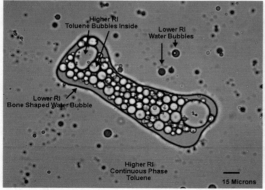 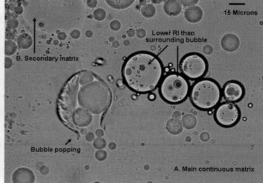

FIGURE 5.2
Optical micrographs of multiple emulsions stabilized with nanoparticles.

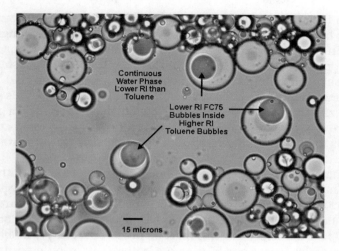

FIGURE 5.3
An optical micrograph of fluorocarbon-in-toluene-in-water emulsion stabilized with nanoparticles.

These multiple emulsions could be made by simple mixing, without any special order of addition of the ingredients. In fact, all three liquid components (the particles were previously dispersed in the respective phases) in the FC-in-toluene-in-water system were combined and emulsified simply by mixing. It is well understood that multiple emulsions are extremely process dependent and are nearly a black art. The ability to create these emulsions with minimal processing is a vast improvement over the current art.[7]

Dispersion stabilization with nanoparticles is also known. A recent example of a dispersion stabilized by nanoparticles was published by Tohver et al.[8] This group used zirconia particles to stabilize an aqueous colloidal system of larger silica particles. The dispersion was stabilized by electrostatic stabilization and thus is essentially applicable only to aqueous systems. Surface modification of the particles changes the stabilization mechanism to steric stabilization, and dispersions in both aqueous and nonaqueous systems have been demonstrated.

Foam, emulsion, and dispersion stabilization can be accomplished with organic molecules. With the proper chemistry, organic molecules also exhibit stabilization capabilities. Because many of these molecules are hydrophobic to begin with, further hydrophobic modification only makes them more compatible with the solvent, resulting in very few, if any, lyophobic/lyophilic regions that would cause them to partition the interface.

Gold Nanoparticle Catalysts

The science of metal nanoparticles has had a significant impact on commerce and industry, most prominently through the chemical transformations of hydrocarbons that lead to fuels and petrochemicals. Additionally, the physics of small metal particles has been fascinating for a long time. A problem considered in latter times has centered on quantum size effects: how electronic, and perhaps chemical, properties change as the particle size is reduced in three (nanoclusters), two (nanowires), or one (rafts or islands) dimensions. Gold nanoparticles are singled out in this chapter because clusters and particles formed from main group and d-block elements comprise too vast a field. The majority of work on gold nanoparticles has been done in the past two decades. This smaller subfield replicates the puzzles and surprises presented by metal nanoparticles in general. Because it is so new, there are many technological opportunities to be explored and many fundamental questions remain to be answered. Gold nanoparticles are at a frontier in catalytic chemistry today.

The surface chemistry of bulk gold is limited in scope compared with its lighter congeners copper and silver, and to elements of the transition series. Oxygen absorption is strong but occurs rapidly only at >500°C.[9,10] Chemisorptive properties are weak although of some importance in electrochemistry. Gold has long been known to catalyze the oxidation of CO to CO_2; the discovery that this reaction proceeds rapidly at ambient temperature on small gold particles was a breakthrough. Catalysis by or on gold nanoparticles has since developed rapidly, and this[11] as well as CO oxidation[12] has been recently reviewed. Broad topics discussed here include a) preparation and characterization of gold nanoparticle catalysts; b) the nature of the active site(s) and the mechanism of CO oxidation; and c) other processes catalyzed by gold nanoparticles.

Catalyst Syntheses and Characterization

Gold catalysts are made through two general methods that differ in terms of the principles involved, the materials produced, and the level of their understanding that has been achieved. In both, a solid support is used to stabilize the particles against sintering and to provide them in a form that is convenient to use.

Chemical Methods

The first method involves chemical reduction of Au(III), usually as the hydrated H_3O^+ salt of $AuCl_4^-$. A preformed support may be impregnated with this "$HAuCl_4$" and then treated with a reducing agent. Alternatively, when the pH of a mixture of $HAuCl_4$ and various metal ions is raised, gold hydroxide and metal hydroxide coprecipitate. Drying, calcination, and reduction of this mixture provide Au(0) supported upon a metal oxide. Suitable reducing agents include H_2, BH_4^-, citrate, and ascorbate. Permutations of support materials and reaction conditions are myriad. Some systematic studies are available,[13–18] but a real understanding of the role of processing variables has not been achieved. Supports can be carbon, metal oxides, or mesoporous oxides such as zeolites. Oxides can be categorized as reducible (TiO_2, Fe_2O_3, CeO_2) or not (ZrO_2, SiO_2, Al_2O_3).

Chemical vapor deposition (CVD) has been used to prepare nanoparticle catalysts, although the technology is quite expensive. Dimethylgold acetylacetonate absorbs on supports such as MgO. Decomposition at >100°C produces an efficient CO oxidation catalyst. X-ray absorption near edge (XANES) and extended X-ray absorption fine structure spectroscopies (EXAFS) showed that, under steady-state CO oxidation conditions, the catalyst contained Au(0) in the form of (on average) Au_6 clusters and also additional gold as Au(I).[19–21]

Preformed gold clusters, obtained by the reduction of $(Ph_3P)AuCl$ followed by treatment with O_2 to remove capping ligands and then with H_2, have been used as catalysts.[22]

Usually, these catalysts are characterized in terms of their efficiency in some test reaction, usually CO oxidation, under arbitrary conditions (O_2 and CO concentrations, humidity, flow rates, and catalyst time on stream) that would be difficult to reproduce in another laboratory. Characterization on any length scale, from molecular to morphological, is extremely difficult. In many cases, it is uncertain whether all the reactants and by-products have been removed from the catalyst surface. It is reported that chlorides poison the active sites.[23] This raises the question of a role for gold ions, and these, rather than Au(0), are reported to be the active sites in the water-gas shift reaction.[24]

Transmission electron microscopy (TEM) and scanning tunneling microscopy (STM) have been used to visualize small particles.[25–27] Catalysts can contain large amounts of large particles and crystalline bulk gold. One study of Au/Al_2O_3, prepared by sputtering, found that some gold was unaccounted for in the mass balance and in size regime below TEM resolution limits (~1 nm).[28] It is not clear whether all the catalytically active gold has (or can be) viewed by microscopy. Gold particles < 2 nm in size have structureless absorption, and larger ones show a surface plasmon peak in the visible.[29,30] Colors of gold catalysts thus can provide a crude, qualitative measure of gold aggregation. On MgO, they vary from light blue to blue, purple, and then pink with increasing particle size.[11] Individual characterization techniques have different limitations. For example, in x-ray photoelectron spectroscopy (ESCA) and TEM, a sample prepared (and used) under ambient conditions is studied under high vacuum conditions. In ESCA, photoelectrons can cause spurious reduction of cationic gold.

Incisive characterization of gold nanoparticle catalysts, however prepared, faces formidable obstacles. The root cause is this: virtually all experimental probes average over or "see" all (or almost all) of the gold in the sample. But it is not clear whether all the gold is responsible for all the catalytic activity. The effect of a small fraction of the total gold can be greatly amplified in catalysis. The same problem exists with particles and clusters of other metals. However, for gold, the effect on activity of particle size is extremely large, going from high at < 10 nm diameter to zero for the bulk-like metal. In contrast, Ru and Pt are active under all degrees of dispersion (allowing for scaling for surface area).

Sputtering Methods

This process is conducted under vacuum and deposition is line-of-sight. It is expensive and best suited to planar surfaces having relatively small areas. Impediments to commercial applications are aids to fundamental research. Catalysts can be prepared, studied, and tested under UHV conditions and in the absence of chemical contaminants. Most of our detailed understanding of gold nanoparticle catalysts comes from work on sputter-coated model catalysts.

Basic Studies

It is difficult to deduce what gold particle morphologies arise from heterogeneous chemical reduction of $HAuCl_4$. Understanding of the model catalysts is much easier. In brief, a) nucleation of gold clusters occurs at surface defects that act as traps; b) on Al_2O_3, there are two kinds of traps at < 0.8 and >1.6 eV; c) the defect density is ca. 3×10^{12} sites per cm^2 (10^{-3} monolayer); and d) when the clusters grow to >600 atoms, they leave the traps. This can explain the bimodal size distribution of the clusters.[31] Atomistic definition of these traps is needed.

Fundamental studies of metal nanoparticle growth on oxide surfaces indicate that, on a larger length scale, particles tend to collect at dislocations and grain boundaries,[32–34] and mobility on rough surfaces is lower than on smooth ones.[35] There is a significant size dependence,[36] and oxygen can enhance lateral diffusion.[37] Many theoretical studies have explored stability–size–geometry relationships in small gold clusters.[38–40] There are "magic numbers": Au_{20} in T_d form is particularly stable.[41] Metal–support interactions (which need to be explored further) and relativistic effects[42] tend to stabilize flat 2-D structures over polyhedral 3-D ones.

Ionized gold clusters generated in the gas phase can be mass sorted, deposited with low kinetic energy onto oxide surfaces, and then charge neutralized. For Au_n ($n > 2$) on MgO, Au_8 was found to be the smallest cluster that catalyzed CO oxidation and then only on defect-rich surfaces. Adsorption is accompanied by partial electron transfer into the cluster from surface defects (possibly oxygen vacancies). This may play a key role in enhancing catalytic activity. On Au_8, CO is a weakly bound η^1 ligand.[43,44] Oxidation by small gold clusters has been studied experimentally[45] and theoretically.[46,47] Interestingly, in the gas phase, only Au_n^- ($n \le 20$, $\neq 16$ and even) reacts with O_2 to form $Au_n(O_2)^-$.

Scanning tunneling microscopy studies of gold nanoparticles on TiO_2 suggested that clusters may have unusual catalytic properties as one cluster dimension becomes smaller than three atomic spacings. Two-atom thick clusters show maximum activity for catalytic CO oxidation at diameters 2.5–3.0 nm. A metal-to-nonmetal transition occurs as cluster size decreases below 3.5 nm in diameter and 1 nm in height (ca. 300 atoms).[48,49] These results raise fundamental questions about the electronic nature of gold nanoparticles, particularly when they are small and thin. Also, O_2 is reported to bind more strongly to ultrathin Au islands than to thick ones.[50] Two-dimensional Au islands are thermodynamically less stable than three-dimensional aggregates and sintering occurs on annealing or in the presence of O_2.

A different view of CO oxidation comes from computational studies of extended Au(1 1 1). Steps and kinks play a critical role in reducing the activation energy for O_2 dissociation.[51] The enhanced activity of thin islands may be related to step density (geometric effects) rather than to quantum size effects.[52] The literature suggests a key role for such defects. Gold is an effective catalyst because it can bind CO and O_2 but weakly enough so that subsequent processes have achievable activation barriers.

Hydrogenation

Use of gold nanoparticles as hydrogenation catalysts is a recent development.[53] In hydrogenation of acrolein using Au/ZnO, edge sites effected reduction of carbonyl groups, whereas reduction of the C = C bonds occurred on facial sites.[54] Thiophene doping of supported gold increased the rate of carbonyl hydrogenation of crotonaldehyde,[55] a remarkable example of promotion of gold by sulfur.

Silane Chemistry

Alkylsilanes, $RSiH_3$, are absorbed on gold with loss of hydrogen to form monolayers of RSi $\equiv$ moieties.[56] Reaction of $RSiH_3$ with water is catalyzed by gold nanoparticles. Products include siloxane nanowires, filaments, and tubes.[57] These results indicate that gold nanoparticles have considerable potential in the synthesis of novel materials and in developing new chemistry of p-block hydrides.

Other Applications

Gold nanoparticles have been used to catalyze the liquid-phase oxidation of *vic*-diols, e.g., ethylene glycol to glycolic acid.[58] This may presage an extensive liquid-phase catalytic chemistry. In the gas phase, $Au_n(OH)^-$ clusters form adducts with O_2.[59] Propylene oxide is formed over Au/TiO_2 from a mixture of propylene, H_2, and O_2. Very small gold particles tend instead to form propane.[60–63] Numerous other applications will surely soon be published.

Nanophase Glass-Ceramics

Nanophase glass-ceramics are an important class of materials that incorporate a nanoscale, particulate phase in a continuous matrix. These nanoscale grains are formed by controlled nucleation and growth from a precursor glass during a separate heat-treatment step that follows higher-temperature forming. A nanophase microstructure can impart unique optical and mechanical properties to glass-ceramic materials, as demonstrated through some commercially important examples.

Glass-ceramic materials were invented in the 1950s at Corning Glass Works by S. Donald Stookey[64] who discovered that when small amounts (e.g., < 5 wt.%) of nucleating agents (e.g., TiO_2, ZrO_2) were added to certain glass-forming compositions, a separate heat-treatment step after forming could generate highly crystalline bodies. The devitrified materials displayed superior mechanical and thermal properties, including exceptional thermal shock resistance and strength. Glass-ceramic materials with many-fold increases in strength, as well as elimination of thermal expansion, as compared with their parent glasses, can be engineered. Through development efforts at Corning, Schott, Nippon Electric Glass, and IBM, among others, glass-ceramic materials have found useful application in cookware, cooktops, dental restorations, electronic packaging substrates, building materials, and machined structural components. Commercial glass-ceramics are sold under such trade names as MACOR®, DICOR®, Vitronit™, Corning Ware®, Vision®, Neoceram®, CERAN®, EMPRESS®, and others.[65]

Although a complete treatment is beyond the scope of this chapter, it is worth considering the source of improved mechanical properties of glass-ceramic materials over simple glass materials. There are many aspects of mechanical performance that combine to define the utility of a material, including numerous modes of strength and fracture toughness, as well as surface hardness and thermomechanical attributes. At the same time, the variety of glass-ceramic compositions has become enormous in the past half century. Accordingly, there is no way to efficiently capture the full breadth of what is known about the development of mechanical properties of glass-ceramic materials in general. However, there are some dominant themes that have emerged. Glass-ceramic materials, as brittle solids, display superior strength mainly by virtue of improved abrasion resistance and flaw tolerance, which develop when crystalline grains form regions with some inherently superior properties. However, those grains also frequently serve to create unique localized stress conditions for crack suppression and barriers for crack deflection in the microstructure. More detailed analyses of specific materials systems can follow numerous lines of explanation that include considerations of such factors as differences in thermal expansion coefficient, anisotropy in thermal expansion, and chemical substitution effects. Glass-ceramic materials, as composite materials, develop properties that depend on their different constituent phases as well as interactions between them. An example is the elimination or reversal of thermal expansion of some glass-ceramics that results from the crystallization of low or net-negative thermal expansion phases in a higher net-positive thermal expansion residual glass matrix. Finally, effects of grain size on strength of ceramics, mainly the effect that decreasing grain size almost always leads to higher strength, are manifest in glass-ceramics as well. Thus, it is not uncommon to observe eventual reductions in strength when glass-ceramics are heat treated to increase the grain size.

Recently, glass-ceramic materials with exceptionally fine-scale microstructures have been reported at an increasing rate. In most of the original work by Stookey, crystal sizes in the range of 0.1–20 μm were reported. Although such materials have great utility in a wide range of applications, certain advantages can be realized when the crystal size is further reduced to the nanoscale (< 100 nm). Primarily, with a nanoscale crystal size, one can access advantageous combinations of high mechanical performance and improved optical transparency for many compositions. As mentioned, glass-ceramic materials in general are characterized by much higher strength, fracture toughness, and thermal shock resistance than their glass counterparts. However, glass-ceramics take on even more value when their optical properties can be tailored for certain applications. The rapid growth in the use of glass ceramic materials in cooktops, following a migration from opaque β-spodumene to transparent β-quartz glass-ceramics, is an example. In the latter case, optical transparency is achieved mainly through index of refraction matching and low birefringence for the crystalline phase, although a small crystal size of around 100 nm is achieved.

A good example where a unique combination of mechanical and optical properties has been achieved in a nanophase glass-ceramic material relates to high index of refraction beads or microspheres. 3M has recently patented transparent nanophase glass-ceramic microspheres that include comparatively large amounts of ZrO_2 and TiO_2, with index of refraction values that can exceed 2.0.[66–68] The materials are particularly useful as exposed lens elements in durable retroreflective pavement markings. Without index matching between the crystal grains and the surrounding glass phase, within the microspheres, these materials rely on ultrafine crystal size for transparency. Useful beads comprise crystal grains, with a size usually smaller than 20 nm. Other new nanophase glass-ceramic materials from 3M have been reported with ternary rare-earth-oxide-Al_2O_3–ZrO_2 compositions.[69] The glasses are formed with near-eutectic compositions, and can be devitrified with nanoscale grain structure and exceptional mechanical properties, including very high hardness. At the same time, optical transparency is preserved.

There are other examples of glass-ceramic materials that are crystallized to develop a particular property, but crystallized specifically with nanoscale grain structure to preserve optical transparency. Transparent oxyfluoride glass-ceramics with exceptional optical frequency upconversion were first reported by researchers at Sumita Optical Glass and NTT Opto-Electronics Laboratories in 1993.[70] Since that time, research at Corning and elsewhere has further proven the potential utility of the materials, which comprise a nanoscale dispersion of doped lead or rare earth fluoride crystals in an aluminosilicate matrix.[71] The nanophase glass-ceramics are preferred to competing materials, like simple glasses and single crystals. Nanophase glass-ceramics offer the outstanding fluorescence properties of an active rare earth doped fluoride, with their low phonon coupling losses, together with mechanical and chemical durability of an aluminosilicate glass, which derives from the residual matrix after devitrification.

Nanophase glass-ceramics, being nanocomposite materials, offer the potential to achieve unique combinations of properties that have not been demonstrated in other materials systems. Most significantly, small feature sizes in the multiphase microstructures can lead to optical transparency. At the same time, an interaction between phases, or a functional phase itself, can impart additional properties, such as high strength. Just as glass-ceramics in general have experienced extensive technical development and commercial application since their invention half a century ago, the advancement and the use of nanophase glass-ceramics in particular can be expected to grow substantially over the coming years.

Conclusions

The capability to controllably engineer discrete nanoscale particulate domains in a material is a powerful tool for accessing new and unique material properties that enable innovative industrial applications. It is expected that research and development work on nanomaterials will continue to increase and numerous new applications will be identified.

References

1. Scarlett, A.J., Morgan, W.L., Hilderbrand, J.H. Emulsification by solid powder. J. Phys. Chem. **1927**, *31*, 1566–1571.
2. Binks, B.P. Particles as surfactants—similarities and differences. Curr. Opin. Coll. Surf. Sci. **2002**, *7* (1–2), 21–41.
3. Nikolov, A.D., Wasan, D.T. Ordered micelle structuring in thin films formed from anionic surfactant solutions. J. Coll. Interface Sci. **1989**, *133* (1), 1–12.
4. Asakura, S., Oosawa, F. Interaction between particles suspended in solutions of macromolecules. J. Polym. Sci. **1958**, *33* (126), 183–192.
5. Kralchevsky, P.A., Nikolov, A.D., Wasan, D.T., Ivanov, I.B. Formation and expansion of dark spots in stratifying foam films. Langmuir **1990**, *6* (6), 1180–1189.
6. Hug, J.E., von Swol, F., Zukowski, C.F. The freezing of colloidal suspensions in confined spaces. Langmuir **1995**, *11* (1), 111–118.
7. Silva Chunha, A., Grossiord, J.L., Seiller, M. The formulations and industrial applications of multiple emulsions: an area of fast development. In *New Products and Applications in Surfactant Technology*. Karsa, D.R., Ed. CRC Press LLC: Boca Raton, FL, 1998; 1, 205–226.
8. Tohver, V., Chan, A., Sakurada, O., Lewis, J.A. Nanoparticle engineering of complex fluid behavior. Langmuir **2001**, *17* (12), 8414–8421.
9. Chesters, M.A., Somorjai, G.A. Chemisorption of oxygen on stepped surfaces. Surf. Sci. **1975**, *52*, 21–28.
10. Jugnet, Y., SantosAires, F.J., Deranlot, C.V., Piccolo, L., Bertolini, J.C. Chemisorption on Au(1 1 0). Surf. Sci. **2002**, *521* (1,2), L639–L634.
11. Bond, G.C., Thompson, D.T. Catalysis by gold. Catal. Rev. Sci. Eng. **1999**, *41* (3,5), 319–388.
12. Bond, G.C., Thompson, D.T. Gold catalyzed oxidation of carbon monoxide. Gold Bull. **2000**, *33* (2), 41–51.
13. Schumacher, B, Plzak, V., Kinne, M., Behm, R.J. Highly active Au/TiO_2 catalysts. Catal. Lett. **2003**, *89* (1,2), 109–113.
14. Kung, H.H., Kung, M.C., Costello, C.K. Supported catalysts for CO oxidation. J. Catal. **2003**, *126* (1,2), 425–432.
15. Wolf, A., Schüth, F. Synthesis of highly active gold catalysts. Appl. Catal. A. **2002**, *226*, 1–13.

16. Grisel, R.J., Kooyman, P.J., Nieuenhuys, B.E. Influence of the preparation of Au/Al$_2$O$_3$ on oxidation activity. J. Catal. **2000**, *191* (2), 430–437.
17. Schubert, M., Hackenberg, S., van Veen, A.C., Muhler, M., Plzak, V., Behm, R.J. CO oxidation over gold catalysts. J. Catal. **2001**, *197* (1), 113–122.
18. Zanella, R., Giorgio, S., Henry, C.R., Louis, C. Methods for preparation of gold nanoparticles. J. Phys. Chem. B. **2002**, *106* (31), 7634–7642.
19. Guzman, J., Gates, B.C. Cationic and reduced gold in oxidation catalysts. J. Phys. Chem. B. **2002**, *106* (31), 7659–7665.
20. Guzman, J., Gates, B.C. Mononuclear gold-complex catalyst. Angew. Chem. Int. Ed. **2002**, *42* (6), 690–693.
21. Okumura, M., Tsubota, S., Haruta, M. Preparation of supported gold catalysts. J. Mol. Catal. A. **2003**, *199* (1,2), 73–84.
22. Martra, G., Prati, L., Manfredi, C., Biella, R.M., Coluccia, S. Deposition of gold sols from (Ph$_3$P)$_3$AuCl. J. Phys. Chem. B. **2003**, *107* (23), 5343–5459.
23. Ohm, H.S., Yang, J.H., Costello, C.K., Wang, Y.M., Bare, S.R., Kung, H.H., Kung, M.C. Effect of chloride on supported catalysts. J. Catal. **2002**, *210* (2), 375–386.
24. Fu, Q., Saltsburg, H., Flytzani-Stephanopoulos, M. Ceria-based water gas shift catalysts. Science **2003**, *301* (5635), 935–938.
25. Boccuzzi, F., Cerrato, G., Pinna, F., Strukul, G. Reduced reactivity on used gold samples. J. Phys. Chem. B. **1998**, *102* (30), 5733–5736.
26. Guczi, L., Horváth, D., Pászti, Z., Toth, L., Horváth, A., Karacs, A., Petõ, G.J. Modeling gold nanoparticles. J. Phys. Chem. B. **2000**, *104* (14), 3183–3193.
27. Akita, T., Lu, P., Ichikawa, S., Tanaka, K., Haruta, M. Dispersion of nanoparticles in Au/TiO$_2$. Surf. Interface Anal. **2001**, *31* (2), 73–78.
28. Carrey, J., Maurice, J.-L., Petroff, F., Vaurès, A. Evidence of cluster mobility. Surf. Sci. **2002**, *504*, 75–82.
29. Alvarez, M.M., Khoury, J.T., Schaaf, T.G., Shafigullin, M.N. Spectra of nanocrystal gold molecules. J. Phys. Chem. **1997**, *101* (19), 3706–3712.
30. Link, S., El-Sayed, M.A. Surface plasmon electronic oscillations. J. Phys. Chem. B. **1999**, *103* (40), 8410–8426.
31. Carrey, J., Maurice, J.-L., Varuès, A. Evidence of carrier mobility. Phys. Rev. Lett. **2001**, *86* (20), 4600–4603.
32. Frank, M, Bäumer, M. From atoms to crystallites. Phys. Chem. Chem. Phys. **2000**, *2* (17), 3723–3738.
33. Freund, H.-J., Baumer, M., Kuhlenback, H.K. Supported cluster model systems. Adv. Catal. **2000**, *45*, 333–378.
34. Henry, C.R. Studies of supported catalysts. Surf. Sci. Reports. **1998**, *31* (7,8), 235–325.
35. Yoon, B., Luedtke, W.D., Gao, J., Landman, U. Gold on defective graphite clusters. J. Phys. Chem. B. **2003**, *107* (24), 5882–5891.
36. Campbell, S.C., Parker, S.C., Starr, D.E. Size-dependent nanoparticle sintering. Science **2002**, *298* (5594), 811–814.
37. Sykes, E.H.C., Williams, F.J., Tikhov, M., Lambert, R.M. Nucleation, growth mobility and absorption. J. Phys. Chem. B. **2002**, *106* (21), 5390–5394.
38. Wang, J., Wang, G., Zhuo, J. Density functional study of Au$_n$ clusters. Phys. Rev. B **2002**, *66* (3), 035418–035424.
39. Häkkinen, H., Landman, U. Clusters and their anions. Phys. Rev. B. **2000**, *62* (7), R2287–R2290.
40. Häberlen, O., Chung, S.-C., Stener, M., Rösch, N. From clusters to bulk. J. Chem. Phys. **1997**, *106* (12), 5189–5201.
41. Li, J., Li, X., Zhai, H.-J., Wang, L.-S. Au$_{20}$: tetrahedral cluster. Science **2003**, *299* (5608), 864–867.
42. Häkkinen, H., Moesler, M., Landman, U. Bonding in gold: relativistic effects. Phys. Rev. Lett. **2002**, *89* (3)033401.
43. Sanchez, A., Abbet, S., Heiz, U., Schneider, W.-D., Häkkinen, H., Barnett, R.N., Landman, U. Why gold is not noble. J. Phys. Chem. B. **1999**, *103* (48), 9573–9578.

44. Heiz, U., Sanchez, A., Abbet, A., Schneider, W.-D. Using nanoassembled model catalysts. Chem. Phys. **2000**, *262* (1), 189–200.
45. Stolcic, D., Fischer, M., Ganteför, G., Kim, Y.D., Sun, Q., Jena, P. Observation of key reaction intermediates. J. Am. Chem. Soc. **2003**, *125* (10), 2848–2849.
46. Lopez, N., Nørskov, J.K. Catalytic oxidation by gold nanoparticles. J. Am. Chem. Soc. **2002**, *124* (38), 11262–11263.
47. Hammer, B., Nørskov, J.K. Theoretical surface science and catalysis. Adv. Catal. **2000**, *45*, 71–129.
48. Valden, M., Lai, X., Goodman, D.W. Onset of activity of gold on titania. Science **1998**, *281* (5383), 1647–1650.
49. Lai, X., St. Clair, T.P., Valden, M.; Goodman, D.W. Scanning tunneling microscopy of clusters. Prog. Surf. Sci. **1998**, *59* (1-4), 25–52.
50. Bondzie, V.A., Parker, S.C., Campbell, C.T. Kinetics of CO oxidation. Catal. Lett. **1999**, *63* (3,4), 143–151.
51. Liu, Z.-P., Hu, P., Alavi, A. DFT study of CO oxidation. J. Am. Chem. Soc. **2002**, *124* (49), 14770–14779.
52. Mavrikitis, M., Stoltze, P., Nørskov, J.K. Making gold less noble. Catal. Lett. **2000**, *64* (2-4), 101–106.
53. Mohr, C., Claus, P. Hydrogenation properties of gold. Sci. Prog. **2001**, *84* (4), 311–334.
54. Mohr, C., Hofmeister, H.; Radnik, J.; Claus, P. Gold-catalyzed hydrogenation of acrolein. J. Am. Chem. Soc. **2003**, *125* (7), 1905–1911.
55. Bailie, J.E., Hutchings, G.J. Promotion by sulfur of gold catalysts. Chem. Commun. **1999**, *21*, 2151–2152.
56. Owens, T.M., Nicholson, K.T., Banaszak Holl, M., Süzer, S. Alkylsilane-based monolayers on gold. J. Am. Chem. Soc. **2002**, *124* (24), 6800–6801.
57. Prasad, B.L.V., Stoeva, S., Sorensen, C., Zaikovski, V., Klabunde, K.J. Polymerization of alkylsilanes to nanowires. J. Am. Chem. Soc. **2003**, *125* (35), 10,488–10,489.
58. Prati, L., Rossi, M.J. Selective liquid phase oxidation. J. Catal. **1998**, *176* (2), 552–560.
59. Wallace, W.T., Wyrwas, R.B., Whetten, R.L., Mitrić, R., Bonačyić-Koutecký, V. Absorption on gold cluster anions. J. Am. Chem. Soc. **2003**, *125* (27), 8408–8414.
60. Hayashi, T., Tanaka, K., Haruta, M. J. Catalysis **1998**, *178* (2), 566–575.
61. Ajo, H.M., Bondzie, V.A., Campbell, C.T. Propene absorption on gold. Catal. Lett. **2002**, *78* (1–4), 359–368.
62. Zwijnenburg, A., Goosens, A., Sloof, W.G., Crajé, W.J.; van der Kraan, A.M.; de Jongh, L.J.; Makkee, M., Moulijn, J. Mossbauer characterization of Au/TiO$_2$. J. Phys. Chem. B. **2002**, *106* (38), 9853–9862.
63. Kapoor, M.P., Sinha, A.K., Seelan, S., Inagaki, S., Tsubota, S., Yoshida, H.; Haruta, M. Hydrophobicity induced vapor phase oxidation. Chem. Commun. **2002**, *23*, 2902–2903.
64. Stookey, S.D. Ceramic Body and Method of Making US Patent 2,971,853, 14 February 1961.
65. Höland, W.; Beall, G. *Glass-Ceramic Technology*. The American Ceramic Society: Westerville, OH, 2002.
66. Kasai, T., Budd, K.D., Lieder, S.L., Laird, J.A., Yokoyama, C., Naruse, T., Matsumoto, K., Ono, H. Transparent beads and their production. US Patent 6,335,083, 1 January 2002.
67. Budd, K.D., Kasai, T., Roscoe, S.B., Yokoyama, C., Bailey, J.E. Transparent Microspheres US Patent 6,245,7001, 2 June 2001.
68. Frey, M.H., Studiner, C.J., Budd, K.D., Kasai, T., Roscoe, S.B., Yokoyama, C., Bailey, J.E. Glass-ceramic Microspheres Impart Yellow Color to Retroreflected Light US Patent 6,479,417, 12 November 2002.
69. Rosenflanz, A. Al$_2$O$_3$-Y$_2$O$_3$-ZrO$_2$/HfO$_2$ Materials, and Methods of Making and Using U.S. Patent Application 20,030,110,708, 19 June 2003.
70. Wang, Y, Ohwaki, J. New transparent vitroceramics codoped with Er^{3+} and Yb^{3+} for efficient frequency upconversion. Appl. Phys. Lett. **1993**, *63* (24), 3268–3270.
71. Beall, G.H., Pinckney, L.R. Nanophase glass-ceramics. J. Am. Ceram. Soc. **1999**, *82* (1), 5–16.

6

Nanostructured Materials

Vikrant N. Urade and Hugh W. Hillhouse

CONTENTS

Introduction 67
General Strategies to Synthesize Nanostructured Materials from the Bottom-Up
Approach 69
    Templating Route to Synthesize Nanostructured Materials 69
    Types of Templates 70
Use of Single Molecules as Templates to Synthesize Microporous Materials 70
Use of Self-Assembled Arrays of Molecules as Templates for Mesoporous Materials 71
    Self-Assembled Surfactant Templates 71
    Use of Functional Dendrimers for the Self-Assembly of Nanobuilding Blocks 72
    Synthesis Strategies 72
        Cooperative Self-Assembly 73
        Ligand-Assisted Templating 75
    Processing of Mesostructured Materials: Effect of Synthesis Parameters 76
        Control of Reactivity of Inorganic Precursor 76
        Surfactant to Inorganic Precursor Ratio 77
        Effect of Additives 77
        Effect of Environmental Variables 78
    Strategies for Stabilization of the Inorganic Framework and Removal of
    the Template 79
    Morphology Control of Mesostructured Materials 80
Use of Rigid Templates 80
    Rigid vs. Nonrigid Templates 80
    Materials Synthesized 80
Conclusions 81
References 82

Introduction

Profound changes may occur in the electronic properties of materials as their characteristic length scale is reduced to the nanoscale. These changes affect the optical properties, mechanical response, adsorption behavior, and catalytic properties of the material. The possibility of manipulating the structure of materials at the nanometer length scale to alter these properties has opened up an array of opportunities for the materials science, chemistry, and the engineering communities. Synergistic efforts by researchers in these areas have led to the discovery of a large number of new nanostructured materials. Many of these materials have promising applications in heterogeneous catalysts, molecular-sieve adsorbents, sensors, hydrogen and methane storage materials, solar energy conversion, thermoelectric devices, and magnetic data storage. However, in many cases, these properties are sensitively dependent on the processing conditions.

Because of the explosion in new materials, we introduce a nomenclature scheme to help clarify the field. In general, we define a nanomaterial as any material with a characteristic length scale that is less than 100 nm. This characteristic length scale may be the overall size of the sample (such as a 10 nm diameter particle), or it may be an internal length scale over which the structure varies. The realm of nanomaterials may then be divided into four subcategories depending on the dimensionality of the material (Figure 6.1). Thus we reserve the term "nanostructured materials" for the materials that are macroscopic in three dimensions but contain internal structure, on the nanometer length scale, of less than 100 nm but greater than 1nm (this excludes normal crystal structures such as fcc, rocksalt, zinc blende, etc.). This classification thus encompasses an array of materials including zeolites, self-assembled inorganic materials such as mesoporous metal oxides, and self-assembled block-copolymers. If the material is porous and has pores of diameter less than 100 nm, then we classify the material as nanoporous. Historically, porous materials have been classified by the IUPAC as being microporous if the pore dimensions are below 2 nm, mesoporous if they are between 2 and 50 nm, and macroporous if the pore dimensions exceed 50 nm. While the choice of the term "microporous" for a material with pores less than 2 nm is unfortunate, this terminology is still heavily used.

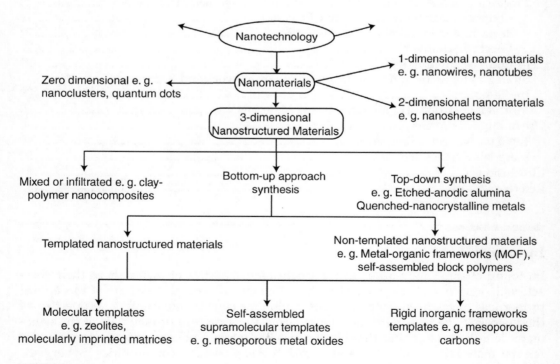

FIGURE 6.1

A schematic illustrating various classes of nanostructured materials based on the synthesis strategies.

There have been many breakthroughs in the synthesis of nanostructured materials in the last 10–15 years, including the synthesis of large pore zeolites with pores larger than 1 nm,[1,2] the discovery of surfactant templated mesoporous materials by the scientists at Mobil,[3] the extension of this templating method to nonsilica systems including transition metal oxides, metals, and carbons,[4–6] and the synthesis of nanoporous metal organic frameworks (MOFs).[7] This chapter focuses on the synthesis strategies and the mechanisms involved in the synthesis of nanostructured materials and the processing parameters that are vital for the successful synthesis of these nanostructures. Special emphasis is placed on the synthesis of materials that are closer to applications and are therefore of more immediate technological importance to the process engineer. Further, within the branch of nanostructured materials, materials may be further categorized depending on the synthesis procedures, as illustrated in Figure 6.1. This chapter also focuses on templated nanostructured materials assembled by the bottom-up approach.

General Strategies to Synthesize Nanostructured Materials from the Bottom-Up Approach

Nanostructured materials can be synthesized from the so-called "top down" or "bottom-up" approach. In the first approach, features at the micron (or submicron) length scale are created on a substrate by masking and exposing selected regions of a radiation sensitive layer (typically a polymeric photoresist) to a UV source. This exposure is followed by various chemical treatments and mechanical steps to obtain the desired spatial pattern on a substrate. However, the feature sizes that can be obtained with this approach are limited to the length scale of the wavelength of the radiation employed. If features at the nanometer scale are desired, one must start from the bottom (i.e., use individual molecules or clusters) and assemble templates that will impart the nanostructure to the desired material.

Templating Route to Synthesize Nanostructured Materials

The general strategy for the synthesis of nanostructured materials involves the use of templates, which can be ionic, molecular, or supramolecular structures and act as molds. These molds guide the formation of the structure and leave behind porosity when removed by suitable means (if desired). The templating approach allows control over the size and the shape of the resulting pores. In a typical synthesis of a nanoporous metal oxide material, a suitable template molecule (typically an amphiphile) is added to the synthesis mixture. The synthesis mixture contains inorganic precursors that are capable of interacting with the template molecule either via electrostatic or entropic pathways to create an ordered nanostructured assembly of the template and the inorganic material. The composition of the mixture is adjusted such that the desired material self-assembles and phase separates from the mixture. At this stage, the template molecule is occluded in the inorganic framework, directing the nanoscale structure. The removal of the template creates ordered voids in the inorganic framework, whose size and topology are determined by the template used. Template removal is essential for applications of these materials as catalyst supports, adsorbents, or molecular sieves.

The inorganic framework must be structurally rigid and capable of supporting itself in the absence of the template.

Types of Templates

Depending on the size and the shape of the pore system desired, the nature of template molecules also varies. Microporous materials are typically synthesized using individual molecules or ions as templates. In this case, these molecules may not constitute the entire template, but may be combined with solvent or water molecules to form the templating species. When larger pore sizes are desired, the templates are supramolecular assemblies of surfactants arranged in various configurations depending on the synthesis parameters. It is also possible to use other porous inorganic materials as templates and create new porous materials inside them. These templates are termed as "rigid templates," as opposed to "soft" surfactant templates, which are labile and undergo phase transition in the solution. The words "endotemplating" and "exotemplating" are also used to indicate the use of surfactant micelles and inorganic porous solids as templates, respectively.

Use of Single Molecules as Templates to Synthesize Microporous Materials

The use of single molecules or ions as templates results in materials with the smallest pore sizes (typically less than 1 nm), and hence they are called microporous according to the IUPAC definitions. Most microporous materials synthesized to date are zeolites. Zeolites are a subclass of microporous materials in which the crystalline inorganic framework is composed of four-coordinated species interconnected by twocoordinated species. Traditionally these materials are aluminosilicates; however, many different compositions have been synthesized.[8] The templates used in the synthesis of microporous materials are typically small ionic or neutral molecular species. The function of the template in the synthesis of microporous materials is little understood, and there are at least four different modes by which an additive can operate in a zeolite synthesis:[9] a) It may act as a space filler occupying the voids in the structure, thereby energetically stabilizing less dense inorganic framework; b) the additive may control the equilibria in the synthesis mixture, such as solution pH or complexation equilibria; c) it may preorganize the solution species to favor the nucleation of a specific structure; d) it may act as a true template determining the size and the shape of the voids in the structure.

Noncrystalline microporous materials may also be synthesized by a technique called molecular imprinting. In this process, a molecular template, called a "print molecule," is used to direct the arrangement of functional monomers around the template, which can then be fixed by chemical polymerization with a crosslinking monomer. This results in the formation of a rigid matrix with the template embedded in it. The removal of the template exposes the functional sites, which can specifically recognize the print molecule or molecules similar to it. This technique can be effectively used to synthesize catalytic and enzymatic hosts having specific interactions with a particular kind of molecule and can be used for separations as well. For a review of molecular imprinting, the reader is referred to Ekberg and Mosbach,[10] Collinson,[11] and Wulff.[12]

Use of Self-Assembled Arrays of Molecules as Templates for Mesoporous Materials

The pore sizes obtained using molecular or ionic templates are restricted because of the small sizes of the templates used (typically less than 15 Å), restricting the accessibility of the internal surface to small molecules. Hence, a great deal of effort has been devoted to the synthesis of materials having pore sizes larger than those in zeolites. In the last decade, progress toward this goal has been made following the discovery of surfactant templated silica by the scientists at Mobil.[3] A variety of materials (silica and nonsilica oxides, carbons, metals, chalcogenides, phosphates, borates, and sulfides) have been synthesized in highly ordered forms having extremely high surface areas (>1000 m²/g) and narrow pore size distributions, with the pore size tunable all the way from 2 to 30 nm by varying the size of the surfactant and the synthesis conditions. These materials, with pore diameters between 2 and 50 nm, are called mesoporous or mesostructured materials, depending on whether the template has been removed or not. The most well known of these materials is the Mobil family of mesoporous materials (M41S), including a two-dimensional hexagonal mesoporous silica (MCM-41) and a cubic phase mesoporous silica (MCM-48).

The templates used in the synthesis of mesostructured and mesoporous materials can be classified into two categories: The first class of templates includes "soft" templates, which are ordered arrays of self-assembled surfactant micelles, similar to the ones used by the researchers at Mobil. Alternately, mesoporous materials can themselves be used as templates to synthesize new mesostructured materials, and such templates can be termed as "rigid" templates. In the following sections, we focus on the use of supramolecular assemblies of surfactants as well as on the use of rigid templates as the templates for the synthesis of mesostructured materials.

Self-Assembled Surfactant Templates

Surfactants are amphiphiles and as such have both hydrophilic and hydrophobic portions. If the contrast in the hydrophobicity of the hydrophilic and hydrophobic moieties is strong, then above a certain concentration (critical micelle concentration, cmc) these molecules may self-assemble in solution to form aggregates called micelles. If the concentration is increased further, the micelles may themselves self-assemble into a tertiary structure and form a lyotropic liquid crystalline phase to minimize the free energy of the system. There are three important parameters that characterize the self-assembly process of the surfactant: cmc, aggregation number, and molecular packing factor. Below the cmc, surfactant molecules predominantly exist as monomers in solution. The cmc is experimentally identified as the concentration at which added surfactant starts to enter into an aggregate.[13] The aggregation number is the dominant number of monomeric units in an aggregate. The geometry and the topology of the assembly are affected by the aggregation number, the geometry of the surfactant molecule, concentration of the surfactant, temperature, and the presence of other species in solution. For the effect of the aggregation number on the microstructure of some block copolymers that are commonly used as templating agents, the reader is referred to Vasilescu, Caragheorgheopol, and Caldararu.[14] The effect of surfactant geometry can be qualitatively, and in many cases quantitatively, described by making use of the concept of molecular packing factor, *g*.

Surfactant molecules aggregate into different shapes in solution (spherical, cylindrical, lamellar, spherical bilayer, vesicular, etc.), which result in minimum interaction between the hydrophobic moieties of the surfactant and the polar solvent. The concept of molecular

TABLE 6.1

Molecular Packing Factor Values Corresponding to Various Micellar Structures

Packing Factor, g	Structure Obtained
1/3	Spherical
1/2	Infinite cylinders
1	Planar bilayers

Between $g = 1/3$ and $g = 1/2$ or between $g = 1/2$ and $g = 1$, complex phenomena such as formation of aggregates of lower or higher symmetry or phase separation may occur.

packing factor is useful in visualizing and qualitatively predicting the geometry of the micelles from the known geometry of individual surfactant molecules. Israelachvili et al.[15] proposed the concept of molecular packing parameter and demonstrated how the size and the shape of the aggregate at equilibrium can be predicted from a combination of molecular packing considerations and general thermodynamic principles. The molecular packing factor is defined as:

$$g = \frac{V_0}{al_0}$$

where V_0 is the volume of the surfactant tail, l_0 the critical length of the tail (not the actual length of a fully extended tail), and a the effective area of the head group at the surface of the micelle. Thus, low values of g result in structures of high curvature, while a value of $g = 1$ has no net curvature. Table 6.1 summarizes the range of values of the packing factor for which various structures are obtained. The key point to keep in mind is that solution conditions can drastically affect the effective area of the head group. For instance, highly charged metal oxide ions can effectively screen the charge of adjacent head groups, reducing their effective area, and thus increase the packing parameter.

Use of Functional Dendrimers for the Self-Assembly of Nanobuilding Blocks

Chemically well-defined inorganic clusters (often termed as "nanobuilding blocks") can be assembled into an ordered array using dendrimers capped with functional groups that can chemically bond with the inorganic clusters. Various metal oxo-based hybrid materials, including Ti- and Ce-based gels, have been synthesized using this synthetic strategy.[16,17]

Synthesis Strategies

Self-assembled mesostructured materials are typically synthesized from dilute solutions of inorganic precursors and surfactant molecules wherein the self-assembly process occurs through electrostatic and entropic interactions and is a cooperative one. However, in some cases, mesostructured materials are synthesized at much higher surfactant concentrations such that a liquid crystalline phase actually pre-exists in solution, and the inorganic precursor is preferentially accommodated in the hydrophilic domains. This approach, called "true liquid crystal templating" (TLCT, route II in Figure 6.2), was used by Attard et al. for the synthesis of mesoporous silica, metals, and alloys.[18,19] In addition, there have been several novel synthesis routes developed that utilize inorganic precursors with covalently bonded hydrophobic ligands or take advantage of specific solution interactions such as hydrogen bonding.

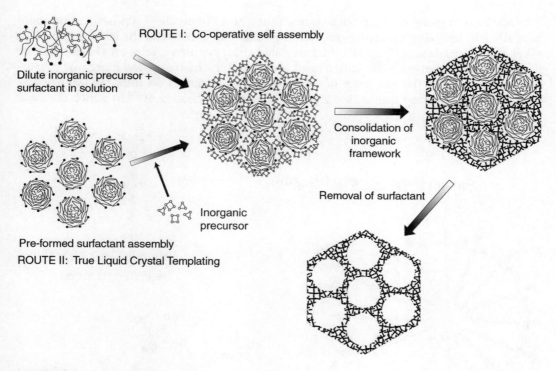

ROUTE I: Co-operative self assembly

Dilute inorganic precursor + surfactant in solution

Consolidation of inorganic framework

Removal of surfactant

Inorganic precursor

Pre-formed surfactant assembly

ROUTE II: True Liquid Crystal Templating

FIGURE 6.2

A schematic illustrating various routes used to synthesize mesostructured materials. Route I: A cooperative selfassembly route relying on the interaction between the surfactant molecules and the inorganic precursors. Route II: TLCT route. The inorganic framework condenses around preformed surfactant micelles in this case.

Cooperative Self-Assembly

After the addition of the inorganic precursor, cooperative assembly between the surfactant and the inorganic precursor takes place, leading to the formation of the mesostructure. This approach is similar to the one used by the Mobil scientists and is also the most commonly used one for the synthesis of nonsilica systems, where the control of precursor reactivity demands lower concentrations of the inorganic precursors and hence of the surfactant. In Figure 6.2, the formation of a mesostructure by this approach is illustrated (route I).

To help explain this mechanism of the self-assembly process, Stucky and coworkers[20,21] proposed a very useful concept of charge density matching. This concept refers simply to the fact that electroneutrality is maintained locally at the interfaces. At the isoelectric point (IEP) of the inorganic material, the inorganic framework is charge neutral. When the pH is below the IEP, the framework carries a positive charge, and when it is greater, it is negatively charged. The electrostatic interactions between the charged inorganic moieties (I^- or I^+ depending on pH) and the surfactant head groups (S^+ and S^- depending on the type of surfactant, cationic or anionic) lead to the formation of a hybrid interface. The interaction can be direct (S^+I^- or S^-I^+) if the surfactant head groups and inorganic moieties carry opposite charges or they can be mediated by counterions associated with the surfactant ($S^+H^-I^+$ or $S^-H^+I^-$). The shape of the resultant hybrid structure is determined by the balance of charge density between the inorganic moieties and the surfactant head groups. If the charge density of the inorganic framework is reduced, the average head group area of the surfactant assembly increases and the packing factor g decreases, resulting in the formation of more curved structures.

In addition to using charged surfactants, Tanev and Pinnavaia[22] synthesized mesoporous silica by utilizing the hydrogen bonding interactions between the head group of an alkyl amine surfactant (S[0]) and the hydroxylated silica precursor, tetra ethyl ortho silicate (I[0]). The bonding between the amine head group and the hydroxylated inorganic precursor occurs through the exchange of the lone pair of electrons on nitrogen (Figure 6.3). This work led to a very useful development reported by Zhao et al.[23] in which nonionic

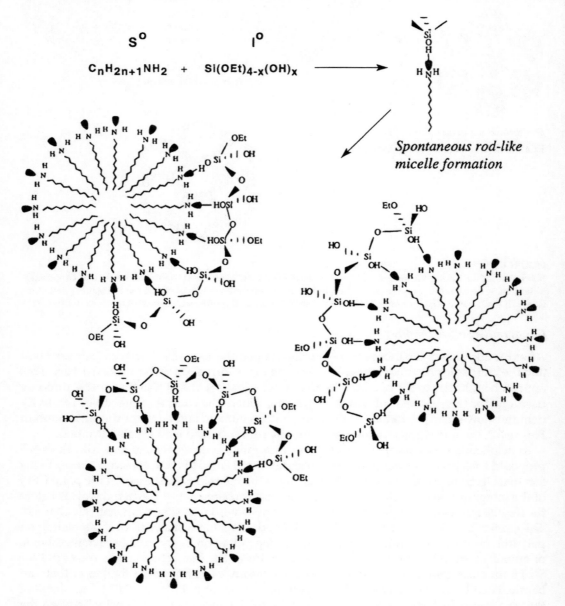

FIGURE 6.3
The neutral templating route proposed by Pinnavaia et al. The electron lone pairs indicated by the shaded lobes on the surfactant head groups participate in the hydrogen bonding process with the framework silanol (Si–OH) groups. (From Tanev and Pinnavaia.[22])

amphiphilic triblock-copolymers were used as templates. These commercially available templates are quite robust in the synthesis of mesoporous materials and produce thicker walled materials than the cationic surfactants. One unique feature of the triblock-copolymer templated materials is that many structures have micropores in the walls that interconnect the mesoporosity.

Ligand-Assisted Templating

As opposed to the liquid crystal or electrostatic templating mechanisms, which rely on coulombic, van der Waals, or hydrogen bonding interactions between the surfactant molecules and the inorganic species, Antonelli et al.[24] proposed a ligand assisted templating mechanism, which relies on the formation of a covalent bond between the surfactant head group and the inorganic moieties. The amine surfactants were pretreated with metal alkoxides in the absence of water to form metal-ligated surfactants. Addition of water, which acts as both a solvent and a reactant, to the alkoxide–surfactant solution initiates the self-assembly and alkoxide hydrolysis–condensation. The reaction scheme following this route is given in Figure 6.4.

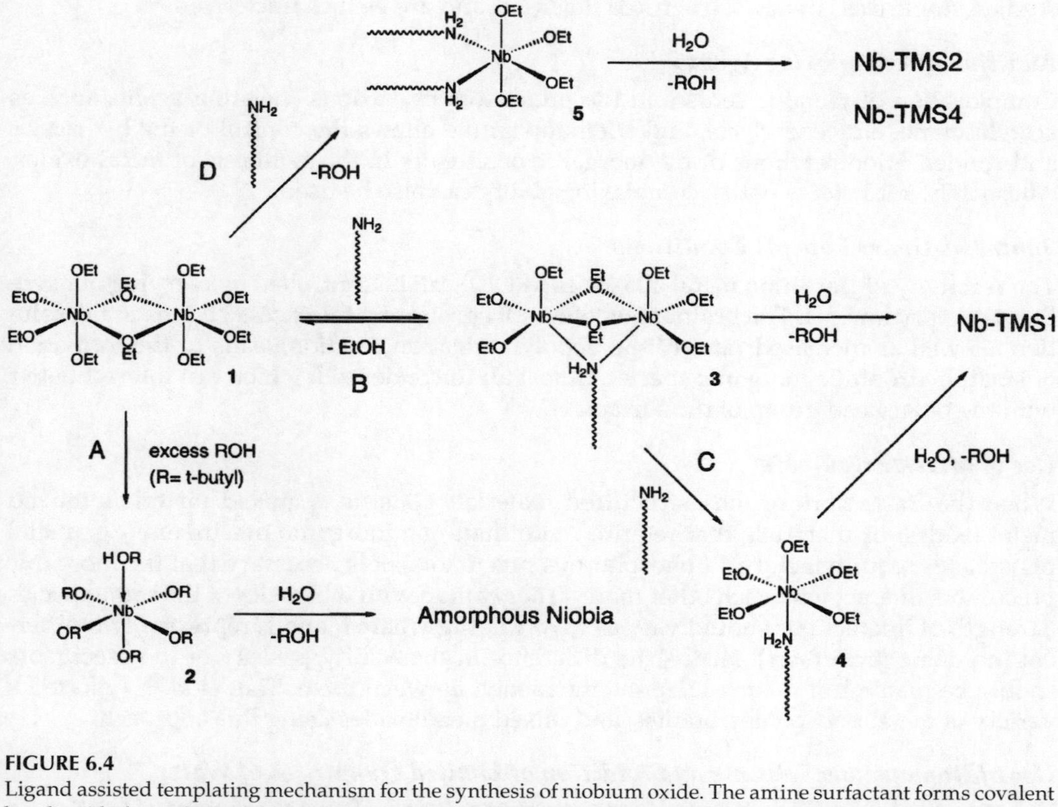

FIGURE 6.4

Ligand assisted templating mechanism for the synthesis of niobium oxide. The amine surfactant forms covalent bonds with the niobium (V) ethoxide, resulting in the formation of a hybrid interface. Upon exposure to water, the system organizes into various structures such as bidimensional hexagonal (Nb-TMS1), three-dimensional hexagonal (Nb-TMS2), lamellar (Nb-TMS4), or a cubic phase. (Nb-TMS3 is not shown in the figure.) (From Antonelli, Nakahira, and Ying.[24])

Processing of Mesostructured Materials: Effect of Synthesis Parameters

The formation of self-assembled materials is governed by a number of experimental parameters: the choice of inorganic precursors and surfactants, the inorganic to surfactant ratio, amount of water and other solvents used, pH during synthesis, additives, reaction time and temperature, treatments used to stiffen the inorganic framework, and treatments to remove surfactant and obtain porosity, all decide the final ordering, porosity and surface area. It is well known that successful synthesis depends on tuning the rates of the reactions that lead to the formation of inorganic framework with that of the self-assembly of the surfactant micelles.

Control of Reactivity of Inorganic Precursor

Silica is the most extensively studied system among the self-assembled materials. Commonly used precursors in the synthesis of silica are tetraalkyl orthosilicates ($Si(OR)_4$). However, in nonsilica systems, including transition metal oxides or chalcogenides, the precursors are typically very reactive, and hence some means must be devised to control the reactivity of these precursors to avoid outright precipitation of solids having no mesostructure. The ways to control the reactivity of the inorganic precursors are summarized here. These are specific mostly to metal oxide systems, as they are the most extensively studied single class of mesostructured materials and are well characterized.

Addition of Complexing Agents

Complexation of metallic centers in the precursors by various chelating agents such as acetylacetone, ethylene glycol, and triethanolamine allows the control of the hydrolysis and condensation reactions of the inorganic precursors in the synthesis of metal oxides. Alternately, surfactants with a complexing ability can also be used.

Synthesis Under Low pH Conditions

The reactivity of transition metal alkoxides $M(OR)_n$ can be controlled by carrying out syntheses at very low pH. The protonation of M–OH groups by the excess H^+ ions in the solution, as well as increased rates of the depolymerization reaction, leads to the formation of small hydrophilic inorganic species almost all uncondensed, which can interact better with the polar head group of the surfactant.

Use of Mixed Precursors

When the framework of mesostructured materials consists of mixed materials (mixed metal oxides) or materials that require more than one inorganic precursors (e.g., metal phosphates require metal and phosphorous precursors), it is necessary that the inorganic precursors interact with each other more strongly than with molecules of the same species (strength of interactions should vary as $I_1I_2 > I_1I_1, I_2I_2$, where I_1 and I_2 represent two different inorganic precursors). Hence, the difference in the acidity/basicity of the precursors should be maximum to maximize the interaction between them. Tian et al.[25] reported a variety of metal phosphates, borates, and mixed metal oxides using this approach.

Use of Nonaqueous Solvents and Addition of Limited Quantities of Water

Use of limited quantities of water in nonaqueous solvents allows the control of the degree of hydrolysis of the transition metal oxide precursors. The resultant hydrolyzed, polar species can have a better interaction with the polar head group of the surfactant, resulting in the formation of ordered structures. Soler-Illia et al. reported a "modulation of hybrid interface" approach,[26] which relies on the addition of controlled quantities of water to the solution of the inorganic precursor and a nonionic surfactant in an organic solvent to obtain ordered mesostructures.

Use of Dilute Solutions

The reactivity of the inorganic precursors can also be controlled by using low concentrations to lower the reaction rates. Consequent evaporation of solvent results in the formation of organized structure. This approach, introduced by Brinker et al.[27] and termed as "evaporation induced self-assembly." (EISA), allows the formation of powders, films, gels, and monoliths.

Use of Nanoparticles

Instead of using reactive precursors that undergo reactions converting them to the final inorganic material, nanometer sized particles of final inorganic materials can themselves be used as the precursors. These particles will be inert, in that they will not interact with each other, but their interaction with the surfactants will allow formation of ordered mesostructure.

Surfactant to Inorganic Precursor Ratio

The type of mesostructure obtained depends strongly on the surfactant to inorganic ratio. In fact, there is a close correlation between the surfactant to solvent ratio in the phase diagram of a surfactant and the surfactant to inorganic ratio in the mesostructured materials obtained. Alberius et al. demonstrated this correlation by the so-called "general predictive synthesis" approach.[28] They used the phase diagrams of the water-surfactant system to guide the synthesis of mesoporous silica and titania films. There was a very close correlation between the values of the volume fraction of the surfactant over which different phases are obtained in the water-surfactant system and in the silica-surfactant and titania-surfactant systems.

Effect of Additives

Additives can have multifold effects on the structure of mesoporous materials: they can affect the structure at the nanometer level, i.e., change the phase obtained, or they can act at the micrometer level, changing the morphology of the resultant materials. The morphology control is discussed separately. When they are acting at the nanometer level, the additives can modulate the rate of the hydrolysis and condensation reactions associated with the inorganic precursors and hence change the range of variables over which ordered materials can be obtained. For example, Kim et al. obtained highly ordered silica mesostructures over a diverse range of pH by controlling the relative rates of hydrolysis and condensation of the silica species through the use of fluoride ions and tetramethyl orthosilicate (TMOS) as the inorganic precursor.[29]

Ionic additives can accumulate at the surfactant–inorganic interface and modulate the interfacial curvature, thereby changing the value of "a," the effective head group area per surfactant molecule. This can change the structure of surfactant micelles and hence the structure of the mesophase obtained. The addition of salts to solution can also lower the critical micelle concentration, thus broadening the synthesis domain in surfactant concentration, as well as increase the range of surfactants that can be used in producing ordered mesostructures. The salt addition can also affect the radius of the micelles and hence the unit cell parameter and porosity of the inorganic material obtained after surfactant removal.[30]

The dramatic effect of salt addition on the mesostructure was demonstrated by Che et al.[31] Starting with solutions with the same surfactant: inorganic ratio and working under the same synthesis condition, and by varying only the type of acid used (H_2SO_4, HCl, HBr, and HNO_3), the authors obtained a three-dimensional hexagonal structure with H_2SO_4 (*P6₃/mmc* space group), a two-dimensional hexagonal with HBr (*p6m*), a bicontinuous cubic with HNO_3 (*Ia3d*), and a cubic (*Pm3n*) with HCl. (For an introduction to the space group nomenclature, which is quite common among the zeolite and mesoporous materials community to describe the crystal structures, the reader is referred to Sands.[32] Various "International Tables for Crystallography" also constitute a useful and handy reference.) This modulation of structure was explained in terms of the adsorption strength of the acid anions on the head groups of the surfactant micelles. The acid anions are more or less hydrated in the surfactant solution. Less strongly hydrated ions have, in general, smaller ionic radii and bind more closely and strongly on the head group of the surfactant. The small anions contribute to the partial reduction in the electrostatic repulsion between the charged surfactant head groups and the decrease in the effective area of surfactant, "*a*," thereby resulting in a significant increase in the *g* value. Ionic radii decrease in the following order: $\frac{1}{2}SO_4^{2-} > Cl^- > Br^- > NO_3^-$. Hence, H_2SO_4 leads to the formation of the 3D hexagonal *P6₃/mmc* mesophase with a smaller *g* parameter and HNO_3 favors the formation of the *Ia3d* mesophase with a larger *g* parameter. Liu et al.[33] found that the addition of organosiloxane or small organic molecules like benzene and its alkyl derivatives to the synthesis mixture containing TEOS, as silica source, and nonionic block-polymeric surfactants results in the formation of cubic bicontinuous *Ia3d* phase at room temperature under acidic conditions. Previous reports of the synthesis of this phase were limited to alkaline media and high temperature synthesis.

Effect of Environmental Variables

Relative Humidity (RH) of Atmosphere

Relative humidity is an extremely important parameter for the synthesis of thin films from solution by dip or spin coating. The dynamics of evaporation is highly dependent on the RH of the atmosphere. The importance of controlling the RH was demonstrated by Cagnol et al.[34] Starting with a solution with a fixed composition, different mesostructures were obtained just by changing the RH of the atmosphere. Structures with higher curvature and lower values of the packing factor g, such as cubic structures, require higher values of RH. More water is retained in the films at higher RH and hence the flexibility of inorganic framework, where the water is mainly located, is enhanced. The inorganic framework, therefore, can conform to the more curved interface resulting in the formation of cubic structure. At lower values of RH, on the other hand, structures requiring less curvature (such as 2D hexagonal mesostructure) are obtained. It was also demonstrated that the final structure is formed through a series of intermediate structures, the sequence of which depends on the relative humidity. It is also possible to change the structure after the dip coating by changing the relative humidity.

Organic Solvent Environment

The partial pressure of the organic solvent in the atmosphere has effects similar to that of water. The organic solvent generally has high volatility and evaporates faster. At high concentrations of the solvent, though, a high quantity of solvent is maintained in the films and the films remain low in viscosity, paving way for phase transitions.[35] Gallis and Landry studied the transformation processes leading to the formation of MCM-48 from MCM-41 and found that the presence of ethanol at the interface was necessary for the phase transition.[36] The authors theorized that hydrolysis of TEOS leads to the production of ethanol in the vicinity of the organic–inorganic interface. Upon heating, ethanol was driven further into the organic region, increasing the surfactant packing parameter and causing the system to transform from MCM-41 into MCM-48.

Strategies for Stabilization of the Inorganic Framework and Removal of the Template

Most of the applications of surfactant templated materials depend on the porosity of the inorganic framework; hence removal of the template is necessary. However, the inorganic framework must be strong enough to retain its fidelity even in the absence of the template. This is particularly important for the synthesis of materials with reactive precursors, as the strategies to control the precursor reactivity result in a framework that has a lower degree of polymerization and, hence, lower structural stability. This necessitates post-treatment of the mesostructured materials to enhance the polymerization of the inorganic framework and removal of template without destroying the mesostructure. The strategies employed to achieve this end are summarized here.

Thermal Treatments to Consolidate the Inorganic Framework

The surfactants are usually removed by heating the organic–inorganic hybrids to temperatures high enough to oxidize the template. Thermal treatment of the hybrids at a temperature that is lower than that required for template removal but high enough to enhance the rate of condensation will lead to consolidation of the inorganic framework in the presence of the template; this will enhance the stability of the inorganic framework for subsequent template removal at higher temperatures.

Mild Template Removal

Instead of using strong thermal treatments to remove the templates, mild template removal techniques, e.g., by refluxing in a solvent that can dissolve the template, use of UV irradiation, or use of mild oxidants, can also be used for template removal without structural collapse.

Doping with Foreign Atoms to Reduce the Size of Crystallites

In the case of transition metal oxides, calcination of the mesostructures is accompanied by crystallization of the inorganic framework. The size of the crystallites formed in the inorganic framework walls grows as the temperature increases, and it has been shown that when the size of these crystallites becomes comparable to the wall thickness, the structure collapses. Doping of the inorganic walls by heteroatoms prevents the growth of the crystallites, and, therefore, retains the structural integrity of the inorganic framework after template removal.

Aging Under Mild Conditions

Longer duration of aging, if it is permitted by the synthesis parameters, leads to a higher degree of condensation, and, therefore, stronger inorganic walls.

Treatment with Vapors of the Inorganic Precursor

The density of the inorganic walls can be increased postsynthesis by subjecting the material to an environment saturated with the vapors of the inorganic precursor. The vapors infiltrate into the wall of the structure, and fill any defects and microporous voids in the framework, which would otherwise lead to the collapse of the structure on template removal. The denser walls that result from the vapor treatment have much higher structural strength than those of untreated materials.

Morphology Control of Mesostructured Materials

The control of morphology of the materials at a macroscopic level is essential from an application point of view. Various morphologies of mesostructured materials, especially silica and nonsilica mesoporous oxides, have been obtained, such as free standing and supported thin films, spheres, fibers, and monoliths, using a combination of sol–gel chemistry and emulsion chemistry. Syntheses in acidic solutions afford better control over themorphology of thematerials. Addition of a cosolvent immiscible with the solvent used for synthesis also leads to the formation of different morphologies of the resultant materials. Completely different morphologies are observed on the oil side and the water side of the resultant emulsion.[37] Morphology can also be affected by the synthesis temperature; polymerization at low temperatures leads to the formation of faceted single crystals under thermodynamically controlled conditions. For a complete discussion of the control of morphology of mesostructured materials, the reader is referred to a review by Zhao et al.[38]

Use of Rigid Templates

Rigid vs. Nonrigid Templates

Although very versatile and well researched, the surfactant templating route has some disadvantages. Surfactants are labile molecules and undergo thermal degradation typically at temperatures lower than ~350°C. Most transition metal oxides undergo amorphous to crystalline transition at a temperature higher than this, which is responsible for the destruction of mesostructure, as discussed earlier. Most applications of the metal oxides, on the other hand, depend on the crystallinity of the wall. This necessitates the use of rigid templates. It has also been demonstrated that the use of mesoporous carbons as rigid templates for the synthesis of mesoporous metal oxides results in the formation of mesostructures that are otherwise not obtained with surfactant templates. A complete infiltration of inorganic precursors into the mesopores of the template is necessary to obtain dense wall with as little volume shrinkage as possible.

Materials Synthesized

The first ordered mesoporous materials synthesized using rigid templates were carbons. Ryoo et al. reported the synthesis of highly ordered cubic mesoporous carbons (termed CMK-1) using MCM-48 silica as the template[6] (and references therein). As a general route for the preparation of mesoporous carbons, the silica template is infiltrated with a solution of carbon precursor or an organic compound. This is followed by pyrolysis, in which the carbon precursor is converted to carbon. The silica template is then removed by treating with HF or NaOH. Mesoporous carbon materials typically have very high

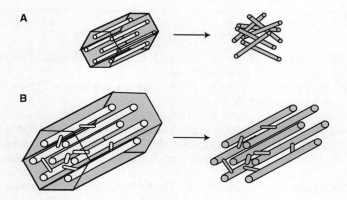

FIGURE 6.5
A schematic illustrating the necessity of an interconnected pore system for the formation of mesoporous carbon. When MCM-41 (A), which does not have interconnections, was used as template, the structure collapsed upon calcination. The use of large pore interconnected SBA-15 (B) led to a stable, highly ordered material. (Adapted from Ryoo, Joo, Kruk, and Jaroniec.[6])

surface areas (1500–1800m²/g). But the carbon prepared using MCM-41 as a template (which has a 2D hexagonal structure) collapses upon the template removal to yield disordered microporous structure with high surface area. The reason behind this is explained in Figure 6.5. It is necessary that when the silica template is 2D hexagonal, there should be some micropores connecting the mesopores to each other. Otherwise, the resultant carbon nanorods are disconnected and they collapse upon one another after template removal. When large pore 2D hexagonal silica, SBA-15, having micropores in the wall was used as template, the synthesized mesoporous carbon, termed CMK-3, was a true replica of the silica structure.

Tian et al. demonstrated the use of microwave digested silica as a template for the synthesis of a variety of metal oxides having crystalline structures in the form of nanowires and nanospheres.[39] The use of microwaves instead of the usual high temperature calcination to remove the surfactant from the silica during its synthesis leaves behind a larg e number of surface silanol groups (Si–OH). This leads to better hydrophilic affinity interactions with the inorganic precursors resulting in higher loading of the pores.

Conclusions

Phenomenal progress has been made in the synthesis of nanostructured materials in the last decade. A deeper understanding of the formation mechanisms has been established and it is now possible to synthesize these materials in a reproducible way. Modification of the properties of these materials should now pave the way for the use of these materials for conventional applications (catalysis, separation, adsorption, etc.) and for novel applications in the fields of solar energy conversion, electronics, hydrogen generation, etc.

References

1. Davis, M.E., Saldarriaga, C., Montes, C., Garces, J., Crowder, C. A molecular-sieve with 18-membered rings. Nature **1988**, *331* (6158), 698–699.
2. Dessau, R.M., Schlenker, J.L., Higgins, J.B. Framework topology of Alpo4-8—the 1st 14-ring molecular-sieve. Zeolites **1990**, *10* (6), 522–524.
3. Kresge, C.T., Leonowicz, M.E., Roth, W.J., Vartuli, J.C., Beck, J.S. Ordered mesoporous molecular-sieves synthesized by a liquid-crystal template mechanism. Nature **1992**, *359* (6397), 710–712.
4. Yang, P.D., Zhao, D.Y., Margolese, D.I., Chmelka, B.F., Stucky, G.D. Generalized syntheses of large-pore mesoporous metal oxides with semicrystalline frameworks. Nature **1998**, *396* (6707), 152–155.
5. Attard, G.S., Coleman, N.R.B., Elliott, J.M. The preparation of mesoporous metals from performed surfactant assemblies. In Mesoporous Molecular Sieves; **1998**; Vol. 117, 89–94.
6. Ryoo, R., Joo, S.H., Kruk, M., Jaroniec, M. Ordered mesoporous carbons. Adv. Mater. **2001**, *13* (9), 677–681.
7. Chen, B.L., Eddaoudi, M., Hyde, S.T., O'Keeffe, M., Yaghi, O.M. Interwoven metal–organic framework on a periodic minimal surface with extra-large pores. Science **2001**, *291* (5506), 1021–1023.
8. Breck, D. *Zeolite Molecular Sieves: Structure, Chemistry, and Use*; Krieger Publishing Company, 1984.
9. Schuth, F. Endo- and exotemplating to create high-surface-area inorganic materials. Angew. Chem. Int. Ed. **2003**, *42* (31), 3604–3622.
10. Ekberg, B., Mosbach, K. Molecular imprinting—a technique for producing specific separation materials. Trends Biotechnol. **1989**, *7* (4), 92–96.
11. Collinson, M.M. Sol–gel strategies for the preparation of selective materials for chemical analysis. Crit. Rev. Anal. Chem. **1999**, *29* (4), 289–311.
12. Wulff, G. Molecular imprinting in cross-linked materials with the aid of molecular templates—a way towards artificial antibodies. Angew. Chem. Int. Ed. (English) **1995**, *34* (17), 1812–1832.
13. Evans, D.F., Wennerström, H. *The Colloidal Domain: Where Physics, Chemistry, Biology, And Technology Meet*, 2nd Ed.; Wiley-VCH: New York, 1999.
14. Vasilescu, M., Caragheorgheopol, A., Caldararu, H. Aggregation numbers and microstructure characterization of self-assembled aggregates of poly(ethylene oxide) surfactants and related blockcopolymers, studied by spectroscopic methods. Adv. Colloid Interf. Sci. **2001**, *89*, 169–194.
15. Israelachvili, J., Mitchell, D.J., Ninham, B.W. Theory of self-assembly of hydrocarbon amphiphiles into micelles and bilayers. J. Chem. Soc. Faraday Trans. **1976**, *2* (72), 1525.
16. Soler-Illia, G., Rozes, L., Boggiano, M.K., Sanchez, C., Turrin, C.O., Caminade, A.M., Majoral, J.P. New mesotextured hybrid materials made from assemblies of dendrimers and titanium(IV)-oxo-organo clusters. Angew. Chem. Int. Ed. **2000**, *39* (23), 4249–4254.
17. Bouchara, A., Rozes, L., Soler-Illia, G.J.D., Sanchez, C., Turrin, C.O., Caminade, A.M., Majoral, J.P. Use of functional dendritic macromolecules for the design of metal oxo based hybrid materials. J. Sol–Gel Sci. Technol. **2003**, *26* (1–3), 629–633.
18. Attard, G.S., Glyde, J.C., Goltner, C.G. Liquidcrystalline phases as templates for the synthesis of mesoporous silica. Nature **1995**, *378* (6555), 366–368.
19. Attard, G.S., Leclerc, S.A.A., Maniguet, S., Russell, A.E., Nandhakumar, I., Bartlett, P.N. Mesoporous Pt/Ru alloy from the hexagonal lyotropic liquid crystalline phase of a nonionic surfactant. Chem. Mater. **2001**, *13* (5), 1444–1446.
20. Huo, Q.S., Margolese, D.I., Ciesla, U., Demuth, D.G., Feng, P.Y., Gier, T.E., Sieger, P., Firouzi, A., Chmelka, B.F., Schuth, F., Stucky, G.D. Organization of organic-molecules with inorganic molecular-species into nanocomposite biphase arrays. Chem. Mater. **1994**, *6* (8), 1176–1191.
21. Monnier, A., Schuth, F., Huo, Q., Kumar, D., Margolese, D., Maxwell, R.S., Stucky, G.D., Krishnamurty, M., Petroff, P., Firouzi, A., Janicke, M., Chmelka, B.F. Cooperative formation of inorganic-organic interfaces in the synthesis of silicate mesostructures. Science **1993**, *261* (5126), 1299–1303.

22. Tanev, P.T., Pinnavaia, T.J. Mesoporous silica molecular sieves prepared by ionic and neutral surfactant templating: a comparison of physical properties. Chem. Mater. **1996**, *8* (8), 2068–2079.
23. Zhao, D.Y., Feng, J.L., Huo, Q.S., Melosh, N., Fredrickson, G.H., Chmelka, B.F., Stucky, G.D. Triblock copolymer syntheses of mesoporous silica with periodic 50 to 300 angstrom pores. Science **1998**, *279* (5350), 548–552.
24. Antonelli, D.M., Nakahira, A., Ying, J.Y. Ligand-assisted liquid crystal templating in mesoporous niobium oxide molecular sieves. Inorg. Chem. **1996**, *35* (11), 3126–3136.
25. Tian, B.Z., Liu, X.Y., Tu, B., Yu, C.Z., Fan, J., Wang, L.M., Xie, S.H., Stucky, G.D., Zhao, D.Y. Self-adjusted synthesis of ordered stable mesoporous minerals by acid–base pairs. Nat. Mater. **2003**, *2* (3), 159–163.
26. Soler-Illia, G.J.D., Sanchez, C., Lebeau, B., Patarin, J. Chemical strategies to design textured materials: from microporous and mesoporous oxides to nanonetworks and hierarchical structures. Chem. Rev. **2002**, *102* (11), 4093–4138.
27. Brinker, C.J., Lu, Y.F., Sellinger, A., Fan, H.Y. Evaporation-induced self-assembly: nanostructures made easy. Adv. Mater. **1999**, *11* (7), 579–585.
28. Alberius, P.C.A., Frindell, K.L., Hayward, R.C., Kramer, E.J., Stucky, G.D., Chmelka, B.F. General predictive syntheses of cubic, hexagonal, and lamellar silica and titania mesostructured thin films. Chem. Mater. **2002**, *14* (8), 3284–3294.
29. Kim, J.M., Han, Y.J., Chmelka, B.F., Stucky, G.D. One-step synthesis of ordered mesocomposites with non-ionic amphiphilic block copolymers: implications of isoelectric point, hydrolysis rate and fluoride. Chem. Commun. **2000**, *24*, 2437–2438.
30. Yu, C.Z., Tian, B.Z., Fan, B., Stucky, G.D., Zhao, D.Y. Salt effect in the synthesis of mesoporous silica templated by non-ionic block copolymers. Chem. Commun. **2001**, *24*, 2726–2727.
31. Che, S.N., Lim, S.Y., Kaneda, M., Yoshitake, H., Terasaki, O., Tatsumi, T. The effect of the counteranion on the formation of mesoporous materials under the acidic synthesis process. J. Am. Chem. Soc. **2002**, *124* (47), 13962–13963.
32. Sands, D.E. *Introduction to Crystallography*; Dover Publications: New York, 1994.
33. Liu, X.Y., Tian, B.Z., Yu, C.Z., Gao, F., Xie, S.H., Tu, B., Che, R.C., Peng, L.M., Zhao, D.Y. Room-temperature synthesis in acidic media of large-pore three-dimensional bicontinuous mesoporous silica with *Ia3d* symmetry. Angew. Chem. Int. Ed. **2002**, *41* (20), 3876–3878.
34. Cagnol, F., Grosso, D., Soler-Illia, G., Crepaldi, E.L., Babonneau, F., Amenitsch, H., Sanchez, C. Humidity-controlled mesostructuration in CTABtemplated silica thin film processing. The existence of a modulable steady state. J. Mater. Chem. **2003**, *13* (1), 61–66.
35. Alonso, B., Albouy, P.A., Durand, D., Babonneau, F. Directing role of pH and ethanol vapour on the formation of 2D or 3D mesostructured silica and hybrid organo-silica thin films. N.J. Chem. **2002**, *26* (10), 1270–1272.
36. Gallis, K.W., Landry, C.C. Synthesis of MCM-48 by a phase transformation process. Chem. Mater. **1997**, *9* (10), 2035–2038.
37. Schacht, S., Huo, Q., VoigtMartin, I.G., Stucky, G.D., Schuth, F. Oil–water interface templating of mesoporous macroscale structures. Science **1996**, *273* (5276), 768–771.
38. Zhao, D.Y., Yang, P.D., Huo, Q.S., Chmelka, B.F., Stucky, G.D. Topological construction of mesoporous materials. Curr. Opin. Solid-State Mater. Sci. **1998**, *3* (1), 111–121.
39. Tian, B.Z., Liu, X.Y., Yang, H.F., Xie, S.H., Yu, C.Z., Tu, B., Zhao, D.Y. General synthesis of ordered crystallized metal oxide nanoarrays replicated by microwave-digested mesoporous silica. Adv. Mater. **2003**, *15* (16), 1370–1374.

Section III

Solids Handling and Processing

7

Adsorption

Shivaji Sircar

CONTENTS

Introduction 87
Adsorption as a Separation Process 88
Adsorbent Materials 90
Key Adsorptive Properties for Separation 92
Adsorption Equilibria...92
Heat of Adsorption 96
Heat of Immersion...97
Adsorption Kinetics..98
Description of Selected Adsorptive Separation Processes 100
    Adsorptive Drying..100
    Air Fractionation 101
    Production of Hydrogen 103
    Separation of Bulk Liquid Mixtures 104
Emerging Adsorptive Separation Processes...104
    Rapid PSA Cycles...105
    Radial and Rotary Bed Adsorbers 105
    Hybrid Gas Separation Using Adsorption 106
    Nanoporous Carbon Membrane 106
    Simultaneous Adsorption and Reaction 107
Conclusions...107
References..108

Introduction

The separation and purification of fluid mixtures (gas or liquid) by adsorption is a major unit operation in the chemical, petrochemical, environmental, pharmaceutical, and electronic gas industries. A list of the key commercial applications of this technology is given in Table 7.1.[1] The phenomenal growth in the development of this technology is demonstrated by Figure 7.1, which shows a year-by-year tally of U.S. patents issued between 1980 and 2000 on five different topics of adsorption.[1] The total number of patents is overwhelming.

TABLE 7.1

Key Commercial Applications of Adsorption Technology

Gas separation
 Gas drying
 Trace impurity removal
 Air separation
 Carbon dioxide–methane separation
 Solvent vapor recovery
 Hydrogen and carbon dioxide recovery from steam-methane reformer off-gas
 Hydrogen recovery from refinery off-gas
 Carbon monoxide–hydrogen separation
 Alcohol dehydration
 Production of ammonia synthesis gas
 Normal–isoparaffin separation
 Ozone enrichment
Liquid separation
 Liquid drying
 Trace impurity removal
 Xylene, cresol, cymene isomer separation
 Fructose–glucose separation
 Fatty chemicals separation
 Breaking azeotropes
 Carbohydrate separation
Environmental separation
 Municipal and industrial waste treatment
 Ground and surface water treatment
 Air pollution control
 VOC removal
 Mercury vapor removal
Bioseparation and pharmaceutical separation
 Recovery of antibiotics
 Purification and recovery of enzymes
 Purification of proteins
 Recovery of vitamins
 Separation of enantiomers of racemic compounds
 Removal of micro-organisms
 Home medical oxygen production
Electronic gas purification
 Production of ultrahigh-purity N_2, Ar. He, H_2, O_2
 Purification of fluorinated gases NF_3, CF_4, C_2F_6, SiF_4
 Purification of hydrides NH_3, PH_3, ASH_3, SIH_4, Si_2H_6

Adsorption as a Separation Process

Adsorption is a surface phenomenon. When a multicomponent fluid mixture is contacted with a solid adsorbent, certain components of the mixture (adsorbates) are preferentially concentrated (selectively adsorbed) near the solid surface creating an adsorbed phase. This is because of the differences in the fluid–solid molecular forces of attraction between the

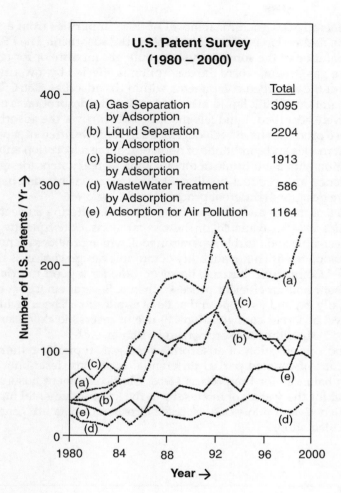

FIGURE 7.1
U.S. patent survey of adsorption topics.

components of the mixture. The difference in the compositions of the adsorbed and the bulk fluid phases forms the basis of separation by adsorption. It is a thermodynamically spontaneous process, which is exothermic in nature. The reverse process by which the adsorbed molecules are removed from the solid surface to the bulk fluid phase is called desorption. Energy must be supplied to carry out the endothermic desorption process. Both adsorption and desorption form two vital and integral steps of a practical adsorptive separation process where the adsorbent is repeatedly used. This concept of regenerative use of the adsorbent is key to the commercial and economic viability of this technology.

Three generic adsorptive process schemes have been commercialized to serve most of the applications shown in Table 7.1. They include 1) temperature swing adsorption (TSA); 2) pressure swing adsorption (PSA); and 3) concentration swing adsorption (CSA).[2–9] The fluid mixture (feed) to be separated is passed over a regenerated adsorbent (contained in an adsorber vessel) to produce a stream enriched in the less strongly adsorbed components of the mixture, followed by desorption of the adsorbed components, which produces a stream enriched in the more strongly adsorbed components of the mixture. The TSA processes are generally designed for removal of trace impurities from a mixture (gas or liquid), where the desorption is effected by heating the adsorbent. The PSA processes are designed for separation of the components of a bulk gas mixture or for removal of dilute impurities from a gas stream, where the desorption is effected by lowering the gas phase partial pressure of the adsorbed components within the adsorber. The CSA processes are designed for separation of bulk liquid mixtures, where the desorption is effected by flowing a less selectively adsorbed liquid (eluent or desorbent) over the adsorbent. Numerous variations of these processes have been developed to achieve different separation goals by using 1) different modes and conditions of operation of the adsorption and the desorption steps in conjunction with a multitude of other complementary steps (designed to improve separation efficiency and product quality); 2) different types of adsorbents; 3) different process hardware designs; 4) different process control logic, etc.

Several families of micro- and mesoporous adsorbents offering a spectrum of adsorption characteristics are also available for these separations. Consequently, the technology has been a very versatile and flexible separation tool, which provides many different paths for a given separation need. This availability of multiple design choices is the driving force for innovations.[10] Commercial success, however, calls for a good marriage between the optimum adsorbent and an efficient process scheme. Several emerging concepts in this field can potentially expand its scope and scale of application. These include 1) rapid PSA processes; 2) novel adsorber configurations; 3) use of reversible chemisorbents; 4) adsorbent membranes; 5) simultaneous sorption and reaction, etc.[10]

The design and optimization of adsorptive processes typically require simultaneous numerical solutions of coupled partial differential equations describing the mass, heat, and momentum balances for the process steps. Multicomponent adsorption equilibria, kinetics, and heat for the system of interest form the key fundamental input variables for the design.[11,12] Bench- and pilot-scale process performance data are generally needed to confirm design calculations.

Adsorbent Materials

A key factor in the development of adsorption technology for the fluid separation has been the availability of appropriate adsorbents. The most frequently used categories include crystalline materials like zeolites, and amorphous materials like activated carbons, silica and alumina gels, polymeric sorbents, and ion-exchange resins. These materials exhibit a large spectrum of pore structures (networks of micro- and mesopores of different shapes and sizes) and surface chemistry (degrees of polarity), which provide a large choice of core adsorptive properties (equilibria, kinetics, and heat) to be utilized in the design of the separation processes. Table 7.2 lists some of the physical properties of common adsorbents.

The microporous alumino-silicate zeolites (Types A, X, and mordenite are frequently used) provide a variety of pore openings (3–10 Å), cavity and channel sizes, and framework Si/Al ratios. They are also available in various cationic exchanged forms (Na, K, Li, Ag, Ca, Ba, Mg), which govern their pore openings and cationic adsorption site polarities. They are highly hydrophilic materials and must be dehydrated before use. The amorphous adsorbents contain an intricate network of micropores and mesopores of various shapes and sizes. The pore size distribution may vary over a wide range. The activated carbons and the polymeric sorbents are relatively hydrophobic in nature. The silica and

TABLE 7.2

Physical Properties of some Adsorbents

	NaX Zeolite (Bayer, Germany)[b]	BPL Carbon (Calgon, U.S.A.)	Molecular Sieve Carbon (Takeda, Japan)	H151 Alumina (Alcoa, U.S.A.)	Silica Gel (Grace, U.S.A.)
BET area (m²/g)	—	1100	—	350	800
Pore volume (cm³/g)	0.54	0.70	0.43	0.43	0.45
Bulk density (g/cm³)	0.65	0.48	0.67	0.85	0.77
Mean pore diameter (Å)	7.4[a]	30	3.5	43	22

[a] Crystal pore aperture size.
[b] Manufacturer given in parentheses.

alumina gels are more hydrophilic (less than zeolites) and they must also be dehydrated before use.

Commercial adsorbents are generally produced in bound forms (0.5–6.0 mm diameters) in regular particle shapes (beads, pellets, extrudates, granules, etc.). The purpose is to reduce pressure drops in adsorbers. Clay, alumina, polymers, pitch, etc. are used as binders, which typically constitute 10–20% (by weight) of the final product. The binder phase usually contains a network (arteries) of meso- and macropores (0.5–50.0 μm diameters) to facilitate the transport of the adsorbate molecules from the bulk fluid phase to the adsorption sites (within zeolite crystals and micropores of amorphous adsorbents) and vice versa. Adsorption of fluid molecules on the binder material is generally very weak. Figure 7.2 shows a schematic drawing of a bound zeolite pellet depicting the pathways for transport of the adsorbate molecules.

The vast majority of fluid separation by adsorption is affected by the thermodynamic selectivity of the adsorbent for certain components of the fluid mixture over others. Physisorption is the dominant mechanism for separation. Thus, it is governed by the surface polarity of the adsorbent and the polarizability and permanent polarity of the adsorbate molecules. All adsorbate molecules, in this case, have access to the adsorption sites. The separation can also be based on a kinetic selectivity by the adsorbent where certain

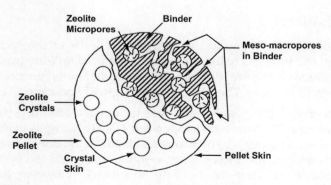

FIGURE 7.2
Schematic drawing of a bound adsorbent particle.

molecules of the fluid mixture diffuse into the adsorbent pores faster than the others because of their relative size differences. Size or steric exclusion of certain components of a fluid mixture from entering the adsorbent pores (typically for zeolites) is also possible. The last case is known as "molecular sieving." Adsorbents may be energetically homogenous, containing adsorption sites of identical adsorption energy (heat of adsorption), or energetically heterogenous, containing a distribution of sites of varying energies. The cause of adsorbent heterogeneity is generally physicochemical in nature. It is created by a distribution of micro- and mesopores of different sizes and shapes within the adsorbent particle as well as by a distribution of adsorption sites of different surface chemistry and polarity within the micropores.

An adsorbent is often tailor-made to suit a separation need or a process can be designed to best fit the properties of an adsorbent. Special adsorbents are also available for specific applications (e.g., removal of mercury vapor, drying of reactive fluids, resistance to acids, etc). More recently, adsorbents have been produced that use reversible chemisorption as the mechanism for gas separation.[13] Creation of new adsorbents and modification of existing adsorbents continue to be an active area of research and development.

Key Adsorptive Properties for Separation

All practical adsorptive separation processes are carried out using a stationary packed bed (adsorber) of the adsorbent particles. Each particle is subjected to the adsorption, the desorption, and the complementary steps of the process in a cyclic fashion. The ad(de)sorption characteristics exhibited by the particle during different periods of the cycle are governed by the multicomponent adsorption equilibria, kinetics, and heat for the fluid mixture of interest under the local conditions (e.g., fluid phase pressure, temperature and composition, adsorbate loadings in the particle, and its temperature) that the particle experiences. As these conditions can vary over a wide range during a process cycle, it is imperative that those adsorptive properties be accurately known over that range for reliable process design.

Adsorption Equilibria

Adsorption equilibria determine the thermodynamic limits of the specific amounts of adsorption (mol/g) of a pure gas or the components of a fluid mixture (gas or liquid) under a given set of conditions [pressure (P), temperature (T), and mole function (y_i or x_i) of component i] of the bulk fluid phase. The simplest way to describe adsorption equilibria of pure gas i is in the form of adsorption isotherms where the amount adsorbed $\left(n_i^0 \right)$ is plotted as a function of gas pressure (P) at a constant temperature (T). The pure gas adsorption isotherms can have various shapes (Types I–V) by Brunauer classification depending on the porosity of the adsorbent (microporous, mesoporous, or nonporous) and the system temperature (below or above the critical temperature of the adsorbate).[9] However, the most common isotherm shape is Type I, which is depicted by most microporous adsorbents of practical use. These isotherms exhibit a linear section in the very low-pressure region (Henry's law region) where the amount adsorbed is proportional to the gas pressure $\left[\left(n_i^0 = K_i P\right)\right]$. The proportionality constant is called Henry's law constant (K_i), which is a function of temperature only. The amount adsorbed monotonically increases with increasing pressure beyond the Henry's law region with a progressively decreasing isotherm slope and finally the amount adsorbed asymptotically approaches the saturation adsorption capacity (m_i) of the adsorbate. Figures 7.3A and 7.3B show examples of Type I isotherms for adsorption of pure N_2 and pure O_2, respectively, on various zeolites at 25°C.[2] The figures demonstrate that N_2 is more strongly adsorbed than O_2 on all zeolites and their adsorption characteristics are significantly affected by the structure of the zeolite, as well as by the nature of the cation present in them. The LSX zeolites in Figure 7.3 represent X zeolite structure with low Si/Al ratio. The amounts adsorbed of a pure gas at any given pressure decrease with increasing temperature because of the exothermic nature of the adsorption process.

The equilibrium amounts adsorbed of component i from a binary gas mixture (n_i) are generally described as functions of gas phase mole fractions (y_i) at a constant system temperature (T) and total gas pressure (P). An example is given in Figure 7.4 for adsorption of binary N_2–O_2 mixtures on Na–mordenite at various temperatures where the total gas pressure was 1.0 atm.[2] These binary isotherm shapes are typical for Type I adsorption systems on microporous adsorbents.

The relative adsorption between components i and j of a gas mixture is expressed in terms of the selectivity of adsorption ($S_{ij} = n_i y_j / n_j y_i$). Component i is more selective than component j if $S_{ij} > 1$. The thermodynamic selectivity decreases with increasing T for any given values of n_i. For adsorption on a homogenous adsorbent at constant T, S_{ij} can be constant, increase, or decrease with adsorbate loading depending on the size differences between the molecules of components i and j.[14] For adsorption on a heterogenous adsorbent, S_{ij} generally decreases with increasing adsorbate loading.[2] Table 7.3 gives a list of Henry's law selectivity ($S_{ij}^* = K_i / K_j$) for several binary gas mixtures at 30°C on a zeolite and an activated carbon.[2] The first-mentioned gas of a pair is the more selectively adsorbed component.

Separation of a gas mixture by a time dependent kinetic selectivity [$S_{ij}(t) = n_i(t) y_j / n_j(t) y_i$] has also been used in practice when there is a difference in the rates of adsorption of the components of the gas mixture. $n_i(t)$, in this case, is the amount of component i adsorbed at time t.

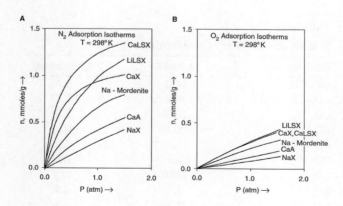

FIGURE 7.3
Pure gas adsorption isotherms for (A) nitrogen and (B) oxygen on various zeolites.

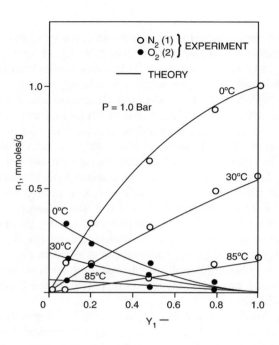

FIGURE 7.4
Binary gas adsorption isotherms for nitrogen (1) and oxygen (2) mixtures on sodium mordenite.

Numerous models have been developed to describe pure and multicomponent gas adsorption on porous adsorbents. The analytical models are, however, most useful for process design. A few analytical models for Type I adsorption systems, which are thermodynamically consistent, are given below[2]:

$$\text{Langmuir:} \quad b_i P y_i = \theta_i / (1 - \theta_i) \tag{1}$$

$$\text{Multisite Langmuir:} \quad b_i P y_i = \theta_i / \left(1 - \sum \theta_i\right)^{a_i} \tag{2}$$

TABLE 7.3

Selectivities of Binary Gas Mixtures

Gas Mixture	5A Zeolite	BPL Carbon
CO_2–CH_4	195.6	2.5
CO_2–CO	59.1	7.5
CO_2–N_2	330.7	11.1
CO_2–H_2	7400.0	90.8
CO–CH_4	3.3	0.33
CO–N_2	5.6	1.48
CO–H_2	125.0	12.1
CH_4–N_2	1.7	4.5
CH_4–H_2	37.8	36.6
N_2–H_2	22.3	8.2

Martinez-Basmadjian: $\quad b_i P y_i = \left\{ \theta_i \Big/ \left(1 - \sum \theta_i\right)^{a_i} \right\} \exp\left(-\sum a_i w_i \theta_i\right)$ (3)

Toth: $\quad b_i P y_i = \theta_i \Big/ \left\{ 1 - \left(\sum \theta_i\right)^k \right\}^{1/k}$ (4)

The frequently used Langmuir model describes adsorption of equal-sized adsorbates ($m_i = m_j$) on an energetically homogenous adsorbent. The multisite Langmuir model is an extension to include the effects of dissimilar adsorbate sizes ($m_i \neq m_j$). The Martinez–Basmadjian model is a further extension to include lateral interactions in the adsorbed phase. The Toth model is developed to describe adsorption of equal-sized molecules on an energetically heterogenous adsorbent. The variables of Eqs. (1)–(4) are the fractional coverage of component i of the gas mixture ($\theta_i = n_i/m_i$) at P, T, and y_i, the number of adsorption sites occupied by the adsorbate type i (a_i), the energy of lateral interactions between i molecules in the adsorbed phase (w_i), the gas–solid interaction parameter for component i (b_i), and the adsorbent heterogeneity parameter for all adsorbates ($k < 1$). The temperature coefficient of the parameter b_i is given by:

$$b_i = b_i^0 \exp\left(q_i^*/RT\right)$$ (5)

where b_i^0 is a constant. q_i^* is the isosteric heat of adsorption of pure gas i in the Henry's law region. R is the gas constant. The pure gas adsorption isotherms $\left(\theta_i^0 \text{ vs. } P\right)$ for these models can be obtained by setting $y_i = 1$.

The extent of specific equilibrium adsorption of component i from a liquid mixture having mole fraction x_i for that component is expressed in terms of a variable called the Gibbsian surface excess $\left[n_i^e \text{ (mol/g)}\right]$, which is related to the actual amounts adsorbed by[15]

$$n_i^e = n_i - \left(\sum n_i\right) x_i$$ (6)

The surface excess of component i is equal to its actual amount adsorbed $\left(n_i^e \sim n_i\right)$ only when $x_i \ll 1$ and the component i is very selectively adsorbed ($S_{ij} \gg 1$) over other components of the mixture.[15] The binary liquid phase surface excess adsorption isotherm $\left(n_1^e \text{ vs. } x_1\right)$ at a constant temperature (pressure is not a variable) on a microporous adsorbent is often U-shaped, as shown in Figure 7.5A. Component 1 is selectively adsorbed if $n_1^e > 0$. For adsorption of a dilute solute from a liquid mixture, on a microporous solid, the excess isotherm is similar in shape as Type I isotherm for vapor adsorption (as shown in Figure 7.5B).

Numerous analytical models have also been developed for binary liquid phase surface excess isotherms. A model equation that accounts for adsorbate size differences, bulk liquid phase nonideality, as well as a simplified description of adsorbent heterogeneity is given below[16]:

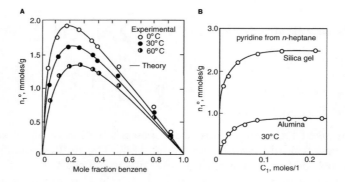

FIGURE 7.5
Binary surface excess isotherms for adsorption of liquid mixtures: (A) benzene (1) + cyclohexane (2) on silica gel and (B) pyridine (1) + n-heptane (2) on silica gel and alumina.

$$n_1^e = \frac{m_1}{\left(S_{0H} - S_{0L}\right)} \left\{ \frac{x_2}{a_1} \left[\left(\frac{S_H}{S_{0H}}\right)^{1/(\beta-1)} - \left(\frac{S_L}{S_{0L}}\right)^{1/(\beta-1)} \right] + \frac{a_2}{a_1(\beta-1)} \left(\frac{S_{0H}}{S_H} - \frac{S_{0L}}{S_L}\right) \right\} \tag{7}$$

$$\left(\frac{S_H}{S_{0H}}\right) = \left(S_H a_1 + a_2\right)^{(\beta-1)/\beta} ; \left(\frac{S_L}{S_{0L}}\right) = \left(S_L a_1 + a_2\right)^{(\beta-1)/\beta} \tag{8}$$

where the variable $\beta = (m_1/m_2)$, a_i is the activity of component i in the bulk liquid phase, and S_{0L} and S_{0H} are the selectivities of adsorption of component 1 over component 2 at the limit of $x_1 \to 0$ at the lowest and the highest energy sites of the adsorbent.

Heat of Adsorption

The pertinent thermodynamic variable to quantitatively describe the thermal effects in the exo(endo) thermic gas ad(de)sorption process is called the isosteric heat of adsorption.[17] The isosteric heat of adsorption of component i of an ideal gas mixture (q_i) at adsorbate loading of n_i and temperature T is given by the following thermodynamic relationship[17]:

$$q_i\left(n_i\right) = RT^2 \left\{ \frac{\delta \ln\left(Py_i\right)}{\delta T} \right\}_{n_i} \tag{9}$$

Eq. (9) is frequently used to obtain the isosteric heat of adsorption of a pure gas $i\left(q_i^0\right)$ as a function of n_i^0 and T from measured isotherms at different temperatures (y_i is equal to unity in that case). Estimation of isosteric heat of adsorption of component i of a gas mixture by using Eq. (9) is, however, not practical and they must be obtained by calorimetric

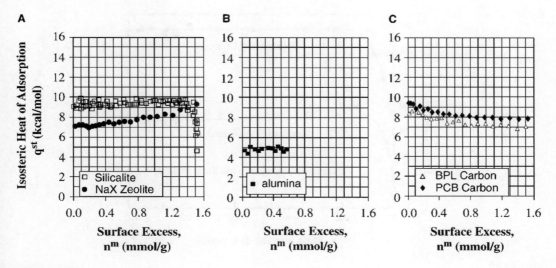

FIGURE 7.6
Isosteric heat of pure SF_6 on various absorbents: (A) zeolites, (B) alumina, and (C) activated carbons.

measurements. In the absence of lateral interactions, the isosteric heat of adsorption of a pure gas on an energetically homogenous adsorbent is independent of adsorbate loadings. It remains constant at its value at the Henry's law region $\left(q_i^*\right)$ at all coverages. The presence of lateral interactions between adsorbed molecules (pronounced at higher coverages) can increase q_i^0 with increasing n_i^0. The isosteric heat decreases with increasing adsorbate loading when the adsorbent is energetically heterogeneous. The isosteric heat is generally a very weak function of T. Figure 7.6 shows several calorimetrically measured examples of these behaviors for adsorption of pure SF_6 on various micro- and mesoporous adsorbents. [18] The isosteric heat of adsorption of a component of a gas mixture is equal to that of the pure gas for adsorption on a homogenous adsorbent. However, the component isosteric heat of a mixture can be substantially different from that of the pure gas at the same loadings when the adsorbent is energetically heterogeneous.[17] Furthermore, the dependence of component isosteric heat on the adsorbate loadings can be very complex in that situation.[17]

Heat of Immersion

The thermodynamic property to quantify the heat effects for ad(de)sorption of a liquid mixture is the heat of immersion. Figure 7.7 shows examples of heat of immersion of binary benzene–cycloxane mixtures on two activated carbons at 30°C.[15] The corresponding surface excess isotherms are U-shaped (e.g., see Figure 7.5A). Benzene is the more selectively adsorbed species. The temperature coefficients of liquid phase adsorption isotherms are generally much smaller than those for gas phase adsorption. The temperature changes in a liquid phase adsorber because of the ad(de)sorption processes are also small owing to the high heat capacity of the liquid.

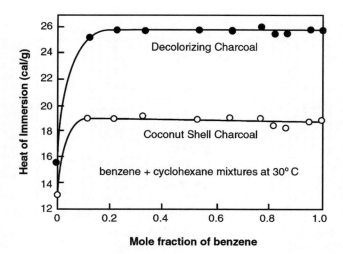

FIGURE 7.7
Heat of immersion of binary liquid mixtures of benzene (1) + cyclohexane (2) on activated carbons.

Adsorption Kinetics

The actual kinetics of the ad(de)sorption process is generally very fast (in the order of microseconds). However, a significant resistance may exist for transfer of the adsorbate molecules from the bulk fluid phase to the adsorption sites inside the micropores of the adsorbent (see Figure 7.2). For gas adsorption, these resistances may be caused by 1) molecular diffusion through a fluid film outside the adsorbent particles (for mixtures only); 2) Knudsen diffusion through the meso- and macropores of the adsorbent and the binder material in parallel with surface diffusion of adsorbed molecules on the walls of those pores (if any); and 3) activated (hopping) diffusion of adsorbed molecules inside the micropores. The same resistances exist for adsorption of liquid mixtures except that the flow through the meso-macropores of the adsorbent and the binder is controlled by molecular and surface diffusion through liquid filled pores. Additional mass transfer resistance called "skin resistances" at the surface of the adsorbent particles or zeolite crystals has also been observed.[2] Some or all of these processes are strongly influenced by the local fluid phase concentration and temperature within the adsorbent particle.

The magnitudes of gas diffusivities by different mechanisms follow the order: meso-macropore gas diffusivity $(D_p) \gg$ surface diffusivity (D_s) in those pores $\geq$ micropore diffusivity (D_m). For example, D_p and D_s for adsorption of water vapor into the mesopores of an activated alumina sample were ~5 – 10 × 10^{-2} cm^2/sec and ~3 × 10^{-6} cm^2/sec, respectively.[4] The micropore diffusivity of water vapor into the crystals of NaA zeolite was ~1 × 10^{-6} cm^2/sec.[4]

The diffusivity of a liquid adsorbate through the meso-macropores is much slower than that for a gas. For example, the diffusivity of bulk liquid phase water from ethanol into an alumina sample was ~6 × 10^{-7} cm^2/sec.[4]

The time constant for meso-macropore diffusion into an adsorbent particle of radius R_p is given by $\left(D_p/R_p^2\right)$. The time constant for micropore diffusion into a pore (having a characteristic diffusional distance of R_m) is given by $\left(D_m/R_m^2\right)$. Because R_m (1–2 μm) is typically much smaller than R_p (1–2 mm), the meso-macropore diffusion generally controls the overall mass transfer into a practical adsorbent particle, even though $D_m \ll D_p$. An exception will be the case where the diameter of the adsorbate molecule is very close to that of the micropore, so that D_m is extraordinarily small, and the micropore diffusion becomes the controlling mechanism. Table 7.4 shows the micropore diffusivities of various gases in the Henry's law region into the crystals of several zeolites at 300 K.[2]

The table shows the remarkable decrease in the micropore diffusivity of a gas when its molecular diameter approaches that of the zeolite pore. The temperature coefficients of D_s and D_m are given by the Arrhenius relationship [D_s, $D_m = D^0 \exp(-E/RT)$] because these diffusions are activated processes. E is the activation energy for the diffusion process and D^0 is a constant. These diffusivities can also be complex functions of adsorbate loadings and compositions.[19]

The most rigorous formulation to describe adsorbate transport inside the adsorbent particle is the chemical potential driving force model. A special case of this model for an isothermal adsorption system is the Fickian diffusion (FD), model which is frequently used to estimate an effective diffusivity for adsorption of component i (D_i) from experimental uptake data for pure gases.[2,19] The FD model, however, is not generally used for process design because of mathematical complexity. A simpler analytical model called linear driving force (LDF) model is often used.[20] According to this model, the rate of adsorption of component i of a gas mixture into an adsorbent particle of radius R is given by[11]:

$$\frac{d\bar{n}_i}{dt} = k_{ii}\left(\bar{n}_i^* - \bar{n}_i\right) + \sum k_{ij}\left(\bar{n}_j^* - \bar{n}_j\right) \tag{10}$$

where $\bar{n}_i$ is the average adsorbate loading of component i in the particle at time t. The variable $\bar{n}_i^*$ represents the adsorbate loading of component i that would be in equilibrium with the superincumbent gas phase conditions at time t. k_{ii} and k_{ij} are straight and cross mass transfer coefficients for component i. Further simplifications are often made by assuming that $k_{ij} = 0$ and k_{ii} is a function of T only. The relation between k_{ii} and D_i is generally given by $K_{ii} = \Omega(D_i/R^2)$, where Ω is a constant.[20] A value of $\Omega = 15$ is often used even though other values are also possible.[20] A parallel formulation called the surface excess linear driving force (SELDF) model for describing adsorption kinetics from liquid mixtures using Gibbsian surface excess as the variable has also been used successfully.[4]

Table 7.5 shows examples of LDF mass transfer coefficients for adsorption of several binary gas mixtures on BPL activated carbon particles (6–16 mesh) at ~23–30°C.[21] The data show that the mass transfer coefficients are relatively large for these systems. There is a

TABLE 7.4

Examples of Pure gas Diffusivities into Zeolites at 300 K[a]

Gas	Kinetic Diameter (Å)	NaA Zeolite (4A) (m²/sec)	Na-CaA Zeolite (5A) (m²/sec)	Nax Zeolite (13X) (m²/sec)
N_2	3.70	3.1×10^{-14}	1.1×10^{-10}	Fast
Kr	3.65	1.2×10^{-17}	—	—
CH_4	3.76	1.2×10^{-15}	6.0×10^{-10}	—
n-C_4H_{10}	4.69	9.6×10^{-20}	1.5×10^{-13}	2.4×10^{-11}

[a] Effective crystal pore openings of 4A, 5A, and 13X zeolites are 4.0, 4.9, and 7.6 Å, respectively.

TABLE 7.5

Examples of Binary LDF Mass Transfer Coefficients on BPL Carbon

Gas Mixtures	k_{11} (sec^{-1})	k_{22} (sec^{-1})
CO_2 (1) + He (2)	0.44	—
CH_4 (1) + He (2)	1.42	—
N_2 (1) + He (2)	3.33	—
CO_2 (1) + CH_4 (2)	0.35	0.76
CO_2 (1) + N_2 (2)	0.35	0.66

scarcity of multicomponent adsorption equilibria, kinetics, and heat data in the published literature. This often restricts extensive testing of theoretical models for prediction of multicomponent behavior.

Description of Selected Adsorptive Separation Processes

Adsorption has become the state of the art technology for many separation applications as listed in Table 7.1. The more prolific areas include:

a. Drying of gaseous and liquid mixtures.

b. Production of oxygen and nitrogen enriched gases from air.

c. Production of ultrapure hydrogen from various gas sources.

d. Separation of bulk liquid mixtures where distillation is not convenient.

There has been extensive research and development in all of these areas during the last 30 yr. For example, topics (a)–(c) alone have generated more than 600 U.S. patents during the period 1980–2000.[1] They have been assigned to 160 different corporations around the world. For the sake of brevity, only a selected few commercial processes will be discussed here.

Adsorptive Drying

Both TSA and PSA processes are commercially used for removal of trace or dilute water contamination from a gas. They are commercially designed to handle 1–40,000 ft³ of feed gas per minute. The basic steps of a conventional TSA process consists of: 1) flowing the contaminated gas over a packed bed of a desiccant (silica gel, alumina, zeolite) at a near-ambient temperature and withdrawing a dry product gas until the moisture concentration in the product gas rises to a preset level; 2) heating the adsorbent to ~150–300°C by flowing a hot dry gas countercurrently through the bed and rejecting the water laden effluent gas; and 3) cooling the bed to feed gas temperature by countercurrently flowing a dry gas through the bed at feed gas temperature while rejecting the effluent gas. The cycle is then repeated. A part of the dry gas (~10–30%) produced in step 1 is generally used to supply the gas for steps 2 and 3. The effluent gas from step 3 is often heated to supply the gas for step 2. Figure 7.8A is a schematic diagram of a three-column TSA gas drying unit. The total cycle time (all steps) for TSA processes generally varies between 2 and 8 hr. A typical

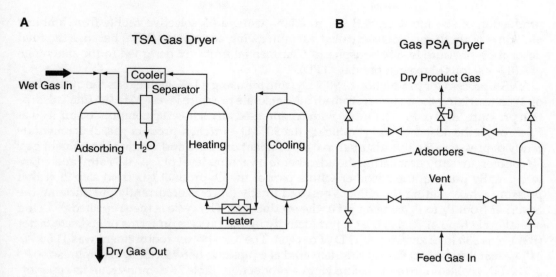

FIGURE 7.8
Schematic drawings of (A) three-column TSA and (B) two-column PSA processes.

dynamic water removal capacity of an alumina dryer is ~5–15% by weight. Product gas dew points of less than −40°C can be easily obtained. Many different process modifications like thermal pulsing, elimination of the cooling step, lower temperature regeneration, etc. are also used for decreasing the costs of drying.[4]

The TSA dryers for liquid mixtures use similar process steps to those used for gas dryers except that the adsorbers are first drained to remove the void liquid before they are heated countercurrently using a hot gas. Heating vaporizes the adsorbed water as well as liquid films adhering to the adsorbent particles. The hot effluent gas is cooled to condense out the components of the feed liquid mixture. The adsorber is then cooled by countercurrently flowing a cold gas and refilled with dry liquid before starting a new cycle.[4]

The basic steps of a conventional PSA gas drying process (Skarstrom cycle) consists of: 1) adsorption of water vapor from the feed gas by flowing the gas over a desiccant bed (silica gel, alumina, zeolite) at an elevated pressure (say 5–15 atm) and withdrawing a dry product gas at feed pressure; 2) countercurrently depressurizing the adsorber to near-ambient pressure and venting the effluent; 3) countercurrently purging the adsorber with a part of the dry gas produced by step 1) at near-ambient pressure while venting the effluent; and 4) countercurrently pressurizing the adsorber with a part of the product gas from step 1). Adsorption at a relatively lower feed gas pressure (1.3–1.7 atm) and desorption under vacuum (both depressurization and purge steps) are also practiced. Figure 7.8B is a schematic diagram of a two-bed PSA dryer. The process can be used to obtain a very dry product (say −60°C dew point). A typical process uses ~15–30% of the dry product gas as purge. The total cycle times for PSA processes generally vary between 2 and 6 min.[4]

Air Fractionation

A large variety of PSA process concepts have been commercialized for: 1) production of 23–95 mol% oxygen using a N_2 selective (thermodynamic) zeolite; 2) production of 99+ mol% nitrogen using an O_2 selective (kinetic) carbon molecular sieve; and 3) simultaneous production of 90+ mol% O_2 and 98+ mol% N_2 using a N_2 selective zeolite from ambient air. Some of these concepts are called vacuum swing adsorption (VSA) because the final desorption pressure is subatmospheric. Commercial units are designed in the size range of 0.012–100 tons of oxygen per day (TPD).[2]

A VSA process for production of 90% O_2 from air using a LiX zeolite uses the cyclic steps of: 1) pressurizing the adsorber from an intermediate pressure level (P_I) to the final adsorption pressure level of P_A (~1.43 atm) with compressed air; 2) flowing compressed air feed at P_A through the adsorber and producing the 90% O_2 enriched product gas; 3) countercurrently depressurizing the adsorber to near-ambient pressure and venting the effluent gas; (4) countercurrently evacuating the adsorber to a vacuum level of P_D (~0.34 atm) and countercurrently purging the adsorber with a part of the O_2 product gas from step 2) at that pressure while venting the effluent gas; and finally (5) countercurrently pressurizing the adsorber from P_D to P_I using a part of the product gas. The cycle is then repeated.[22] Using a total cycle time of 70 sec, the O_2 productivity by the process, in terms of bed size factor (BSF), was 830 lb of zeolite per TPD of oxygen. The total power requirement was 11.6 kW/TPD when the product O_2 gas was delivered at a pressure of 6.45 atm by recompression.[22]

The LiX zeolite is currently favored for O_2 production. Table 7.6 demonstrates its superiority over other zeolites for making 90 mol% O_2 from air using a specific VSA cycle operating between final adsorption and desorption pressures of 1.48 and 0.25 atm, respectively.[23]

The Carbon Molecular Sieves (CMS) contain constricted pore mouths that permit the slightly smaller O_2 molecules to diffuse into the pores of the carbon faster than the N_2 molecules from air. This produces an O_2 enriched adsorbed phase based on kinetic selectively when the CMS is contacted with air for a short period of time. The material has practically no thermodynamic selectivity for either gas.[24] A simple four-step PSA process using a CMS consists of: 1) passing compressed air into the adsorber to pressurize it to P_A, and then withdrawing a N_2 enriched product gas at that pressure; 2) connecting the adsorber with another adsorber that has completed step 3) below, to pressure equalize the adsorbers; 3) countercurrently depressurizing the adsorber to near-ambient pressure and venting the O_2 enriched gas, and finally 4) pressure equalizing the adsorber with the companion adsorber, which has completed step 1). The cycle is then repeated. Using a feed air pressure of ~8.1 atm ($=P_A$), the process could produce a 99.0 mol% N_2 product gas with a N_2 recovery of 49.4% from the feed air. The N_2 productivity was ~92 ft^3/hr per cubic foot of adsorbent.[24] The O_2 enriched waste gas contained 33.8% O_2. The N_2 productivity and recovery were reduced to 73% and 39.2%, respectively, when its purity was raised to 99.5 mol%. Other process cycles have been designed using CMS adsorbents to raise the N_2 product gas purity above 99.9 mol%.[25]

TABLE 7.6

Comparative Performance of Various Zeolites for O_2 Production by a VSA Process

Zeolite	BSF	Energy of Separation
NaX	1.00	1.00
CaX	1.28	1.78
LiX	0.51	0.88

Production of Hydrogen

The most common industrial method to make ultrapure hydrogen is by steam-methane reforming (SMR) using a catalyst at the temperature 890–950°C. The reformed gas is then subjected to a high temperature water gas shift (WGS) reaction at 300–400°C. The WGS reactor effluent typically contains 70–80% H_2, 15–25% CO_2, 1–3% CO, 3–6% CH_4, and trace N_2 (dry basis), which is fed to a PSA system at a pressure of 8–28 atm and a temperature of 20–40°C for production of an ultrapure (99.99+ mol%) hydrogen gas at the feed pressure. Various PSA systems have been designed for this purpose to produce 1–120 million cubic feet of H_2 per day.

A popular PSA cycle called polybed process consists of 11 cyclic steps.[26] They include: 1) passing the feed gas at pressure P_F through a packed adsorbent column and withdrawing the high-purity H_2 product gas; 2) cocurrently depressurizing the adsorber from pressure P_F to P_I while producing a stream of essentially pure H_2; 3) further cocurrently depressurizing the adsorber from pressure P_I and P_{II} and withdrawing another stream of H_2 enriched gas; 4) even further depressurizing the adsorber cocurrently from pressure P_{II} and P_{III} and again withdrawing a stream of H_2 enriched gas; 5) countercurrently depressurizing the adsorber from pressure P_{III} to a near-ambient pressure (P_D) and venting the effluent gas; 6) countercurrently purging the adsorber with a stream of essentially pure H_2 produced by a companion adsorber undergoing step 4); 7) countercurrently pressurizing the adsorber from P_D to P_{II} by introducing the effluent gas from another adsorber carrying out step 3; 8) further pressurizing the adsorber countercurrently from pressure P_{II} and P_I using the gas produced by another adsorber undergoing step 2; and 9) finally pressurizing the adsorber countercurrently from pressure P_I to P_F using a part of the product gas produced by another adsorber carrying out step 1. The cycle is then repeated. Increasing the pressure ratio between the feed gas and the purge gas (P_F/P_D) generally improves the separation performance of the PSA process by: 1) providing higher specific adsorption capacities for the selectively adsorbed components of the feed gas during the adsorption step (thus reducing adsorbent inventory) and 2) lowering the quantity of the purge gas required for adsorbent regeneration (thus increasing product recovery). However, increased feed gas pressure also increases the void gas quantity inside the adsorber at the end of the adsorption step, which: 1) increases the amount of adsorbent needed to contain the impurities during the subsequent cocurrent depressurization steps and 2) increases product loss during the countercurrent depressurization step. Consequently, there is an upper feed gas pressure limit (typically <40 atm) for optimum operation of the PSA cycle.

The most distinguishing feature of the polybed process is that step 1) is terminated when there is a substantial adsorption capacity left for the feed gas impurities inside the column so that these impurities are removed from the expanding void gases in steps 2–4 to produce high-purity H_2 effluents. The adsorbers are generally packed with a layer of an activated carbon at the feed end primarily for selective removal of H_2O and CO_2. A layer of 5A zeolite at the product end removes any remaining CO, CH_4, and N_2. These adsorbents are chosen due to the ease of desorption of the impurities from them. Figure 7.9A shows a schematic diagram of a polybed system containing nine parallel adsorbers. Using a total cycle time of 13.3 min, the polybed process could produce a 99.999% pure H_2 product from a WGS reactor effluent gas at 20.7 atm and 21°C with a H_2 recovery of 86.0%. The feed processing capacity was ~35 ft³ of feed gas per cubic foot of total adsorbent in the system/cycle.[26]

Pressure swing adsorption processes are also designed to produce high-purity (99.95+ %) H_2 products from refinery-off gases containing H_2 (65–90%) and C_1–C_5 hydrocarbon

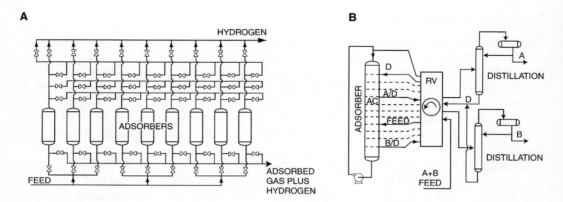

FIGURE 7.9
Schematic drawings of (A) nine column polybed PSA and (B) SMB process with two distillation columns.

impurities with high H_2 recoveries (~86+ %). Silica gel and activated carbons are used as adsorbents.[27]

Separation of Bulk Liquid Mixtures

A variety of simulated moving bed (SMB) processes has been commercialized for adsorptive separation of bulk liquid mixtures where distillation is not cost-effective. They are designed to process 100–120,000 tons of bulk chemicals per year. Key examples include separation of: 1) n-paraffins from branched and cyclic compounds; 2) olefins from paraffins; 3) xylene isomers; 4) glucose–fructose mixture; and 5) enantiomers of racemic compounds. The SMB concept is based on the principles of liquid phase chromatography. It uses a stationary bed of an adsorbent with multiple inlet and outlet ports along its length as shown by Figure 7.9B. Continuous countercurrent flow of the solid and the liquid phases inside the adsorber is mimicked by periodically moving the feed introduction and product withdrawal points to and from the adsorber using a rotary valve (RV). A weakly adsorbed desorbent liquid (D) is continuously circulated through the column with a pump. The pump flow rate is periodically changed with the change in the position of the RV. The process splits a feed liquid mixture (A + B) into two easy to separate liquid mixtures (A + D and B + D), which can then be further separated by distillation, if needed. For example, an SMB process could separate a n-paraffin + n-olefin feed mixture of C_{11}–C_{14} hydrocarbons containing ~9.0 wt% olefins to produce an olefin enriched stream containing ~96.2% olefins with an olefin recovery of ~94.0%. The paraffin enriched stream contained ~98.5 wt% paraffins. A hydrocarbon mixture of lower boiling range was used as the desorbent liquid.[6] The preferred adsorbent was NaX zeolite (caustic treated).[3,6]

Emerging Adsorptive Separation Processes

Design of new adsorptive process cycles using new (or modified) and existing adsorbents continues to grow. The following areas have attracted considerable attention in recent years.

Rapid PSA Cycles

The concept involves the use of faster cycle times (seconds) than those for conventional PSA cycles (minutes) to increase the productivity rate (volume of product gas/volume of adsorbent/hour) by several folds. This is achieved by operating an existing PSA cycle faster using modified hardware or by designing new process cycles using conventional hardware. An example of the latter case is a rapid PSA (RPSA) process for production of ~27.5% O_2 from air.[2] It uses multiple layers of a N_2 selective zeolite stacked in a single adsorber vessel. The layers are cyclically subjected to: 1) simultaneous air pressurization and adsorption and 2) simultaneous depressurization and back purging with O_2 product gas. Using a total cycle time of 10 sec, the O_2 productivity could be raised to ~2300 ft³/hr per cubic foot of adsorbent with an O_2 recovery of ~64%. This productivity is an order of magnitude higher than conventional PSA processes.[2]

Radial and Rotary Bed Adsorbers

The maximum permissible gas flow rate through a conventional packed bed adsorber is governed by pressure drop, local fluidization, and channeling within the column. Radial bed adsorbers are designed to circumvent some of these problems. The adsorbent is placed in an annular section between two coaxial cylindrical chambers with perforated walls, and the gas flows radially through the arrangement. Figure 7.10A is a schematic drawing of a radial bed adsorber.[28] The adsorber allows lower pressure drops, faster cycle times, higher throughputs, and elimination of fluidization, but they are more expensive to build. Radial beds have been used for both PSA and TSA processes.

The rotary bed adsorber (also called adsorption wheel) is designed to provide a truly continuous TSA system.[29] It uses a shallow wheel-shaped adsorption bed that continuously turns about an axis inside a fixed supporting frame. A section of the wheel is used to adsorb impurities from a feed gas while the other section is used to thermally regenerate

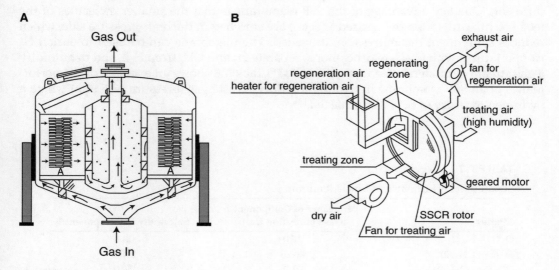

FIGURE 7.10
Schematic drawings of (A) radial bed adsorber and (B) rotary bed adsorber.

the adsorbent. Figure 7.10B is a schematic drawing of the adsorption wheel arrangement. The adsorbent is made from a honeycomb-shaped alumina substrate, which can be coated with layers of different adsorbents. The rotary adsorber has been designed for dehumidification of gases, solvent vapor recovery, volatile organic compounds (VOC) removal, gas deodorization, etc.

Hybrid Gas Separation Using Adsorption

Hybrid concepts combining the principles of selective adsorption with those of membrane technology and reaction engineering have been developed to enhance the overall separation or reaction efficiency.[10] Two examples are given below.

Nanoporous Carbon Membrane

The nanoporous carbon membrane consists of a thin layer (< 10 μm) of a nanoporous (3–7 Å) carbon film supported on a meso-macroporous solid such as alumina or a carbonized polymeric structure. They are produced by judicious pyrolysis of polymeric films. Two types of membranes can be produced. A molecular sieve carbon (MSC) membrane contains pores (3–5 Å diameters), which permits the smaller molecules of a gas mixture to enter the pores at the high-pressure side. These molecules adsorb on the pore walls and then they diffuse to the low-pressure side of the membrane where they desorb to the gas phase. Thus, separation is primarily based on differences in the size of the feed gas molecules. Table 7.7 gives a few examples of separation performance of MSC membranes.[30] Component 1 is the smaller component of the feed gas mixture.

The pores of a selective surface flow (SSF) membrane are large enough (6–7 Å) to permit all molecules of a gas mixture to enter the pores. However, the larger and more polar molecules are selectively adsorbed at the high-pressure side. Then, they selectively diffuse on the pore walls to the low-pressure side, where they desorb into the gas phase. The adsorbed molecules also block the passage of smaller molecules through the void space within the pores. Thus, separation is primarily based on selective adsorption and surface diffusion. One key advantage of the SSF membrane is that the smaller molecules of the feed gas mixture (often the desired species) are enriched in the high-pressure side, which reduces subsequent recompression, if needed. The membrane can be used to enrich H_2 from mixtures with C_1–C_4 hydrocarbons, separate H_2 and CH_4 from H_2S, and dehumidify a gas stream at moderate feed gas pressures.[30] Table 7.8 shows an example of the performance of an SSF membrane in treating a refinery waste gas (pressure ~3.5 atm) where a hydrogen recovery of 60% was achieved.[30]

TABLE 7.7

Separation Performance of MSC Membranes

Mixture	Permeance of Component 1 (Gas Permeation Unit)	Selectivity of Component 1
O_2 (1) + N_2 (2)	16.0	13.6
H_2 (1) + CH_4 (2)	365.5	500.0
CO_2 (1) + CH_4 (2)	91.0	50.0
C_3H_6 (1) + C_3H_8 (2)	183.0	12–15

TABLE 7.8

Hydrocarbon Rejections by SSF Membrane

Hydrocarbons	Feed gas (mol%)	Rejections (%)
C_3H_8	2–7	94
C_3H_6	2–7	94.5
C_2H_6	7–15	84–91
C_2H_4	5–7	86–91
CH_4	35–50	47–55
H_2	14–30	—

TABLE 9

Performance of SERP Concept for H_2 Production[a]

	Product Composition (dry)				Conversion (%)
	H_2	CH_4	CO_2	CO	
SERP concept	94.4%	5.6%	40 ppm	<30 ppm	73.0
Conventional SMR	67.2%	15.7%	15.9%	1.2%	52.6

[a] $T = 490°C$, $P = 26.1$ psia, $H_2O/CH_4 = 6.1$.

Simultaneous Adsorption and Reaction

The concept is based on Le Chatelier's principle that the conversion of an equilibrium controlled reaction as well as the rate of the forward reaction can be significantly enhanced by selectively removing one of the reaction products from the reaction zone. A selective adsorbent can be used for this purpose by mixing it with the catalyst in a reactor. The adsorbent, however, must be periodically regenerated. Such concepts are often called "pressure swing reactors" when the principles of PSA are used for regeneration.[10] One example is a cyclic process called sorption enhanced reaction process (SERP), which was developed for producing H_2 by SMR ($CH_4 + 2H_2O \leftrightarrow CO_2 + 4H_2$) using an admixture of a conventional SMR catalyst and a novel chemisorbent (K_2CO_3 promoted hydrotalcite) for selective removal of CO_2 from the reaction zone. The process used steam purge under vacuum for regeneration of the chemisorbent. The process could directly produce an essentially CO_x-free H_2 product containing ~95% H_2 and 5% CH_4 at a feed pressure of ~26 psia using a H_2O/CH_4 ratio of 6 : 1 as feed. A CH_4 to H_2 conversion of ~73% could be achieved at a reaction temperature of only ~500°C, which was much lower than the temperature of a conventional SMR reaction (~850°C). Table 7.9 describes the pilot test performance of the SERP concept.[13] It also shows the equilibrium limitation of an SMR reactor operated without using the concept. The advantages are obvious.

Conclusions

The range of applications for gas and liquid separation and purification by adsorption is large and growing. The strong research and development activity in this area is facilitated by the flexibility of practical adsorptive process designs such as pressure and thermal swing adsorption, and SMB adsorption, as well as by the availability of a large spectrum of new and old micro- and mesoporous adsorbents.

The fundamental adsorptive properties governing the performance of the separation processes are the multicomponent equilibria, kinetics, and heat. A large volume of data, as well as models to describe them, exist in the published literature only for adsorption of pure gases and binary liquid mixtures. Binary gas adsorption data are sporadic. Multicomponent data are rare. Existence of adsorbent heterogeneity can introduce severe complexity in the multicomponent adsorption behavior.

Despite these limitations, adsorption has become the state of the art technology for many commercial separations. Examples include: 1) gas and liquid drying; 2) production of oxygen and nitrogen enriched air; 3) hydrogen purification; and 4) several bulk liquid phase separations of close boiling compounds. Emerging topics on adsorption research and development include: 1) rapid pressure swing adsorption; 2) novel adsorber configurations; 3) adsorbent membranes; and 4) simultaneous reaction and adsorption.

References

1. Sircar, S. Publications on adsorption science and technology. Adsorption **2000**, *6*, 359–365.
2. Sircar, S., Myers, A.L. Gas separation by zeolites. In *Handbook of Zeolite Catalysts and Microporous Materials*; Aurbach, S.M., Carrado, K.A., Dutta, P.K., Eds., Marcel Dekker, Inc.: New York, 2003; Chapter 22, 1063–1105.
3. Kulaprathipanja, S., Johnson, J.A. Liquid separations. In *Handbook of Porous Solids*; Schüth, F., Sing, K.S.W., Weitkamp, J., Eds., Wiley-VCH: Weinheim, Germany, 2002; Vol. 5, Chapter 6.4, 2568–2622.
4. Sircar, S. Drying processes. In *Handbook of Porous Solids*; Schüth, F., Sing, K.S.W., Weitkamp, J., Eds., Wiley-VCH: Weinheim, Germany, 2002; Vol. 5; Chapter 6.3, 2533–2567.
5. Crittenden, B., Thomas, W.J. *Adsorption Technology and Design*; Butterworth-Heinemann: Oxford, U.K., 1998.
6. Gembicki, S.A., Oroskar, A.R., Johnson, J.A. Adsorption, liquid separation. In *Encyclopedia of Separation Technology*; Ruthven, D.M., Ed., John Wiley: New York, 1997; Vol. 1, 172–199.
7. Ruthven, D.M., Farooq, S., Knaebel, K.S. *Pressure Swing Adsorption*; VCH Publishers: Boca Raton, FL, 1994.
8. Sircar, S. Pressure swing adsorption technology. In *Adsorption: Science and Technology*; NATO ASI Series E; Rodrigues, A.E., Levan, M.D., Tondeur, D., Eds., Kluwer Academic Publishers: London, 1989; Vol. 158, 275–321.
9. Yang, R.T. *Gas Separation by Adsorption Processes*; Butterworths: London, 1987.
10. Sircar, S. Application of gas separation by adsorption for the future. Adsorpt. Sci. Technol. **2001**, *19* (5), 347–366.
11. Sircar, S. Pressure swing adsorption: research needs by industry. Proceedings of the Third International Conference on Fundamentals of Adsorption, Sonthofen, Germany, May 5–9, 1989; Mersmann, A.B., Scholl, S.E., Eds., Engineering Foundation: New York, 1991; 815–843.
12. Hartzog, D.G., Sircar, S. Sensitivity of PSA process performance to input variables. Adsorption **1995**, *1*, 133–151.
13. Waldron, W.E., Hufton, J.R., Sircar, S. Production of hydrogen by cyclic sorption enhanced reaction process. AIChE J. **2001**, *47* (6), 1477–1479.
14. Sircar, S., Rao, M.B. Effect of adsorbate size on adsorption of gas mixtures on homogeneous adsorbents. AIChE J. **1999**, *45*, 2657–2661.

15. Sircar, S., Myers, A.L. Liquid adsorption operations: equilibrium, kinetics, column dynamics and applications. Sep. Sci. Technol. **1986**, *21*, 535–562.
16. Sircar, S. Adsorption from binary liquid mixtures of unequal adsorbate sizes on heterogeneous adsorbents. Surf. Sci. **1984**, *148*, 489–498.
17. Sircar, S., Rao, M.B. Heat of adsorption of pure gas and multicomponent gas mixtures on microporous adsorbents. In *Surfaces of Nanoparticles and Porous Materials*; Schwarz, J.A., Contescu, C.I., Eds., Marcel Dekker Inc.: New York, 1999; Chapter 19, 501–528.
18. Cao, D.V., Sircar, S. Heat of adsorption of SF_6 on various adsorbents. Adsorption **2001**, *7*, 73–80.
19. Karger, J., Ruthven, D.M. *Diffusion in Zeolites and Other Porous Solids*; John Wiley: New York, 1992.
20. Sircar, S., Hufton, J.R. Why does linear driving force model for adsorption kinetics work? Adsorption **2000**, *6*, 137–147.
21. Sircar, S., Kumar, R. Column dynamics for adsorption of bulk binary gas mixtures on activated carbons. Sep. Sci. Technol. **1986**, *21* (9), 919–939.
22. Leavitt, F.W. Vacuum Pressure Swing Adsorption ProcessU.S. Patent 5,415,693, May 16, 1995.
23. Leavitt, F.W. Vacuum Pressure Swing Adsorption ProcessU.S. Patent 5,074,892, Dec 24, 1991.
24. Sircar, S. Adsorption technology—a versatile separation tool. In *Separation Technology—The Next Ten Years*; Garside, J., Ed., Institute of Chemical Engineers: Rugby, U.K., 1994; 47–72.
25. Lemcoff, N.C., Gmelin, R.C. Pressure Swing Adsorption Method for Separating Gaseous Mixtures. U.S. Patent 5,176,722, Jan 5, 1993.
26. Fuderer, A., Rudelsdorfer, E Selective Adsorption Process. U.S. Patent 3,986,849, Oct 19, 1976.
27. Yamaguchi, T., Kobayashi, Y. Gas Separation Process. U.S. Patent 5,250,088, Oct 5, 1993.
28. Smoralek, J., Leavitt, F.W., Nowobiliski, J.J., Bergsten, E., Fassbaugh, J.H. Radial Bed Vacuum/Pressure Swing Adsorber Vessel. U.S. Patent 5,759,242, Jun 2, 1998.
29. Hirose, T., Kuma, T. Honeycomb rotor continuous adsorber for solvent recovery and dehumidification. 2nd Korea–Japan Symposium on Separation Technology, 1990.
30. Sircar, S., Rao, M.B. Nanoporous carbon membranes for gas separation. In *Recent Advances on Gas Separation by Microporous Membranes*; Kanellopoulos, N., Ed., Elsevier: Amsterdam, The Netherlands, 2000; 473–496.

8

Coal–Water Slurries

S. Komar Kawatra

CONTENTS

Introduction 111
Rheology of Coal Slurries 112
    Solid Loading 113
    Particle Size Distribution 114
    Surfactants and Dispersants 115
    Other Factors Affecting Rheology 118
Applications of Coal Slurries 118
    Slurry Fuels 118
    Pipeline Transport 120
    Coal Gasification 121
    Coal Slurry Wastes 122
Conclusions 122
References 122

Introduction

A slurry is defined as a fluid mixture of a pulverized solid with a liquid, usually water.

The term "coal slurry," therefore, includes all mixtures of coal and water that can flow as a fluid. There are four main types of high-percent-solid coal slurries discussed in the literature, with different characteristics depending on the application:

1. *Coal–water slurry fuels*: High-percent-solid slurries that are stable against settling, so they can be pumped, stored, handled, and burned much like heavy fuel oil. Coal–water fuels have about half the heating value of fuel oil, and because of their water content they burn cooler than pulverized coal, leading to lower production of nitrogen oxides.

2. *Coal slurry pipelines*: Slurry is used for transporting the coal through pipelines, but the coal is dewatered at the end before utilization. The slurry is unstable, and the particles are kept in suspension by the turbulence of the flow as they travel through the pipeline. The particles rapidly settle out of suspension once the slurry stops moving.

3. *Gasifier feedstock*: Integrated gasification/combined cycle (IGCC) plants use coal slurry as their feedstock. This simplifies transport of the coal and provides the water needed along with the coal in the gasification reaction. The slurries used

must flow and atomize well in the reactor, but do not need to have long-term stability against settling.

4. *Coal slurry waste*: This is fine coal processing tailings that contain a significant
   amount of coal and enough water so that they can flow. These slurries can be processed to recover fine coal particulates, or disposed of in tailings impoundments.
   These slurries can easily flow over large areas and cause considerable damage if
   the impoundment dam breaks. Coal slurry wastes are very high-percent solids
   and are described as having "about the consistency of molasses."

The properties of coal slurries that are of most interest are: 1) the viscosity/rheology,
which controls the ease of pumping and atomization; 2) the stability against settling,
which controls the ability of the slurry to be stored in tanks without the solids forming a
hard mass that clogs the system; and 3) the solid concentration and coal type/composition,
which determines the heating value per unit volume of slurry.[1]

Of the four applications for coal slurries given here, coal slurry fuels have the most stringent requirements. Hence, the following discussion of slurry properties is with reference
to coal slurry fuels.

Rheology of Coal Slurries

There are two competing considerations for the flow properties (rheology) of coal-water
slurry fuels. The first is that the slurry must have a low viscosity (preferably less than 2000
centipoise) at high shear rates so that it can be pumped and atomized readily.[1] The second
is that it should be sufficiently viscous so that the particles will not settle out during storage, as otherwise the settled solids will clog the pipes leading from the storage tank. There
are two factors that contribute to prevention of slurry settling: development of a yield
stress and thixotropy.

In a slurry that develops a yield stress, the slurry does not begin to flow until a particular
level of shear stress is applied to it. These types of slurries are classified as either pseudoplastic with yield, or Bingham plastic, as shown in Figure 8.1. Pseudoplastic and Bingham
plastic slurries have a high apparent viscosity (ratio of the shear rate to the shear stress)
at low shear rates, but at higher shear rates the apparent viscosity decreases. Coal slurry
rheology can generally be modeled using Herschel and Buckley's yield-power law:[2]

$$\tau = \tau_0 + K\gamma^n$$

where τ is the shear stress (in Pa), τ_0 is the yield stress required to initiate slurry flow (in
Pa), γ is the shear rate (sec^{-1}), n is the flow behavior index (equals 1 for Newtonian slurries,
<1 for pseudoplastic slurries, and >1 for dilatant slurries), and K is the consistency number
(equal to the apparent viscosity for Newtonian slurries).

Slurries that exhibit a yield stress will resist settling when stationary, which is a low
shear condition, but will flow readily when a higher shear force is applied, such as by a
pump or an atomizer.

In a thixotropic slurry, the slurry will, when left undisturbed, develop a three-dimensional network structure where particles attach loosely to each other in an extremely weak

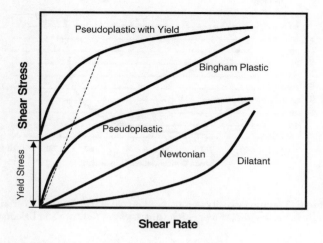

FIGURE 8.1
Classes of rheological behavior that can be shown by coal slurries, as they appear when plotted on a shear rate/ shear stress graph. It is desirable for coal slurries to be Bingham plastic or pseudoplastic with yield, as such slurries flow readily at high shear rates (such as during pumping or atomization), while remaining stable against settling at low shear rates because of their yield stress. Dilatant slurries are completely unsuitable for coal slurry applications because they are extremely difficult to pump.

gel. This holds the solid particles suspended indefinitely. The shear stress required to cause the thixotropic slurry to flow can then be broken into two parts:[3]

$$\tau = \tau_e + \tau_s$$

where τ is the total shear stress (in Pa) required to produce slurry flow, τ_e is the equilibrium shear stress (in Pa) reached after agitating the slurry for an extended time, and τ_s is the structural shear stress (in Pa) resulting from the formation of the structural network.

The structure can then be broken up by stirring, so that the slurry can be refluidized to a free-flowing state. The structural shear stress of the thixotropic slurry decreases with time during agitation according to the following expression:[3]

$$d\tau_s/dt = [\tau_{s\infty} - \tau_s]/\lambda_s - K\eta^{1/2}\gamma\tau_s$$

where $d\tau_s/dt$ is the rate of change of structural shear stress with respect to time, τ_s is the structural shear stress (in Pa), η is the apparent viscosity (in Pa sec), γ is the shear rate (sec^{-1}), $\tau_{s\infty}$ is the final structural shear stress achieved under shear-free conditions, λ_s is the time constant for buildup of structural stress, and K is the proportional constant for the rate of structure breakdown.

Solid Loading

The rheological properties of coal slurries depend on the quantity of solids suspended in the liquid and on the physical and surface chemical properties of the solid particles. The viscosity of the slurry will typically increase gradually with increasing solid loading until a critical point is reached where interparticle friction becomes important. Beyond this point, the viscosity will rapidly increase until the slurry ceases to flow, as shown in Figure 8.2. The value of the

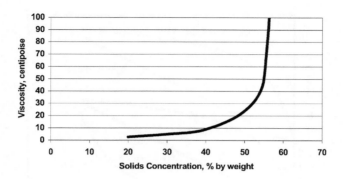

FIGURE 8.2
Effects of solid concentration on the viscosity of a particular coal slurry. Beyond a certain limiting concentration, the viscosity rapidly increases to extremely high values. (From Kawatra and Bakshi.[4])

critical solid loading will vary considerably between coal slurries, depending on the physical and chemical properties of the particles. For most coal slurries, and particularly for coal–water slurry fuels, it is important to be able to make the solid loading of the slurry as high as possible so that the energy content per liter of fuel is maximized, while keeping the viscosity low enough so that the slurry can be pumped economically.

Particle Size Distribution

The size distribution of particles will control the amount of liquid needed to fluidize a given quantity of coal. In general, a fine size distribution will produce a more viscous slurry than a coarse size distribution at the same wt% solids, and the fine particles will produce a more non-Newtonian rheological curve. This can be seen in the laboratory results shown in Figure 8.3, which compares a coarse coal slurry to a fine coal slurry. It is

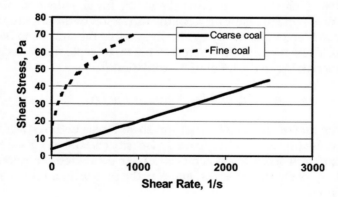

FIGURE 8.3
Comparison of the rheological curves for a fine coal slurry (80% passing 34 µm, top size 100 µm, 52 wt% solids) and for a coarse coal slurry (58 wt% solids). Neither slurry used any additives. Because it is extremely difficult to measure the rheology of unstable slurries with conventional rheometers, these results were obtained using a continuous-pressure-vessel rheometer, which was specially designed for this purpose. The fine coal curve is the average of 10 measurements and the coarse coal curve is the average of 5 measurements, and the standard error of the shear rate measurements was approximately 1.0 Pa for these slurries. The fine coal slurry is clearly pseudoplastic with a yield value of approximately 18 Pa, while the coarse coal slurry is Bingham plastic with an estimated yield value of 4 Pa.

clearly seen that the fine slurry is much more viscous, its pseudoplastic character is very pronounced, and its yield value is high, while the coarse coal slurry is clearly a Bingham plastic.[5]

A graded size distribution, where fine particles fill the interstices between coarse particles as shown schematically in Figure 8.4, will minimize the amount of void space that must be filled by fluid, and so will reduce the quantity of liquid needed to produce a flowable slurry. The best size distributions for this purpose have proven to be multimodal distributions, made up of several fairly narrow size fractions. An example of such a size distribution is shown in Table 8.1.[1] To achieve such a multimodal size distribution, coal slurry production facilities are designed to generate several coal streams, which are each ground and sized to the desired particle sizes and then combined to give the proper size distribution.

An example of how slurry rheology is affected by combining a "coarse" size distribution (208–279 μm) with a "fine" size distribution (smaller than 45 μm) can be seen in Figures 8.5 and 8.6. Adding increasing amounts of coarse particles made the suspensions examined less viscous until approximately 40% of the solids were coarse particles. Increasing the proportion of coarse particles beyond that level produced no further viscosity decrease and tended to destabilize the slurry so that the particles would rapidly settle out.[6]

Surfactants and Dispersants

The interparticle friction in the slurry can be reduced through the use of surface-active reagents (surfactants). The surfactants, therefore, alter the viscosity and overall rheology of the slurry. It is important to remember that coal is not a single tightly defined substance. It covers a wide range of fossilized high-carbon solids, which vary a great deal in composition, heating value, and chemistry. High-rank, high-volatile coals tend to be more hydrophobic than low-rank or low-volatile coals, which contain larger quantities of moisture and oxygen in their structure. It is therefore critical that the surfactants used be selected to match

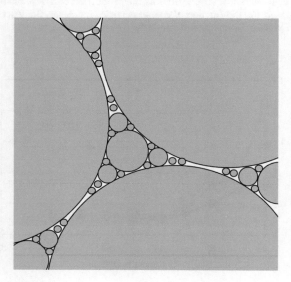

FIGURE 8.4
Graded size distribution to maximize the packing fraction of ideal spherical particles, and thereby minimize the amount of water needed to fluidize the slurry. In this illustration, there are four different sizes of particles present. A similar principle will maximize the packing fraction of real, irregularly shaped particles.

TABLE 8.1

Example of a Multimodal Size Distribution to Produce a
Very High Solid Loading in a Coal Slurry (This Results in
a Powder with Only 24% Pore Volume)

Size Fraction (µm)	Total Solids by Weight (%)
701–833	60
88–104	30
38–43	8
<25	2

(From Mishra and Kanungo.[1])

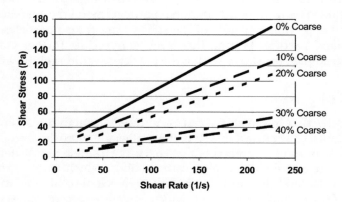

FIGURE 8.5
Changes in the rheology of 40 wt% coal suspensions as the proportion of coarse particles (208–279 µm) to fine particles (<45 µm) is increased. As the fraction of coarse particles increases, the viscosity drops and the yield stress is decreased. (From Mishra, Senapati, and Panda.[6])

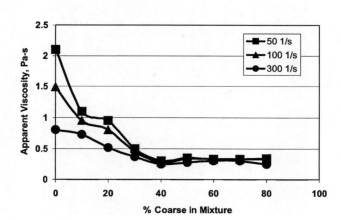

FIGURE 8.6
Apparent viscosity of a 50% solid coal slurry as the proportion of coarse particles (208–279 µm) to fine particles (<45 µm) is increased, at three different shear rates. The viscosity was minimized at 40% coarse particles. (From Mishra, Senapati, and Panda.[6])

the properties of the coal being used for the production of the slurry, as surfactants that are highly successful in one application may be largely ineffective in another.

Because of their structure and chemistry, surfactants are chemicals that tend to concentrate at surfaces in preference to dissolving in the volume of the liquid. They can, therefore, be used to selectively alter the surfaces of particles, changing the way they interact with each other.

In the case of coal particles suspended in water, there are two effects that control interparticle interactions: hydrophobic properties and electrical charge effects. Strongly hydrophobic coal particles will tend to flocculate because of the reduced surface energy of a coal–coal interface compared to a coal–water interface. Surfactants that increase hydrophobicity will therefore tend to increase viscosity owing to increased interparticle interaction. The electrical charge on the particle surfaces, which arises by ion exchange with the fluid, will also affect the viscosity. A low magnitude of charge, within approximately ±5 mV of zero net charge, will allow particles to interact and flocculate readily, resulting in a high viscosity and significantly non-Newtonian behavior. A high magnitude of charge, exceeding –20 mV, will cause particles to repel each other, minimizing flocculation and reducing viscosity.

There are hundreds of different surfactants that have been considered for control of coal slurry rheology. The majority of these reagents fall broadly into the three classes described in Table 8.2: anionic surfactants, nonionic surfactants, and polysaccharides. Anionic surfactants are those that have a negative charge in solution. When they adsorb on the coal

TABLE 8.2

Types of Surfactants that have been Used for Preparing Highly Concentrated Coal-Water Slurries: Many of These Surfactants are Used in Combination with Formulations that Depend on the Characteristics of the Coal

Anionic Surfactants	Nonionic Surfactants	Polysaccharides
Ethylene diamine tetra-acetic acid (EDTA)	Polyethylene glycol ethers	Carboxymethyl cellulose salts
Calcium lignosulfonate	Polypropylene glycol	Cellulose
Ammonium lignosulfonate	Ethoxylated/propoxylated alcohols	Wood particles
Sodium lignosulfonate	Alkylphenyl decaethylene glycol ether	Uronic acid
Sodium tripolyphosphate	Polyethyleneimine	Hydroxypropyl cellulose
Polyalkylene oxide	Propylene oxide polymers	Polysaccharose (produced by
Polysodium styrene sulfonate	Ethylene oxide polymers	lipomyces micro-organisms)
Polyacrylic acid	Phenol butyl naphthalene polymers	Cellulose ether
Sodium allyl sulfonate	Methacrylic acid polymers	
Polyacrylamide	Butyric acid lauryl ethylene diamine salt	
Polyoxyethylene nonyl phenol ether		
Polyvinyl sulfonic acid		
Sulfonated benzene sodium salt		
Naphthalene sulfonic acid		
Tallow soap		
Isobutylene		
Maleic anhydride		
Sodium humate (extracted from brown coal)		

(From Mishra and Kanungo.[1])

surface, they provide it a net negative charge, which increases interparticle repulsion. Nonionic surfactants and polysaccharides have no net charge in solution and are believed to form a more physical barrier that prevents particles from interacting with each other.

Other Factors Affecting Rheology

The rheology of a coal-water slurry is also affected by the composition of the coal. It has been determined that increases in coal ash content lead to increasing viscosity and that slurries are more viscous at pH 6 than at pH 8.[7] Many coals, particularly low-rank or partially oxidized coals, contain humic acids, which act as surfactants and can affect the properties of the slurry.[8] The presence of these humic acids helps to decrease slurry viscosity and increase the maximum particle loading. Additionally, the viscosity is affected by the temperature of the slurry, with increasing temperature tending to decrease the slurry viscosity. It has also been reported that high shear agitation can reduce the viscosity of the coal slurry, apparently by improving the dispersion of surfactants.[9]

Applications of Coal Slurries

Slurry Fuels

As petroleum-based fuels are being depleted, it is expected that coal-derived fuels such as coal slurry fuels will become more important. These fuels are attractive mainly because they are made from coal, which is of low cost compared to petroleum products, and can be pumped, stored, and burned much like fuel oil. The demonstrated coal reserve base in the United States is 508 billion tons, with predicted reserves as high as 1.7 trillion tons.[10] At the current rate of consumption (approximately 1 billion tons/year) this would last for 500–1700 yr, which means that coal, and therefore coal slurry fuels, will be available long after petroleum deposits are depleted. Most of the activity in coal slurry fuel development has used highly cleaned high-rank coals, which can be produced using advanced physical, chemical, and biological techniques to remove ash-forming minerals and sulfur.[11] In addition, processes have been developed to upgrade low-rank coals so that they can be used to produce coal slurries with a high heating value.[12]

Coal slurry fuel production plants have been constructed for producing coal-water slurry fuels. These plants consist of facilities for grinding highly cleaned coal to the desired fineness, adding stabilizing reagents, and increasing the solid content of the slurry to approximately 65–70% solids by weight. A schematic of a typical plant is shown in Figure 8.7. The key design features of such a plant are:

1. *Particle size control*: The plant pictured produces a multimodal size distribution, consisting of relatively large particles from the first stage of grinding, an intermediate size of particles from the second stage of grinding, and particles from the hydrocyclone oversize product that are reground to a very fine size when they are returned to the second grinding stage.
2. *Water content control*: Grinding and size classification of the coal require a slurry with considerably more water than is desired in the final fuel. Hence, filters must be used to remove this excess water once the particles reach their target size.

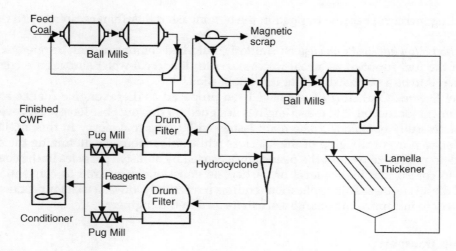

FIGURE 8.7
Simplified schematic of the Beijing Coal Water Fuel Plant, China. Clean coal is first ground in multiple stages to produce a graded size distribution that will fluidize with the minimum amount of water. Drum filters are used to remove excess water from the slurry, and surfactant reagents are combined with the slurry using pug mills. This plant processes 35 metric tons of coal per hour and produces a slurry that is 65% coal by weight. (From Metso Minerals.[22])

3. *Reagent addition*: Before the reagents are added, the material coming from the drum filter will be extremely viscous. Pug mills, which can handle a very high solid content, are, therefore, used to combine the reagents with the coal slurry both to combine the desired particle size fractions and to uniformly disperse the reagents through the coal.

Coal slurry fuels are of interest for the following markets:

Replacement of Heavy Fuel Oil: The original concept for coal-water slurry fuels was as a direct replacement for heavy #5 and #6 fuel oils. The intent was that power plants and industrial heating operations that were using these fuel oils could switch directly to a coal-water slurry with a minimum of modification to their combustors. This requires a very low-ash coal to prevent fouling of the burners and is most attractive when fuel oil prices abruptly increase. The disadvantage of coal-water slurry fuels in this application is that they only provide about half the energy per unit volume as the fuel oil does. Currently, there do not appear to be any industries that see these fuels as an attractive alternative, and they are not being sold commercially for this purpose.

Supplement for Pulverized Coal Combustors: The water content of coal-water slurry fuels has made them attractive for control of nitrogen oxide emissions from pulverized coal combustors at power plants. If the slurry fuel is injected as a supplement for the main fuel (dry coal), the water lowers the combustion temperature and prevents nitrogen oxides from forming, while still allowing complete combustion of the coal. As nitrogen oxides are troublesome to remove or destroy once they are formed, it is generally preferable to prevent their formation at the source. This technology is currently in use for low-NO_x coal combustors. It has also been determined that, when coal-water slurry is added in pulverized coal combustors, a greater fraction of the inorganic hazardous air pollutants (including mercury) became trapped in the bottom ash rather than escaping with the flue gases.[13]

Injection into Fluid-Bed Combustors: Similarly, fluid-bed combustors can be operated with part of the fuel supplied as a coal-water slurry. In Italy, coal-water slurries are currently considered to be a standard fuel for this application.[14]

Diesel Engines: Coal slurry fuels have been proposed as diesel engine fuel as a direct replacement for fuel oil, and diesel engines designed for this purpose have been developed and successfully operated.[15] The main issues with using coal slurry in this application are: first, to prevent clogging of the injectors while still adequately atomizing the slurry; second, to minimize wear of the piston rings caused by abrasive particles in the coal; and third, to control the higher level of particulate emissions compared to conventionally-fueled diesel engines. This application requires particularly low-ash coal for production of the slurry to minimize the quantity of ash formed in the engine.

Pipeline Transport

A coal slurry pipeline is a system for transporting solid coal particles in a liquid carrier. Long-distance coal slurry pipelines are an alternative to railroad transport, and their practicality and economics are largely dependent on land ownership, terrain, water availability and water contamination concerns, political considerations, and coal demand.[16] An example of pipeline transport is the 440 km Black Mesa pipeline, which was completed in 1970 and is currently the only long-distance coal slurry pipeline operating in the United States.[17] Slurry pipelines are also used over shorter distances to transport material within a processing facility.

Unlike coal slurry fuels, which are utilized in a fluid form, coal transported by pipeline is used as a solid after dewatering. As a result, it does not need to be stable when stored in tanks, and it is not so critical that its solid loading be maximized. The following variables are important for a pipeline system:

- Solid concentration range
- Flow velocity range
- Pipe diameter
- Friction losses

These are, in turn, controlled by the size distribution and density of the coal particles. To maintain homogenous flow, which is necessary to minimize wear of the pipeline, a coal slurry should have the following properties, assuming a coal specific gravity of 1.4:[4]

- Maximum particle size: 2400 μm
- Percentage finer than 44 μm: 20%
- Slurry concentration: 50 wt%

To keep the particles in suspension, the flow should be at least 0.15 m/sec faster than either: 1) the critical deposition velocity of the coarsest particles, or 2) the laminar/turbulent flow transition velocity. The flow rate should also be kept below approximately 3 m/sec to minimize pipe wear. The critical deposition velocity is the fluid flow rate that will just keep the coarsest particles suspended, and is dependent on the particle diameter, the effective slurry density, and the slurry viscosity. It is best determined experimentally by slurry loop testing, and for typical slurries it will lie in the range from 1 m/s to 4.5 m/sec. Many empirical models exist for estimating the value of the deposition velocity, such as the following relations, which are valid over the ranges of slurry characteristics typical for coal slurries:[18]

$$F_L = V_c / [2gD(\rho_s - \rho z_t)/\rho_f]^{1/2}$$

where

$F_L = \exp[0.165 - 0.073C_D - 12.5K_2]$
$K_2 = [K_1 - 0.14]^2$
$K_1 = [\mu_L/\rho_L]^{2/3}/g^{1/3}d_{50}$
C_D = particle drag coefficient
d_{50} = mass median diameter of + 74 μm particles (m)
μ_L = viscosity of liquid (Pa sec)
V_c = deposition velocity (m/sec)
g = gravitational acceleration (m/sec^2)
D = pipe diameter (m)
ρ_s = density of solid particles coarser than 74 μm (kg/m^3)
ρ_L = density of fluid (kg/m^3)
$\rho_f = [\rho_s C_f + (1 - C_t)\rho_L]/[1 - C_t + C_f]$
$C_t = C_f + C_r$
C_f = volume fraction of −74 μm particles in slurry
C_r = volume fraction of + 74 μm particles in slurry

Once the coal slurry has been transported to its destination by pipeline, it is dewatered for utilization.[19] Chemical additives are not normally used in slurry pipelines, as they would be an unnecessary added cost and a possible source of water contamination. It is important to monitor the quality of the water removed from the slurry, as it could contaminate waterways if it dissolves harmful materials before being released.

Coal Gasification

For integrated gasification combined cycle (IGCC) plants, coal slurries are pumped under pressure into the reactor, where they react with water to produce combustible gases, particularly methane. There are two reasons for using coal slurries for this application rather than dry coal: 1) a fluid slurry can be easily pumped into a pressurized reactor, whereas it is much more difficult to convey dry solids against a pressure gradient, and 2) the gasification reaction requires water in any case; hence, it is simplest to introduce water along with the coal particles.

The requirements for coal slurry intended to be used in IGCC are similar to those for pipeline transport, with the added requirement that, as for coal-water fuels, the slurry needs to atomize easily in the reactor. Atomization has been found to be easier (producing smaller-diameter droplets at lower jet pressures) as the size of the coal particles in the slurry increases, because of weaker capillary forces holding the slurry together and decreased interparticle friction.[20] As the capillary forces and interparticle friction are also responsible for increasing the viscosity in coal slurries, it is clear that low-viscosity slurries will atomize more readily than high-viscosity slurries.

Coal Slurry Wastes

The bulk of the fine coal that makes up coal slurry wastes is produced in the mine by the miner cutting heads. These fine particles are typically mixed with large quantities of mineral fine substances, particularly clays, to the extent that the fine coal has a very high ash content. Additional fine particles are produced by rotary breakers and other size reduction equipment that is in use in most coal-washing plants. Processing methods exist that can be used to separate coal from the mineral matter in these fine coal slurries, specifically, froth flotation techniques. However, not all plants have the facilities to recover this fine coal, either because their coal is not suited to froth flotation processing, or because they have difficulty meeting ash and moisture specifications with fine coal produced by this means. As a result, many plants must dispose of their fine coal slurries as waste.

Large volumes of coal waste slurry are disposed of in surface impoundments. As the slurry remains easily liquefied long after being impounded, it represents a serious hazard if the impoundment fails. An example of this occurred in Martin County, KY, in 2000, when a breakthrough in a coal slurry impoundment released 245 million gallons of slurry into 75 mi of streams, and caused in excess of $37 million in damages.[21]

Unlike other coal slurry applications where the slurry must be kept fluid and pumpable, coal slurry in impoundments needs to be stabilized to prevent flow. Treatments of coal waste slurry, therefore, concentrate on promoting complete settling, strong interparticle interactions and high viscosities so that impoundment breaks will not release the material as a damaging flow.

Conclusions

Coal slurries are of great importance in a wide variety of coal-handling situations, as the slurry form makes the coal easy to handle by pumping with minimal dust. Highly concentrated slurries that can be burned directly as fuel make it possible to eliminate the need for handling dry coal in the combustion process. The primary considerations in producing and handling coal-water slurries are managing the slurry rheology through manipulation of particle size and surface chemistry and minimizing the water content needed to maintain fluidity. Currently, the most common use of coal-water slurries is as a means for transporting coal, either within a plant or through pipelines. While coal-water slurry fuels have not yet become a widespread replacement for liquid fossil fuels, they are a viable alternative to heavy fuel oils provided that they are made from coal with sufficiently low ash content.

References

1. Mishra, S.K., Kanungo, S.B. Factors affecting the preparation of highly concentrated coal-water slurry (HCCWS). J. Sci. Ind. Res. **2000**, *59*, 765–790.
2. Pulido, J.E., Rojas, C.P., Acero, G., Duran, M., Orozco, M. Rheology of Columbian coal-water slurry fuels: effect of particle-size distribution. Coal Sci. Technol. **1995**, *24*, 1585–1588.
3. Usui, H., Saeki, T., Hayashi, K., Tamura, T. Sedimentation stability and rheology of coal water slurries. Coal Prep. **1997**, *18*, 201–214.

4. Kawatra, S.K., Bakshi, A.K. The on-line pressure vessel rheometer for concentrated coal slurries. Coal Prep. **2002**, *22* (1), 41–56.
5. Nguyen, Q.D., Logos, C., Semmler, T. Rheological properties of South Australian coal-water slurries. Coal Prep. **1997**, *18*, 185–199.
6. Mishra, S.K., Senapati, P.K., Panda, D. Rheological behavior of coal water slurry. Energy Sour. **2002**, *24*, 159–167.
7. Pawlik, M., Laskowski, J.S., Liu, H. Effect of humic acids and coal surface properties on rheology of coal-water slurries. Coal Prep. **1997**, *18*, 129–149.
8. Takao, S., Ozaki, H. The effects of agitation on the rheology of a coal water mixture. Coal Prep. **1997**, *18*, 215–225.
9. EIA. U.S. Coal Reserves. 1997 Update. Energy Information Administration; http://www.eia. doe.gov/cneaf/coal/reserves/front-1.html.
10. Kawatra, S.K., Eisele, T.C. *Coal Desulfurization—High Efficiency Preparation Methods*. Taylor & Francis: New York, 2001.
11. Usui, H., Tasukawa, T., Saeki, T., Katagiri, K. Rheology of low rank coal slurries prepared by an upgrading process. Coal Prep. **1997**, *18*, 119–128.
12. Nodelman, I.G., Pisupati, S.V., Miller, S.F., Scaroni, A.W. Partitioning behavior of trace elements during pilot-scale combustion of pulverized coal and coal-water slurry fuel. J. Hazard. Mater. **2000**, *74*, 47–59.
13. Okasha, F., Miccio, M. Prediction of coal-water slurry dispersion in a fluidized bed combustor, Twenty-Sixth International Symposium on Combustion; The Combustion Institute, 1996; 3277–3285.
14. Hsu, B.D. Coal fueled diesel engine development update at GE transportation systems. In *Coal-Fueled Diesel Engines*. Caton J.A., Webb, H.A., Eds.; ASME, 16-ICE, 1992; 1–9.
15. Cox, C.B. *Comparing the Studies of a Coal Slurry Pipeline*. Virginia Water Resources Research Center, Special Report No. 17; Virginia Polytechnic Institute and State University: Blacksburg, VA, 1983.
16. Anonymous. Black Mesa Pipeline. Black Mesa Pipeline Co.; http://www.blackmesapipeline. com/index.htm, 2003.
17. Gandhi, R.L., Snoek, P.E. Design aspects of slurry pipelines. In *Handbook of Fluids in Motion*. Cheremisinoff, N.P., Gupta, R., Eds.; Ann Arbor Science Publishers: Ann Arbor, MI, 1983; Chapter 36, 945–968.
18. Gilles, R.G., Shook, C.A. A deposition velocity correlation for water slurries. Can. J. Chem. Eng. **1991**, *69*, 1225–1227.
19. Guo, J., Hodges, S., Uhlherr, P.H.T. Dewatering of HTD coal slurry by mechanical expression. Coal Prep. **1997**, *18*, 227–239.
20. Son, S.Y., Kihm, K.D. Effect of coal particle size on coal-water slurry (CWS) atomization. Atomization Sprays **1998**, *8*, 503–519.
21. Office of Surface Mining. *Report on October 2000 Breakthrough at the Big Branch Slurry Impoundment*; U.S. Department of the Interior, 2002; http://www.osmre.gov/pdf/martincounty030402. pdf.
22. Metso Minerals. Coal water fuel (CWF) plants: raw coal transformed into liquid fuel. http:// www.metsominerals.com/inetMinerals/mmcontent2.nsf/WebWID/WTB-030508-2256B-5BA7F/$File/FLUID_CARBON.pdf.

9

Size Reduction

Sunil Kesavan

CONTENTS

Introduction 125
Grinding Energy Requirements 126
Primary and Secondary Crushers 127
   Gyratory Cone Crushers 127
   Jaw Crushers 128
   Roll Crushers 129
Intermediate and Fine Grinders 130
   Impact Mills 130
      Hammer Mills 130
      Centrifugal Pin Mills 130
   Attrition/Disk Mills 131
   Tumbling Mills 132
      Ball Mills 132
      Tube Mills 134
      Rod Mills 134
      Autogenous and Semiautogenous Mills 134
   Rolling Compression Mills 135
Ultrafine Grinders 135
   Fluid-Energy Mills 135
   Agitated Mills 136
      Attritors 137
      Vibratory Mills 139
   Impact Mills with Internal Classification 139
Rotary Cutters 141
Equipment Operation 141
   Open and Closed Circuit Grinding 141
Conclusions 142
References 143

Introduction

The comminution or size reduction process is an important unit operation in the process industries in which solid materials are broken or cut into smaller sizes by the application of mechanical stress. Solids can be reduced in size by crushers and grinders which employ compression, impact, attrition, shear, or combinations thereof. Depending on the degree of particle size reduction desired, the end result is achieved in one or several steps. Devices that are used for size reduction operations can be classified into primary and secondary crushers, grinders, and pulverizers. Explosive blasting is used in many instances for primary size reduction of ore formations into sizes workable by primary crushing machines. The primary crushers are slow-speed machines that reduce the run-of-mine product into 15–25 cm lumps. A secondary crusher reduces these lumps to 5 mm product. Grinders reduce the products of crushing operations into powder. An intermediate grinder typically produces a product that passes a 40-mesh screen. Fine grinders reduce feed into product that passes a 200-mesh screen. Ultrafine grinders can convert the product of secondary crushers into 1–10 μ product. Cutting machines produce particles with a definite size and shape in the 2–10 mm size range.

Size reduction machines can be classified as follows:

    a. Primary and secondary crushers
       1. Gyratory crushers
       2. Jaw crushers
       3. Roll crushers
    b. Intermediate and fine grinders
       1. Impact mills
          a. Hammer mills
          b. Centrifugal pin mill
       2. Attrition mills
       3. Tumbling mills
          a. Ball and pebble mills
          b. Rod mills
          c. Tube mills; compartment mills
       4. Rolling-compression mills
          a. Ring roll mills
          b. Bowl mills
    c. Ultrafine grinders
       1. Fluid-energy mills
       2. Agitated mills
       3. Impact mills with internal classification
    d. Cutting machines
       1. Knife cutters, dicers, slitters

Grinding Energy Requirements

Given the relatively large power consumption accompanying size reduction, it is important to quantify the energy requirements of these unit processes. Comminution theory dates back to the works of Kick[1] and Rittinger[2] that were done in Germany. A generalized differential equation that governs crushing can be written as:

$$dW = -K(dD/D^n) \tag{1}$$

where W is the energy input, K and n are constants, and D is the particle size.

Eq. (1) represents Kick's law when $n = 1$ and Rittinger's law when $n = 2$.

Kick's law[1] essentially states that the work required to obtain a given reduction ratio is the same irrespective of starting size. According to Rittinger's law,[2] work is proportional to surface created. Rittinger's and Kick's laws are only useful over a limited particle size range and are not utilized today.

Bond[3] proposed that work is inversely proportional to the square root of particle diameter. Bond's theory of comminution is represented by Eq. (1) when $n = 1.5$ and can be written as:

$$W = 100W_i\left(1/D_P^{0.5} - 1/D_F^{0.5}\right) \tag{2}$$

where W_i is the Bond work index or work required to reduce a unit weight of product from a theoretical infinite size to a product with 80% passing 100 μ. Extensive data are available on the work index and the Bond law is widely used today.

Primary and Secondary Crushers

Gyratory Cone Crushers

These types of crushers employ a conical crushing element that gyrates in an eccentric manner in a shell resembling an inverted cone (Figure 9.1). The material to be crushed enters the

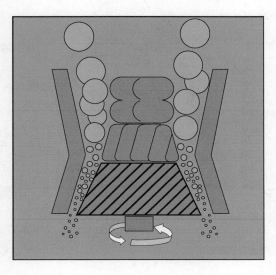

FIGURE 9.1
Schematic representation of the operating principle of a gyratory cone crusher.

top of the crusher where the crushing surfaces are most widely spaced. The product becomes wedged and squeezed between the mantle and the hopper and is progressively broken down until it discharges through the narrow opening at the bottom of the crusher. Gyratory crushers are available in large sizes used for primary crushing and smaller sizes for secondary crushing of soft to medium-hard materials. These continuous-discharge crushers are available in capacities up to 3500 ton/hr and are more cost effective to operate than jaw crushers.

Jaw Crushers

These devices suited for coarse and intermediate crushing of large volumes of hard and semihard materials employ swinging jaws that work against a hardened stationary surface. The jaws, which are essentially flat, form a V-shaped crushing chamber with a wide inlet at the top and a narrow discharge at the bottom. The large feed material gets progressively reduced in size and falls down toward the narrow throat section.

There are three types of jaw crushers (Figure 9.2)—the Blake, the Dodge, and the single-toggle type. In the popular double-toggle Blake jaw crusher (Figure 9.3), the moving jaw

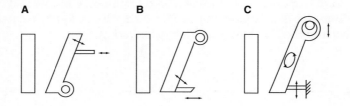

FIGURE 9.2
The essential features of (A) Dodge, (B) Blake, and (C) single-toggle jaw crushers.

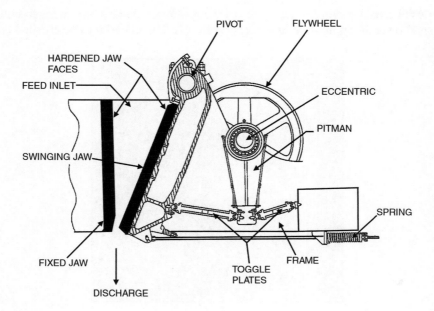

FIGURE 9.3
Cross section of a Blake-type jaw crusher showing the nonchoking design of the swing jaw.

is hinged at the top with the maximum movement at the bottom of this jaw. In the Dodge crusher, the swinging jaw is hinged at the bottom. This gives a fixed discharge opening, giving a more uniform product than the Blake crusher. The greatest movement acts on the large feed at the top of the jaw. The Dodge crusher is not used much in commercial large volume crushing operations. The single-toggle jaw crusher is pivoted at the top like the Blake crusher but has an additional movement in the downward direction. The Blake crusher costs more than the single-toggle machines but the operating costs are less for hard feed materials.

Taggart[4] proposed a rule-of-thumb that a jaw crusher would be the economically preferred size reduction equipment if the required throughput in tons per hour was less than the square of the crusher gap in inches. Depending on size, jaw crushers can handle up to a 48-in. size feed and create product smaller than half an inch. Jaw crushers are available in a variety of sizes up to about 1000 ton/hr and can be stationary, portable, or skid mounted.

Roll Crushers

These tertiary crushers employ smooth or toothed heavy-duty impact and abrasion-resistant steel-rimmed rolls. The rolls are mounted inline in a horizontal manner and turn toward each other at equal speeds to create a nip into which a friable feed material is introduced (Figure 9.4). Heavy-duty compression springs with automatic reset are used to dampen crushing shock and to protect the crusher from tramp iron and oversize material. An adjustable screw that adjusts spring tension changes the crusher opening. A flywheel is used to even out pulses and economize on power consumption. These crushers have a theoretical maximum reduction ratio of 4 : 1 and will only crush materials to about 10 mesh. Roll crushers produce a controlled product size distribution without a lot of fines. The narrow particle size distribution is achieved by controlling a combination of variables including roll speed, gap measure, differential speed, feed rate, and roll surface.

Toothed single roll crushers that crush materials by working on a breaker plate are also available. The crushing teeth are set in segments to facilitate replacement.

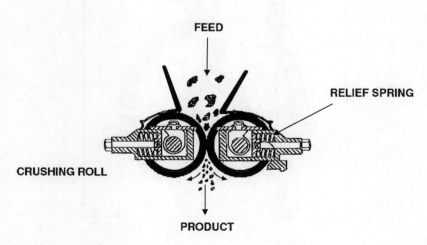

FIGURE 9.4
Schematic of a smooth-roll crusher.

Intermediate and Fine Grinders

Impact Mills

Hammer Mills

Impact mills like the hammer mill use swinging hammers (typically running at 750–1800 rpm) to pulverize the solid feed by impact and attrition. As shown in Figure 9.5, these hammers are mounted by pins on the periphery of a number of disks mounted on a horizontal rotating shaft and are free to swing. The hammers force the product against a rugged breaker plate. The product gets broken down until it is small enough to fall through a discharge grate at the bottom. For some applications like the grinding of dried animal byproducts, a vertical hammer mill that uses a vertical drive shaft with horizontal hammers and screens is more efficient. Hammer mills are used in primary, secondary, and tertiary crushing operations. These mills are relatively inexpensive but tend to produce a lot of fine material. They are best suited for soft or semihard materials, as harder materials tend to rapidly wear out the hammers.

Hammer mills can handle feed sizes of up to 10–20 in. and throughputs up to 500 ton/hr. Air-swept hammer mills use air to help convey the particle product out of the mill.

Centrifugal Pin Mills

An example of a pin mill is shown in Figure 9.6. It applies centrifugal forces to grind feed particles by impact. Feed entering the mill is divided into two streams that drop down on to a rotating plate. Centrifugal force directs the feed outward on to intermeshing pins or blocks mounted on the plate. Pin mills are available in capacities up to 200 ton/hr. These mills are economical, easy to operate, produce a uniform grind, are suitable for wet or dry grinding, and provide high throughput with low energy consumption.

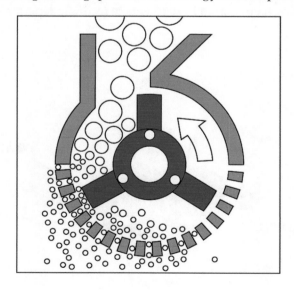

FIGURE 9.5
Schematic representation of the operating principle of a hammer mill. (Courtesy of Sturtevant, Inc., Boston, MA.)

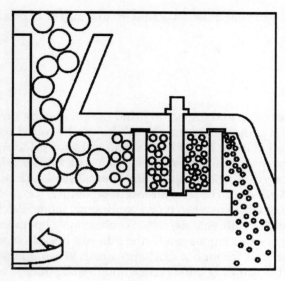

FIGURE 9.6
Simpactor centrifugal pin-type impact mill. (Courtesy of Sturtevant, Inc., Boston, MA.)

Attrition/Disk Mills

These mills use a combination of impact, shear, and cutting action to grind materials between the replaceable wearing surfaces of two grinding disks. One or both of the disks can rotate; if both, they are run counter to each other. The distance between the disks can be adjusted to vary the product particle size. Figure 9.7 is a schematic representation of a typical attrition mill. Liquid cooling of the disks is used with heat-sensitive materials to prevent degradation. Air is sometimes drawn through the mill to help remove product

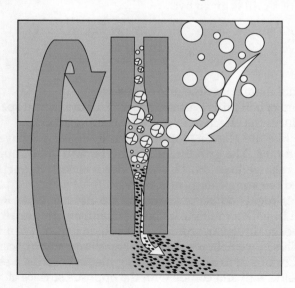

FIGURE 9.7
Schematic of a disk-type attrition mill. (Courtesy of Sturtevant, Inc., Boston, MA.)

and prevent choking. Attrition mills can grind up to 8 ton/hr of product to a particle size passing 200 mesh.

Tumbling Mills

Ball Mills

These mills employ attrition and impact to grind the product by tumbling in cylindrical mills partly filled with grinding media. The media can be round metal balls or nonmetallic pebbles that are 1/2 in. or larger in size. Ball mills for grinding hard materials are usually lined with heavy-duty steel alloy liners. Some ball mills employ internal baffles to prevent slippage of the grinding media over the internal shell surface. Mills for dry grinding are available with either full or semi air-swept capability.

Ball mills can operate either in a batch or a continuous mode. Batch mills typically use a charge of 50–55% of the mill volume, while continuously operating mills use a media charge of 40–45%. In the continuous mode, the mill can operate in such a mode that the product can leave the mill through a discharge grate. When operating in a continuous mode, the effluent of the grinding mill is sent through a classifier to separate out the over-size product to be reprocessed through the mill.

Commonly used types of grinding media are carbon steel, stainless steel, chrome steel, tungsten carbide, ceramic, or zirconia. Ball mills produce up to 50 ton/hr of powder substantially passing a 200-mesh screen. Temperature control can be achieved by the use of jacketed ball mills through which a heat transfer fluid is circulated.

Critical and Operating Speed

In operation, the balls in the mill are carried up in contact with each other and with the walls until the centrifugal force is overcome by the centripetal force. The critical speed N_{cr}, of a ball mill is the speed in rpm above which the grinding media will centrifuge and all milling effectively stops.

The critical speed of a mill is given by:

$$N_{cr} = 42.3/D^{0.5} \tag{3}$$

where D is the internal diameter of the mill in meters.

The operating speed of ball mills is usually 55–75% of the critical speed. Operating close to N_{cr} drastically reduces the effectiveness of the grinding action.

Conical ball mills like the Hardinge mill shown in Figure 9.8 have their larger diameter closer to the feed inlet. The grinding media in the mill fall through different heights depending on their relative location. This provides a classifying action within the mill resulting in increased crushing energy efficiency.

Ball mills with centrifugal and planetary action are also available. Retsch manufactures a planetary ball mill in which multiple grinding chambers are turned around their own axes, and, in the opposite direction, around the common axis of a sun wheel that is driven by a rugged motor. This movement results in the superimposition of centrifugal forces that change constantly (Coriolis motion). The grinding media describe a semicircular motion, separate from the inside wall and collide with the opposite surface at high impact energy. Figure 9.9 illustrates the forces encountered in the operation of this grinding mill that can reduce particles to submicron sizes through the imposition of impact and friction forces.

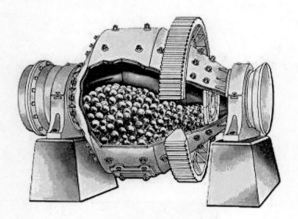

FIGURE 9.8
Hardinge conical ball mill. (Courtesy of Metso Minerals, York, PA.)

The Retsch centrifugal ball mill uses grinding chambers that move in a horizontal plane at speeds of 100–580 rpm. The centrifugal forces that are generated propel the grinding media against the inside walls of the mill where they roll over the product (Figure 9.10). Size reduction is achieved by a combination of impact and friction. The mill is furnished with an automatic reversal system to counter any agglomeration effects and to enhance homogenization.

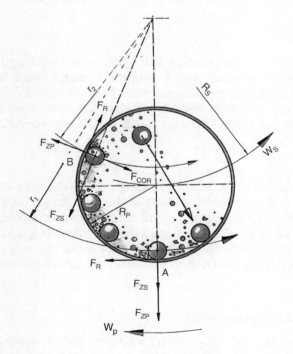

FIGURE 9.9
Schematic of forces operating in a planetary ball mill. (Courtesy of Retsch, Inc., Newtown, PA.)

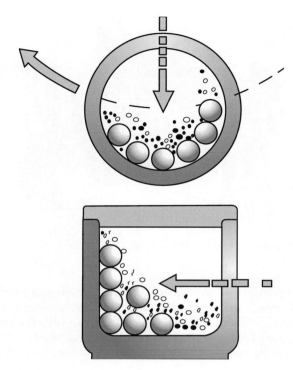

FIGURE 9.10
Schematic of centrifugal ball mill action. (Courtesy of Retsch, Inc., Newtown, PA.)

Tube Mills

Tube mills are basically finish-grinding ball mills with mill length several times the diameter. Tube mills can be made with a uniform diameter throughout the mill length or may be made with several compartments with different section diameters and lengths. Compartment mills are tube mills with slotted transverse partitions that separate grinding media of different sizes with the larger media working on coarser product.

Rod Mills

Rod mills are regarded as intermediate grinding mills and are basically tube mills that employ grinding rods that are about as long as the mill. Rod mills employ rolling compression and attrition and produce very small amounts of oversize or fines. Feed sizes up to 50 mm (2 in.) can be reduced to product in the 5–10 mesh range in open circuit grinding and to a minus 35 mesh range in closed circuit grinding with a classifying device. These mills require operator attention to minimize rod misalignment or entangling during grinding.

Autogenous and Semiautogenous Mills

Autogenous mills are mechanically similar to ball mills but employ the product to be crushed as the grinding medium. These mills can be used for hard, soft, as well as sticky products. The product is discharged through a discharge grate designed to retain oversize product. Autogenous grinding reduces metal wear, eliminates secondary and tertiary crushing stages, and offers savings in capital and operating costs. A variation called air autogenous milling uses air to transport and classify the product and uses no discharge screens or grills that can become blocked. Semiautogenous mills use a combination of the feed, product, and grinding media to achieve the desired size reduction.

Rolling Compression Mills

These mills resemble a mortar and pestle in principle and use a rolling ball or roller member moving against the face of a ring or a casing to grind soft to medium-hard product at rates of up to 50 ton/hr. Pressure is applied by either using heavy springs or by the centrifugal force of the roller. Either the roller or ring may be stationary. Built-in air classification is used to improve the grinding efficiency of these mills. The common types of rolling compression mills are rolling-ring pulverizers, bowl mills, and roller mills. These mills are widely used for grinding coal, cement clinker, and limestone.

Ultrafine Grinders

Fluid-Energy Mills

These energy-intensive mills have no moving parts and use pressurized air, steam, or inert gases to grind particles to ultrafine sizes. The particles to be ground are suspended in a high-velocity fluid stream that carries the particles around in a circular or an elliptical path. The majority of the particle size reduction is accomplished by interparticle attrition with additional grinding caused by particles rubbing against the walls. The particles get classified as they go around the closed path provided by the mill. The larger particles are thrown toward the outer walls and stay in the mill, while the finer particles stay on the inner walls and are removed from the mill through an exit port. The effluent from the mill is sent to a particle separator to collect the ground product. No one jet mill is suitable for all process applications. Several configurations are available to serve specific needs: fluid bed, opposed jet, and multiple-port types.

The Jet-O-Mizer (Figure 9.11) from Fluid Energy Processing can produce product with a 1–50 μ average size and a narrow particle size distribution. This vertical jet mill can also be used for combined operations like grinding/blending and grinding/coating.

For finer finished product down to the submicron size range, the Sturtevant Micronizer jet mill or the Fluid Energy Processing Micro-Jet grinders are suitable. The mills use a circular grinding chamber. The feed metered into the mill enters an air or a gas vortex created by precisely aligned jet nozzles along the mill periphery (Figures 9.12 and 9.13). The tangential angle of the fluid flow causes interparticle impact resulting in size reduction. The desired product and oversize material are separated by air classification in the mill. The mill provides narrow particle size distribution with uniform shape and no heat build-up. Micronizers are available in capacities up to 5 ton/hr.

The Roto-Jet fluid-bed jet mill (Figure 9.14) supplied by Fluid Energy Processing is designed to grind products to the 0.5–40 μ average size range with specific top and/or bottom size requirements. The machine uses a variable-speed rotor for tight control of product size.

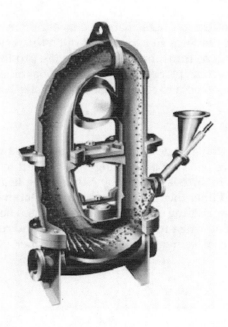

FIGURE 9.11
Cutaway of Jet-O-Mizer showing fluid jets. (Courtesy of Fluid Energy Processing, Telford, PA.) (*View this art in color at www.dekker.com.*)

Agitated Mills

Agitated mills are stirred vertical ball mills in which the grinding media are agitated by vibratory energy or by a rotating impeller that can run at speeds up to 1700 rpm. These mills are suited for both wet and dry grinding in batch or continuous mode. The fast grinding machines are energy efficient, compact, easy to operate, and are best suited for

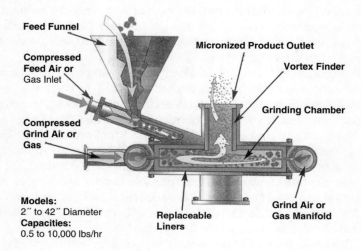

FIGURE 9.12
Operating schematic of the Micronizer jet mill. (Courtesy of Sturtevant, Inc., Boston, MA.) (*View this art in color at www.dekker.com.*)

FIGURE 9.13
Cutaway of the Micro-Jet mill showing fluid jets. (Courtesy of Fluid Energy Processing, Telford, PA.) (*View this art in color at www.dekker.com.*)

reduction of particles to submicron sizes. Particle size reduction is achieved by both shear and impact.

Attritors

The Szegvari attritor uses 1/8–3/8 in. grinding media agitated at speeds up to 350 rpm. Attritors are available that operate in three different modes:

FIGURE 9.14
Cutaway of Roto-Jet showing fluid jets. (Courtesy of Fluid Energy Processing, Telford, PA.) (*View this art in color at www.dekker.com.*)

1. Batch attritor shown in Figure 9.15.
2. Circulation attritors that use an attritor in association with a holding tank that can hold about 10 times the volume of the attritor (Figure 9.16). The contents of the holding tank are passed through the attritor multiple times until the desired reduction in size is achieved. This combination allows the use of small attritors for large grinding jobs through the use of high circulation rates.
3. Continuous attritors with the grinding time controlled by the feed pumping rate (Figure 9.17).

A high-speed attritor that uses 0.5–3 mm grinding media and impeller speeds of up to 1700 rpm has been developed recently for dry grinding.

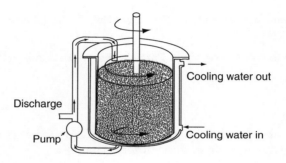

FIGURE 9.15
Batch attritor. (Courtesy of Union Process, Akron, Ohio.)

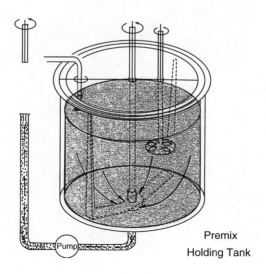

FIGURE 9.16
Circulation attritor. (Courtesy of Union Process, Inc., Akron, Ohio.)

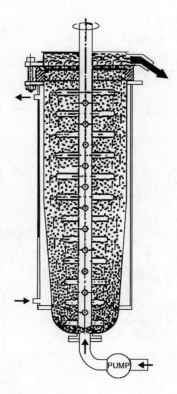

FIGURE 9.17
Continuous attritor. (Courtesy of Union Process, Inc., Akron, Ohio.)

Vibratory Mills

The Sweco vibro-energy mill (Figure 9.18) applies high frequency, three-dimensional vibrations to the grinding chamber that contains small cylindrical grinding media. This helps produce an ultrafine product with a narrow particle size distribution.

The Vibra-Drum vibratory grinding mill supplied by General Kinematics uses a static grinding drum containing grinding media akin to a ball mill. Vibration energy is imparted to the drum from an external mechanism. A subresonant two-mass drive and spring system alternately stores and releases grinding power. Once in motion, the natural frequency design ensures that energy is only needed to move the grinding media as a fluid mass. This mill has lower capital investment, installation, maintenance, and energy costs as compared to conventional rotational mills.

Impact Mills with Internal Classification

An example of a classifying impact mill is the Powderizer marketed by Sturtevant, Inc. (Figure 9.19). The feed to the mill enters the grinding chamber where it is broken by the action of impactors/pins mounted on a rotating disk. A column of air sweeps up the pulverized product through a rotating classifier that rejects oversize products that are returned back for further size reduction. Particle size can be adjusted by changing the

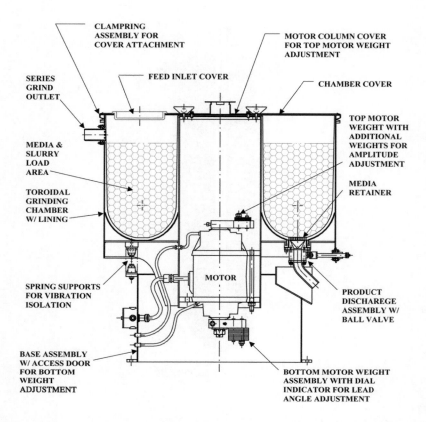

CLAMPRING
ASSEMBLY FOR
COVER ATTACHMENT

MOTOR COLUMN COVER
FOR TOP MOTOR WEIGHT
ADJUSTMENT

SERIES
GRIND
OUTLET

FEED INLET COVER

CHAMBER COVER

TOP MOTOR
WEIGHT WITH
ADDITIONAL
WEIGHTS FOR
AMPLITUDE
ADJUSTMENT

MEDIA &
SLURRY
LOAD
AREA

MEDIA
RETAINER

TOROIDAL
GRINDING
CHAMBER
W/ LINING

MOTOR

SPRING SUPPORTS
FOR VIBRATION
ISOLATION

PRODUCT
DISCHAREGE
ASSEMBLY W/
BALL VALVE

BASE ASSEMBLY
W/ ACCESS DOOR
FOR BOTTOM
WEIGHT
ADJUSTMENT

BOTTOM MOTOR WEIGHT
ASSEMBLY WITH DIAL
INDICATOR FOR LEAD
ANGLE ADJUSTMENT

FIGURE 9.18
Cross section of the Sweco vibro-energy mill. (Courtesy of Sweco, Florence, KY.) (*View this art in color at www*
.dekker.com.)

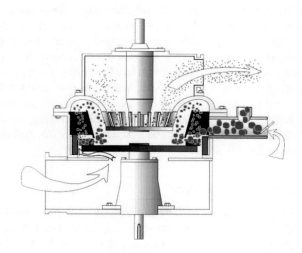

FIGURE 9.19
Schematic of Powderizer air-swept impact mill. (Courtesy of Sturtevant, Inc., Boston, MA.)

classifier speed without shutting down the mill. The product can be pulverized to sizes below 10 μ with narrow size distribution. Pulverizers are available in sizes that can handle 20–7000 kg/hr. The Powderizer is also suitable for heat-sensitive materials.

Rotary Cutters

Cutting mills are used for reducing the size of soft, medium-hard, tough, elastic, fibrous, and temperature-sensitive materials. Examples of products handled in such mills are electronic scrap, film, rubber, foil, foodstuffs, paper and paper products, textiles, and domestic waste. The mills cut, chop, or tear feed using a rotor-mounted set of blades or cutters. Figure 9.20 shows the cutting action of the SM2000 cutting mill supplied by Retsch. The feed material is taken up by the rotor and is crushed by the stainless steel cutting strips inserted in the housing. Helically arranged reversible cutting plates of hard metal operate by successive cutting.

Equipment Operation

Open and Closed Circuit Grinding

Comminution is one of the most inefficient unit operations from the viewpoint of energy consumption. Continuous crushing and grinding equipment can operate in an open circuit mode, where the product being crushed passes through the mill only once. This mode of operation is inefficient as a considerable amount of energy is wasted in regrinding

FIGURE 9.20
Schematic showing the cutting action of the SM2000 cutting mill. (Courtesy of Retsch, Inc., Newtown, PA.)

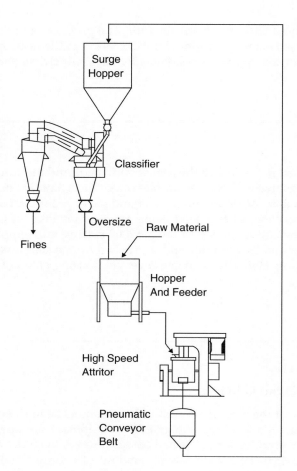

FIGURE 9.21
Representation of a closed-circuit grinding operation employing an attritor. (Courtesy of Union Process, Inc., Akron, Ohio.)

product that has reached the desired product size. A more efficient mode of operation is called closed circuit grinding (Figure 9.21). In closed circuit grinding, the mill discharge is sent to size classification equipment that separates out the product that has reached the desired size. Oversize material is returned to the mill for further size reduction. Closed circuit grinding is suitable for either wet or dry grinding of materials.

Conclusions

The greatest advancements in grinding technology in the recent past have been made in ultrafine grinding equipment like fluid-energy mills that can generate products in the micron and submicron ranges. This trend is expected to continue as comminution processes evolve to address technical issues in the production of ultrafine particles for high technology applications. As comminution is a very important unit operation in the conversion of run-of-mine raw materials, a considerable amount of effort was spent in the early and middle part of the last century to develop efficient size reduction equipment. The basic principles employed in the operation of crushing and grinding machines have not changed substantially over the past few decades. Saving energy, reducing costs, and cutting pollution have been the main areas targeted for continuous improvement in size reduction operations. The technical areas for future improvement of comminution processes fall into the following general categories: 1) Advanced sensors to provide integrated online physical and chemical characterization of feed and product to help in automated process control; 2) Utilization of the knowledge of real-time feed characteristics to optimize comminution processes; 3) Improved modeling of grinding mill operations like three-dimensional simulation of charge motion in tumbling mills; 4) Advanced abrasion-resistant milling media and surfaces; and 5) Novel or improved physical separation processes to be used in conjunction with size reduction equipment. As in the past, it is expected that future developments in conventional comminution will be evolutionary rather than revolutionary.

References

1. Kick, F. *Das Gasetz der propertionalen Widerstande und siene Anwendung*; Arthur Felix: Leipzig, 1885.
2. Rittinger, P.R. *Lehrbuch der Aufbereitungskunde*; Ernst and Korn: Berlin, 1867.
3. Bond, F.C. The theory of comminution—Meeting of AIME, Mexico City, October, 1951. Trans. Am. Inst. Min. Metall. Pet. Eng. **1952**, *193*, 484–494.
4. Taggart, A.F. *Handbook of Mineral Dressing*; Wiley: New York, 1945.

10

Pneumatic Conveying

Kimberly H. Henthorn

CONTENTS

Introduction 145
System Configuration 146
    Vertical vs. Horizontal Transport 146
    Positive vs. Negative Pressure Systems 148
Transport Regimes 149
Experimental Observations 150
    Effect of Particle Properties 150
    System Behavior 151
Modeling 152
Conclusions 153
References 153

Introduction

Pneumatic conveying (pneumatic transport) is the transport of particulate material by a conveying gas stream in a pipe or duct. These systems are popular in industrial practice because of their flexibility and relatively low maintenance requirements. Early pneumatic conveying efforts were mainly focused on the transport of grains, but have since evolved into the movement of a wide range of solid materials including coal, chemicals, cement, catalysts, mining materials, pharmaceuticals, food, and sand. While most solid materials can be transported pneumatically, it can be difficult to transport large, dense, cohesive, brittle, or fiber-like solids due to the increased energy requirement, higher probability of particle attrition, and higher risk of pipeline blockage. Pneumatic conveying systems are currently used within many industrial processes including pneumatic drying operations, dust collection systems, and standpipes in fluidized bed systems.

Although pneumatic conveying is not the most energy-efficient mode of particle transport, it is widely used because it is convenient and the solids can be isolated from the environment, which is important for hazardous or sensitive materials. Many pneumatic conveying systems do not run efficiently, in practice, and this is partly due to a lack of understanding of complex gas–solid flows. It is very difficult to experimentally isolate the effects of phase relationships, which include interparticle, particle–wall, particle–gas, and gas–wall interactions. However, recent progress in experimental techniques and computational modeling approaches has greatly increased our understanding of gas–solid flows and has led to improvements in the way conveying systems are designed. This chapter gives an overview of basic pneumatic conveying system concepts and discusses some of the experimental observations made with respect to system and particle properties.

System Configuration

One feature that makes pneumatic conveying systems so attractive in industrial practice is their flexibility in design. Physical properties of the material being transported, potential explosion or other hazards, and arrangement of pickup and drop-off points contribute to the selection of the best system configuration. When designing a pneumatic conveying system, one must consider a large number of variables and must ultimately determine system properties and operating conditions including gas velocity, number and type of pipeline bends, pipeline dimensions, and pressure drop across the system, to name a few. However, optimization of these parameters can be extremely challenging. For example, using an optimal gas velocity is critical in pneumatic conveying. Gas velocities that are too high can lead to particle attrition, pipe erosion (particularly in bends), and wasted energy. On the other hand, gas velocities that are too low encourage saltation in horizontal flow, where the energy provided by the gas stream is insufficient to maintain particle suspension and the solids begin to "salt out" or deposit along the bottom of the duct or pipe. Saltation typically leads to pipeline blockage, which can be very expensive and time-consuming to remedy. In addition, pressure drop calculations are complicated by variations in particle characteristics such as shape, size, size distribution, and cohesiveness, which can be difficult to quantify.

Conveying systems consist of a "mixing/acceleration zone," where solids are introduced into the flowing gas stream and accelerated to their conveying velocity, a "conveying zone," where the two-phase mixture is transported across a desired distance, and a "separation zone," where the two phases are separated. A variety of technologies have been developed to achieve efficiency for each of these zones, although common configurations include a pump, blower, or compressor to transport the gas phase, a hopper to store and dispense particles, a pipeline in which to transport the two-phase suspension, receiving hoppers or cyclone separator to separate the solids from the gas, and a filtration system to further clean the gas. Air is typically used as the conveying gas unless there is an environmental or health risk, in which case nitrogen or other gases can be used. Because of the additional cost required to use alternative gases, closed-loop systems are typically utilized.

Vertical vs. Horizontal Transport

In most pneumatic conveying systems, solids are transported both horizontally and vertically. Since pipeline blockage and excessive pressure drops must be avoided, it is important to determine the optimal gas velocity for a given system configuration and solid type. In general, systems should be operated at the lowest gas velocity possible while avoiding choking and saltation in vertical and horizontal transport, respectively. While predicting these two critical parameters is difficult due to their complex dependence on particle and system properties, the current literature offers many empirical relationships, a few of which are discussed below.

In vertical pneumatic conveying, low gas velocities can lead to slugging or choking behavior. Figure 10.1 shows a typical pressure gradient profile in dilute transport for varying solid mass flux. At high gas velocities, pressure drop decreases with decreasing flow rate. However, a decrease in gas velocity translates into an increase in solid concentration at constant solid mass flux, and at point *A* in Figure 10.1, total solid entrainment is no longer possible. As gas velocity is further decreased, choking is observed, where solids begin to periodically fall downward and pressure drop is no longer stable. Choking velocity is

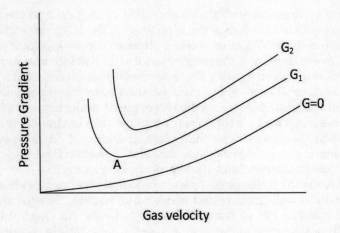

FIGURE 10.1
Pressure drop profiles in dilute vertical pneumatic conveying. G represents solid mass flux, with $G_2>G_1$.

a function of particle properties, and several predictive correlations have been proposed. However, the most widely used correlations include those by Chong and Leung[1] and Punwani et al.[2]

While choking does not occur in horizontal transport, saltation is a common phenomenon that can lead to pipeline blockage. Figure 10.2 shows a typical pressure profile in dilute horizontal transport. At high gas velocities, pressure drop is directly proportional to gas velocity. However, similar to vertical transport, a decrease in gas velocity leads to an increase in solid concentration at constant solid mass flux, and at point A in Figure 10.2, the gas velocity is no longer sufficient to fully suspend the particles and they begin to "salt out" suspension. At very low gas velocities (high solid concentrations), horizontal conveying is possible as dense-phase solid transport, which is discussed later. Several empirical correlations to estimate saltation velocity have been proposed, with the most notable being Matsumoto et al.[3] and Rizk.[4]

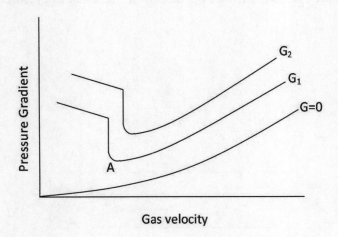

FIGURE 10.2
Pressure drop profiles in dilute horizontal pneumatic conveying. G represents solid mass flux, with $G_2>G_1$.

Most conveying systems consist of a combination of vertical and horizontal sections. Pipeline bends are utilized to change the direction of flow in these systems, but because of the large pressure drop (~7.5 m of vertical pressure drop) associated with changes in momentum and reacceleration of the suspension, it is advised to minimize the number of bends introduced into a given system. The worst transition occurs from a downward vertical flow into a horizontal section; this configuration encourages particle saltation in the horizontal section and greatly increases the risk of pipeline blockage. In addition, the most serious particle breakage and pipeline wear occurs within the bends. It has been shown that saltation velocity is always larger than choking velocity for a given set of materials;[5] as a result, pneumatic conveying velocities should be based on the avoidance of saltation, which would automatically eliminate the risk of choking as well.

Several options exist to minimize particle attrition and pipeline wear in elbows. Gradual bends reduce particle–wall impacts and sudden momentum changes and lead to some improvement in material life as compared to 90° elbows. Although this configuration requires more physical space and the improvement is only slightly incremental, blind tees have been shown to be the best configuration for changes in flow direction. Blind tees are sharp 90° bends that include a short extension of the pipe in the direction of flow beyond the bend (Figure 10.3). A cushion of solids accumulates in the pipe extension so that particles traveling with the suspension experience softer particle–particle impacts rather than particle–wall impacts. Blind tees offer a short turning radius to reduce required space, but greatly reduce erosion in the pipeline. Pressure drop and particle attrition in blind tees are comparable to those in simple bends.[5]

Positive vs. Negative Pressure Systems

Particles can be transported by either or both positive or negative pressure pneumatic conveying systems. Positive systems are the most commonly used and utilize positive gas pressure to push solids through the pipeline. These systems are ideal for situations involving one pickup point and multiple drop-off points. One major difficulty associated with positive pressure systems is effective solid feeding into the pipeline. Rotary airlocks are commonly used to control the solid feed rate against the gas pressure, and screw feeders are utilized to move the solids from bulk storage to the feeder position. In addition, the

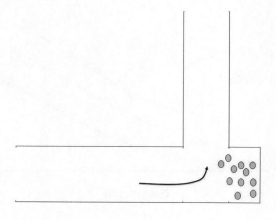

FIGURE 10.3
Blind tee configuration.

risk of leakage is greater with these systems since the pressure within the system is greater than atmospheric pressure. Positive pressure systems are typically limited to a maximum gauge pressure of approximately 1 bar.[5,6]

On the other hand, negative pressure (vacuum) pneumatic conveying systems utilize vacuum to pull particles to their drop-off points. These systems are commonly used to transport materials from open storage, such as unloading truck trailers or ships, and are advantageous for instances where material is to be transported from multiple pickup locations to a single drop-off point. One common example of a negative pressure conveying system is a household central vacuum system, where multiple suction points located within the residence flow into a central dust deposit bin. Negative pressure systems are particularly useful for transporting hazardous materials, because leakage into the environment is minimal compared to positive pressure systems. However, a higher volume of air must be filtered with vacuum systems, so the filtration system is typically more expensive. Negative pressure systems are typically limited to a vacuum of approximately 0.4 bar.[5,6]

Transport Regimes

Pneumatic conveying is typically classified as either dense or dilute phase. Although there is no definitive delineation between the two, dilute-phase flows have been defined as those containing less than 1% solids by volume, pressure drops below 5 mbar/m, and gas velocities greater than 20 m/s, while dense-phase flows contain greater than 30% solids by volume, pressure drops greater than 20 mbar/m, and gas velocities less than 5 m/s.[5] A more widely used definition classifies flows according to the mass ratio of solids to conveying gas, with mass ratios less than 15 corresponding to dilute-phase flows and ratios greater than 15 corresponding to dense-phase flows.[7] The onset of choking and saltation can also be used to define the transition point between dilute- and dense-phase flows.

Dilute-phase flows are the most common mode of transport in industrial pneumatic conveying processes and are characterized by high velocities, low pressure drops, and low solid concentrations. Pressure drop behavior is more stable and more easily predicted in dilute flow than in dense-phase systems. Particulates in dilute flow are fully suspended within the carrier gas and behave as individual entities.

Dense-phase pneumatic conveying systems are much more complex than their dilute-phase counterparts and can be very unstable during operation. However, dense-phase transport is advantageous if the solids are cohesive, brittle, or abrasive. Since dense-phase flow requires lower gas velocities, there is less impact between the solid and pipeline walls, reducing particle attrition and pipe erosion. Dense-phase transport can be categorized as either continuous or discontinuous. In continuous systems, the material is extruded through the pipeline as a single plug, with solids filling the entire pipe cross-section. Since very high pressures are required to operate in this mode, these systems are typically limited to transport over short distances.

Solids in discontinuous dense-phase systems are pushed through the pipeline in plugs, with intermittent portions of the pipe having very low concentrations of particulates. There are different levels of discontinuous flow, ranging from discrete solid plugs that fill the entire pipe cross-section and are separated by sections of very low solid concentration to flows consisting of settled particle dunes at the bottom of the pipe that are eroded downstream by a less concentrated top layer.

Experimental Observations

Experimental investigation of gas–solid flows is abundant in the literature. Because of the complex nature of these flows, a wide range of topics has been studied in an effort to systematically uncover the relationships between relevant parameters. A few key experimental observations from the past several decades are highlighted below.

Effect of Particle Properties

Much of the gas–solid literature focuses on the effect of material properties such as particle size, size distribution, shape, density, and electrostatic behavior. Initial gas–solid work focused on single-particle behavior and gas velocity profiles in single-phase turbulent flow, and many early experiments were hindered by the lack of accurate measurement techniques. Seminal work by Vogt and White[8] investigated the horizontal and vertical transport of non-spherical granular materials and developed an empirical expression for dimensionless pressure drop as a function of particle and system characteristics. They studied particles of varying particle shape, but did not have sufficient data to rigorously evaluate its effect on pressure drop. However, they concluded that particle shape did not play a significant role in the overall pressure drop and could be ignored. Similarly, Reddy and Pei[9] concluded that pressure drop is directly proportional to solid loading and is independent of particle size and air velocity.

Hariu and Molstad[10] investigated the effect of particle shape, solid loading, and solid acceleration on pressure drop in the vertical pneumatic conveying of granular materials. They concluded that the two phases can be considered independent and the overall pressure drop can be estimated by summing the contributions from the gas and solids. Unlike Vogt and White, however, they found that particle shape does in fact have an effect on the overall pressure drop, and its contribution can be corrected in simple models by the addition of an appropriate shape factor. More recent studies support the conclusion of Hariu and Molstad. Pan[11] investigated the importance of fundamental particle properties by collecting experimental data using diverse materials with sizes ranging from 11 μm to 4 mm. The author found a strong relationship between particle properties and pneumatic conveying performance because of the influence of particle size, size distribution, and shape on solid-phase permeability, gas retention capability, and de-aeration behavior.

Henthorn et al.[12] systematically investigated the effects of particle shape and size, mass loading, and Reynolds number on pressure drop in vertical pneumatic conveying. They used a range of particle shapes, with quantified sphericities ranging from 0.59 to 1.0. Particle sphericity is the ratio of surface area of a sphere having the same volume as the particle to the surface area of the particle; thus, the sphericity of a perfect sphere is 1.0. The authors found that particle shape has a strong influence on pressure gradients, and that current empirical correlations and computational models do not adequately capture these effects. Suspensions of non-spherical particles exhibit a higher pressure drop than spherical particles of similar material properties due to increased particle drag. The authors also concluded that pressure drop increases with increasing particle size.

Plasynski et al.[13] also performed a systematic investigation into the effects of particle properties on vertical pneumatic conveying behavior. They measured pressure drop, choking velocity, particle friction factor, pressure fluctuations, and particle velocity, among other factors, for a range of particle types. They concluded that as particle size and density increase, the pressure drop also increases for particles having otherwise similar material properties. They also found that pressure drop increases as mass loading increases due to the increase in solid static head. One interesting conclusion was that for system pressures greater than atmospheric, the gas- and solid-phase friction factors are dependent on each other because of interphase interactions.

Wang et al.[14] studied the transport of cohesive glass spheres in vertical pneumatic conveying and found that interparticle forces and electrostatic behavior of the solid affected the pressure drop differently than larger, non-cohesive glass spheres. Adhesion to pipe walls was much more significant with the cohesive material, but electrostatic charging was more predominant with the larger particles. As a result, the cohesive material experienced a lower minimum pressure gradient at the onset of choking behavior. Ally and Klinzing[15] found that the contribution of electrostatic charging on the pressure drop in vertical conveying was a function of the charging ability of the solid as well as the number density of the particles.

System Behavior

In addition to the effect of particle properties, there has been extensive research into understanding system parameters and the general flow behavior of two-phase mixtures. The Richardson–Zaki equation[16] is one of the most popular empirical equations to relate the slip velocity to voidage fraction in a batch liquid sedimentation system; however, some investigators claim it can be modified to describe the behavior of gas–solid systems as well. Slip velocity is the relative velocity difference between the gas and particle phases and is sometimes estimated to be equivalent to the single particle terminal velocity. However, in high concentration gas–solid flows, the slip velocity is greater than the terminal velocity because of additional factors such as particle recirculation near the walls and solids–wall friction. Capes and Nakamura[17] modified the Richardson–Zaki equation by the addition of an empirical multiplier to account for the increased slip velocity.

Sankar and Smith[18] also found that slip velocity is a strong function of solid concentration and suggested that the Richardson–Zaki equation could not be extended to gas–solid flows. They also claimed that other empirical gas–solid correlations were not applicable across a wide range of operating conditions and omit the effect of critical particle properties. In response, they proposed a new two-condition correlation to relate slip velocity to particle properties, such as size, density, and particle Reynolds number, and found fairly good agreement with experimental data.

Hettiaratchi et al.[19] compared pressure drop data in vertical and horizontal sections of the same system in an attempt to determine their relationship. Their motivation was to develop a method in which the behavior of one type of system could be extrapolated to predict behavior in the other to reduce cost and time acquiring pilot–plant data during system design. They determined that pressure gradients in vertical flow are significantly larger than in horizontal flow. However, the ratios of the two pressure gradients were strongly dependent on particle type, gas velocity, and presence of bends, and their data did not provide an effective empirical relationship between the two.

Early work by Konno and Saito[20] revealed that gas velocity profiles in horizontal and vertical flows are different with the addition of particles. Vertical flows exhibit symmetrical gas velocity profiles with or without the presence of particles. However, the addition of particles led to an asymmetric gas velocity profile in horizontal transport, mainly due to the non-uniform distribution of solids by gravity. The authors also concluded that particle–wall collisions were the major contributor to pressure drop in horizontal conveying.

Satija et al.[21] experimentally investigated the transition from dilute- to dense-phase vertical transport for a range of particle types. Specifically, they measured pressure fluctuations and solid voidage to compare choking velocities for four types of fine powders. They concluded that when a system operates in the dilute regime, pressure fluctuations are negligible, but at the choking velocity, the frequency and intensity of pressure fluctuations sharply increase. The authors used these observations to quantify the onset of choking and found good comparison with the predictive models of Yang.[22,23]

It is generally accepted that pressure drop increases with increased solid mass loading and Reynolds number.[19,20,24–26] However, for finer particles (25–75 μm), pressure drop initially decreases with increasing mass loading at very low solid concentrations (mass loadings between 0.5 and 4.0) and Reynolds numbers between 15,000 and 40,000.[27,28] The gas-phase stress at these conditions is reduced due to particle–particle interactions and the presence of turbulent eddies, which results in the inverse relationship between pressure drop and mass loading.

Modeling

Many pneumatic conveying systems do not operate up to expectations in practice. This is mainly due to the complex nature of two-phase flows and the reliance on simplified empirical correlations during system design. These correlations typically do not take into account particle properties such as cohesion, and many ignore the complex interactions within and between the various phases (solid, gas, pipe). For example, Eq. 1 is a simple empirical equation to predict pressure drop in dilute pneumatic conveying systems:[5,29]

$$\Delta P_T = \frac{1}{2}\varepsilon\rho_g U_g^2 + \frac{1}{2}(1-\varepsilon)\rho_p U_p^2 + \frac{2f_g\rho_g U_g^2 L}{D} + F_{pw}L + \rho_p L(1-\varepsilon)g\sin\theta + \rho_g L\varepsilon g\sin\theta \qquad (1)$$

where θ represents the angle of inclination of the pipe, U_p is the interstitial particle velocity, U_g is the interstitial gas velocity, D is pipe diameter, L is the length over which the pressure measurement is made, ε is the void fraction, and F_{pw} is the solid–pipe friction force [force/length^3]. The terms in this equation represent the pressure drop due to gas acceleration, solid acceleration, gas–pipe friction, solid–pipe friction, static head of the solids, and static head of the gas, respectively. This model treats the two phases independently and does not consider particle shape or particle–particle interactions. Although it has been shown to match fairly well with experimental pressure drop data for spherical particles in vertical pneumatic conveying, the model fails to capture the data for nonspherical or cohesive materials.[12]

Singh[24] further developed the approach leading to Eq. 1 by integrating the effect of particle collisions within vertical binary flow. The author also substituted correlations to describe fluid–particle interactions in place of the friction factor terms in Eq. 1. The model was validated with experimental data and it was found that the model worked relatively well to predict the behavior of small particles, but larger particles tended to recirculate near the pipe walls, which resulted in an overprediction of slip velocity by the model. The terminal velocity of a single particle is sometimes the only parameter used to estimate pressure drop in gas–solid flow, but as Scott[30] claimed, solid–wall friction must also be taken into consideration. In addition, Singh's model still neglected to account for effects of particle cohesion and adhesion, particle shape, and more complex size distributions.

As a result of the lack of effective predictive tools, much effort has been given toward computational modeling to better understand pneumatic conveying systems. Although a complete discussion on computational modeling is beyond the scope of this chapter, interested readers are encouraged to investigate the many publications relating to interphase relationships,[31–36] computational approaches,[37–41] transport regimes,[42–45] population balances within pneumatic conveying,[46] and particle properties.[47–51]

Conclusions

Although pneumatic conveying is ubiquitous in industrial practice, it is very complex and difficult to design an efficient, problem-free system. This is partly because gas–solid systems are not completely understood, and many empirical correlations commonly used for system design do not incorporate complicating parameters such as particle cohesiveness, breakage, shape, or size distribution. In addition, the use of pilot plants to determine optimal operating parameters for the transport of a given material can be very costly and time-consuming, so there is heavy reliance on correlation and theory. Much work is currently being dedicated to the development of computational models, but they still fall short when predicting the behavior of extreme conditions such as very high solid loadings or highly non-spherical materials. However, much progress has been made over the past few decades, and the future direction of pneumatic conveying research looks very promising.

References

1. Chong, Y., Leung, L. Comparison of choking velocity correlations in vertical pneumatic conveying, Powder Tech. **1986**, *47*(1), 43–50.
2. Punwani, D., Modi, M., Tarman, P. A generalized correlation for estimating choking velocity in vertical solids transport. In Proceedings of International of Powder and Bulk Solids Handling and Processing Conference, Powder Advisory Center, Chicago, 1976.
3. Matsumoto, S., Kikuta, M., Maeda, S. Effect of particle size on the minimum transport velocity for horizontal pneumatic conveying of solids, J. Chem. Eng. Japan **1977**, *10*(2), 273.
4. Rizk, F. Dr-Ing. dissertation. Technische Hochschule Karlsruhe, 1973.
5. Rhodes, M. *Introduction to Particle Technology*. Wiley: Chichester, England, 2008.
6. McGlinchey, D. *Bulk Solids Handling: Equipment Selection and Operation*. Blackwell Publishing: Oxford, England, 2008.
7. Klinzing, G., Rizk, F., Marcus, R., Leung, L. S. *Pneumatic Conveying of Solids: A Theoretical and Practical Approach (Particle Technology Series)*. Springer: Dordrecht, 2010.
8. Vogt, E., White, R. Friction in the flow of suspensions: granular solids in gases through pipe. Ind. Eng. Chem. **1948**, *40*, 1731–1738.
9. Reddy, K., Pei, D. Particle dynamics in solids–gas flow in a vertical pipe. I&EC Fundamentals **1969**, *8*(3), 490–497.
10. Hariu, O., Molstad, M. Pressure drop in vertical tubes in transport of solids by gases. Ind. Eng. Chem. **1949**, *41*, 1148–1160.

11. Pan, R. Material properties and flow modes in pneumatic conveying. Powder Tech. **1999**, *104*, 157–163.

12. Henthorn, K., Park, K., Curtis, J. Measurement and prediction of pressure drop in pneumatic conveying: effect of particle characteristics, mass loading, and Reynolds number. Ind. Eng. Chem. Res. **2005**, *44*, 5090–5098.

13. Plasynski, S., Klinzing, G., Mathur, M. High-pressure vertical pneumatic transport investigation. Powder Tech. **1994**, *79*, 95–109.

14. Wang, Y., Williams, K., Jones, M., Chen, B. CFD simulation of gas–solid flow in dense phase bypass pneumatic conveying using the Euler–Euler model. Appl. Mech. Mat. **2010**, *26–28*, 1190–1194.

15. Ally, M., Klinzing, G. Inter-relation of electrostatic charging and pressure drops in pneumatic conveying. Powder Tech. **1985**, *44*, 85–88.

16. Richardson, J., Zaki, W. Sedimentation and fluidization: Part I. Trans. Inst. Chem. Eng. **1954**, *32*, 35–53.

17. Capes, C., Nakamura, K. Vertical pneumatic conveying: An experimental study with particles in the intermediate and turbulent flow regimes. Canadian J. Chem. Eng. **1973**, *51*, 31–38.

18. Sankar, S., Smith, T. Slip velocities in pneumatic transport: Part I. Powder Tech. **1986**, *47*, 167–177.

19. Hettiaratchi, K., Woodhead, S., Reed, A. Comparison between pressure drop in horizontal and vertical pneumatic conveying pipelines. Powder Tech. **1998**, *95*, 67–73.

20. Konno, H., Saito, S. Pneumatic conveying of solids through straight pipes. J. Chem. Eng. Japan **1969**, *2*(2), 211–217.

21. Satija, S., Young, J., Fan, L.-S. Pressure fluctuations and choking criterion for vertical pneumatic conveying of fine particles, Powder Tech. **1985**, *43*, 257–271.

22. Yang, W. Criteria for choking in vertical pneumatic conveying lines. Powder Tech. **1983**, *35*, 143–150.

23. Yang, W. A mathematical definition of choking phenomenon and a mathematical model for predicting choking velocity and choking voidage. AIChE J. **1975**, *21*(5), 1013–1015.

24. Singh, B. Analysis of pressure drop in vertical pneumatic conveying. Generalized approach for gas–particle and liquid–particle systems. Powder Tech. **1982**, *32*, 179–191.

25. Rautiainen, A., Stewart, G., Poikolainen, V., Sarkomaa, P. An experimental study of vertical pneumatic conveying. Powder Tech. **1999**, *104*, 139.

26. Namkung, W., Cho, M. Pressure drop in a vertical pneumatic conveying of iron ore. *Ind. Eng. Chem. Res.* **2002**, *41*, 5316.

27. Marcus, R., Leung, L., Klinzing, G., Rizk, F. *Pneumatic Conveying of Solids*. Chapman and Hall: London, 1990.

28. Klinzing, G. *Gas-Solid Transport*. McGraw-Hill: New York, 1981.

29. Knowlton, T. M. Solids transfer in fluidized systems. In *Gas Fluidization Technology*. Ed. D. Geldart, Wiley and Sons: Chichester, **1986**, 341–414.

30. Scott, A. The influence of particle properties on the pressure drop during the pneumatic transport of granular materials. Pneumotransport 4, Fourth International Conference on the Pneumatic Transport of Solids in Pipes, Carmel-by-the-Sea, CA, 1978.

31. Sinclair, J. Jackson, R. Gas–particle flow in a vertical pipe with particle–particle interactions. AIChE J. **1989**, *35*, 1473.

32. Hadinoto, K., Curtis, J. Effect of interstitial fluid on particle–particle interactions in kinetic theory approach of dilute turbulent fluid–particle flow. Ind. Eng. Chem. Res. **2004**, *43*, 3604–3615.

33. Bolio, E., Yasuna, J., Sinclair, J. Dilute turbulent gas–solid flow in risers with particle–particle interactions. AIChE J. **1995**, *41*(6), 1375–1388.

34. Nieuwland, J., Delnoij, E., Kuipers, J., van Swaaij, W. An engineering model for dilute riser flow. Powder Tech. **1997**, *90*, 115–123.

35. Michaelides, E. Motion of particles in gases: Average velocity and pressure loss. Trans. ASME **1987**, *109*, 172–178.

36. Raczek, J. Palica, M. Selection of the gas working velocity in vertical pneumatic conveying systems for binary mixtures, Chem. Eng. Proc. **1997**, *36*, 167–170.
37. Mezhericher, M., Brosh, T., Levy, A. Modeling of particle pneumatic conveying using DEM and DPM methods. Particulate Sci and Tech. **2011**, *29*(2), 197–208.
38. Singh, V., Lo, S. Predicting pressure drop in pneumatic conveying using the discrete element modeling approach. In Proceedings of the Seventh International Conference of CFD in the Minerals and Process Industries, CSIRO, Melbourne, Australia, 2009.
39. Wang, F., Zhu, J., Beeckmans, J. Pressure gradient and particle adhesion in the pneumatic transport of cohesive fine powders. Int J Multiphase Flow **2000**, *26*, 245–265.
40. Huilin, L., Gidaspow, D., Bouillard, J., Wentie, L. Hydrodynamic simulation of gas–solid flow in a riser using kinetic theory of granular flow. Chem. Eng. J. **2003**, *95*, 1–13.
41. Matsumoto, S., Saito, S. Monte Carlo simulation of horizontal pneumatic conveying based on the rough wall model. J. Chem. Eng. Japan **1970**, *3*(2), 223–230.
42. Sturm, M., Wirtz, S., Scherer, V., Denecke, J. Coupled DEM–CFD simulation of pneumatically conveyed granular media. Chem. Eng. Tech. **2010**, *33*(7), 1184–1192.
43. Li, J., Webb, C., Pandiella, S., Campbell, G., Dyakowski, T., Cowell, A., McGlinchey, D. Solids deposition in low-velocity slug flow pneumatic conveying. *Chem. Eng. Process* **2005**, *44*(2), 167–173.
44. Ratnayake, C., Melaaen, M., Datta, B. Pressure drop prediction in dense phase pneumatic conveying using CFD. In Proceedings of the Fourth International Conference on CFD in the Oil and Gas, Metallurgical & Process Industries, SINTEF/NTNU, Trondheim, Norway, 2005.
45. Fan, J., Jin, J., Zhang, X., Cen, K. A numerical model for dense particle-laden jets. Powder Tech. **2001**, *115*, 256–264.
46. Chapelle, P., Christakis, N., Abou-Chakra, H., Bridle, I., Bradley, M., Patel, M., Cross, M. Computational model for prediction of particle degradation during dilute-phase pneumatic conveying: modeling of dilute-phase pneumatic conveying. Adv. Powder Tech. **2004**, *15*(1), 31–49.
47. Pelegrina, A., Crapiste, G. Modelling the pneumatic drying of food particles. J. Food Eng. **2001**, *48*, 301–310.
48. Pelegrina, A. Analysis of pneumatic conveying of particles. Latin American Applied Res. **2002**, *32*(1), 91–96.
49. Weber, M. Friction of the air and the air/solid mixture in pneumatic conveying. Bulk Solids Handling **1991**, *11*(1), 99–102.
50. Pan, R., Wypych, P. Pressure drop and slug velocity in low-velocity pneumatic conveying of bulk solids. Powder Tech. **1997**, *94*, 123–132.
51. Hilton, J., Cleary, P. The role of particle shape in pneumatic conveying. In Proceedings of the Seventh International Conference of CFD in the Minerals and Process Industries, CSIRO, Melbourne, Australia, 2009.

11

Solid-Liquid Mixing: Numerical Simulation and Physical Experiments

Philippe A. Tanguy, Francis Thibault, Gabriel Ascanio, and
Edmundo Brito-De La Fuente

CONTENTS

Introduction 157
Literature Survey... 157
Description of the Coaxial Mixer Setup... 160
Marine Propeller Mixer 162
   Computational Model 163
Validation of the Numerical Model.. 168
Coaxial Mixer Results and Discussion... 170
Conclusions... 177
References.. 178

Introduction

Solid–liquid mixing processes find numerous applications in industry. The modeling of these mixing operations using first principles is still limited, although some progress has been made in recent years. This study assesses the real capability of the network-of-zone approach in the case of a complex mixing problem involving a coaxial mixer. Coaxial mixers are very popular for the preparation of pastes and slurries in the chemical, food, and coating industries. The mixer considered is composed of two rotating shafts: a fast-driven shaft supporting an open impeller and a slow shaft driving a scraping anchor arm.

Literature Survey

The dispersion of solids in liquids, the preparation of solid suspension in liquid media, and the make-down of pigment slurries in agitated vessels are typical solid–liquid mixing problems that find numerous applications in the process industries. For instance, they are involved in the preparation of paints and coatings, the manufacturing of food products, as well as suspension polymerization.

   In solid–liquid mixing design problems, the main features to be determined are the flow patterns in the vessel, the impeller power draw, and the solid concentration profile versus the solid concentration. In principle, they could be readily obtained by resorting to the CFD (computational fluid dynamics) resolution of the appropriate multiphase fluid mechanics equations. Historically, simplified methods have first been proposed in the literature, which do not use numerical intensive computation. The most common approach is the dispersion–sedimentation phenomenological model. It postulates equilibrium between the particle flux due to sedimentation and the particle flux resuspended by the turbulent diffusion created by the rotating impeller. Based on this concept and assuming a one-dimensional distribution of particles along the vessel height, an analytical expression for the concentration profile can be derived:

$$\frac{\phi}{\bar{\phi}} = \frac{Pe_s}{1 - \exp(-Pe_s)} \exp(-Pe_s z/H) \tag{1}$$

where $\bar{\phi}$ represents the average volume fraction of the suspension, z the vertical coordinate, H the fluid height in the vessel, and Pe_s the Peclet number of the solid particles. The Peclet number is defined as

$$Pe_s = v_t \times H/D_{e,p} \tag{2}$$

where v_t is the settling velocity at equilibrium and $D_{e,p}$ the turbulent diffusion coefficient. This dimensionless number is a model fitting parameter, and it is generally correlated with the operating conditions and the physical properties of the suspension.[1] The model has been verified experimentally in a tank provided with four Rushton turbines (radial discharge impeller) at a very low solids concentration.[2,3] Several improvements have been proposed in the literature to make the model more general. Ferreira, Rasteiro, and Figueiredo[4] introduced new terms in Eq. (1) to take into account the radial variation of the particle concentration. Rasteiro, Figueiredo, and Friere[5] suggested the use of the Richardson and Zaki's expression[6] for the settling velocity, as this velocity is strongly dependent on the solid volume fraction in the suspension.

The application of CFD in the modeling of solid–liquid mixing is fairly recent. In 1994, Bakker et al.[1] developed a two-dimensional computational approach to predict the particle concentration distribution in stirred vessels. In their model, the velocity field of the liquid phase is first simulated taking into account the flow turbulence. Then, using a finite volume approach, the diffusion–sedimentation equation along with the convective terms is solved, which includes D_s, a turbulent diffusion coefficient of the particles, defined as:

$$D_s = \frac{\sqrt{k_t}}{3\pi n_p D_p^2} \tag{3}$$

where n_p is the overall volume of the particles, D_p their diameter, and k_t the turbulent kinetic energy density. As the authors did not use iterative coupling between the computation of the flow field and that of the solids concentration, the effect of the solids on the fluid mechanics in the vessel was not taken into account. As an example, they investigated the suspension of 20 μm particles with several radial discharge impellers. They showed that their approach was capable of predicting the solid–liquid interface in the vessel and the effect of the position and number of turbines on the solid distribution. Unfortunately, no experimental validation of this approach was carried out.

All the above models do not consider the particle–particle interactions, although these interactions influence the settling velocity and ignore the effect of the solid phase on the hydrodynamics in the vessel. As a consequence, the practical range of application is restricted to low solids concentrations.

A particle migration model was proposed by Gadala-Maria and Acrivos[7] to describe experimental shear-induced migration observations. This model allows for a better understanding of the shear effects on particle diffusion for concentrated suspensions.[8–11] Based on these studies, a conservation equation for the solid phase was established by Phillips, Amstrong, and Brown,[12] which takes into account convective transport, diffusion due to particle–particle interactions, and the variation of viscosity within the suspension, namely:

$$\frac{\partial \phi}{\partial t} + \upsilon \cdot \nabla \phi = \nabla \cdot \left(K_c a^2 \phi \nabla \left(|\dot{\gamma}| \phi \right) \right) + \nabla \cdot \left(K_\eta \frac{a^2}{\eta} |\dot{\gamma}| \phi^2 \frac{\partial \eta}{\partial \phi} \nabla \phi \right) \qquad (4)$$

where ϕ is the volume fraction, υ the suspension velocity, a the particle radius, $|\dot{\gamma}|$ the magnitude of the rate-of-strain tensor. K_c and K_η are empirical constants equal to 0.41 and 0.62, respectively. In order to describe the variation of the viscosity η with the particle volume fraction, the authors suggested the use of the Krieger–Dougherty phenomenological model.[13] This approach was applied to investigate the diffusion of suspensions consisting of poly(methylmethacrylate) monodisperse particles ($D_p = 675$ μm) at high concentration ($\phi > 45\%$) in Newtonian silicon oil. Two geometries were tested: a Couette flow (flow between parallel plates with one plate in motion) and a Poiseuille flow (flow in a cylindrical channel). For these two cases, as the concentration varies only in the radial direction, several analytical expressions could be established for the solid volume fraction and the suspension velocity profile. The computational results were compared to concentration measurements based on nuclear magnetic resonance and a qualitative agreement was obtained. It should be noted here that in principle Eq. (4) can be applied at any solids concentration, but it is however restricted to noncolloidal systems.

A different numerical strategy to simulate multiphase mixing was introduced by Mann[14] and Mann and Hackett.[15] The idea of the method, called the network-of-zone, is to subdivide the flow domain in a set of small cells assumed to be mixed perfectly. The cells are allowed to exchange momentum and mass with their neighboring cells by convective and diffusive fluxes. Brucato and Rizzuti[16] and Brucato et al.[17] applied this idea to the modeling of solid–liquid mixing. An unsteady mass balance for the particles was derived to estimate the solid distribution in the vessel, namely:

$$V_c \frac{dC_c}{dt} = Q\left(C_p - C_c \right) + \sum_{i=1}^{2} \alpha Q \left(C_i - C_c \right)$$

$$+ \sum_{i=3}^{5} \left(\upsilon_t S_i C_i - \upsilon_t S_c C_c \right) \qquad (5)$$

where the subscript c is the cell on which the balance is applied, the subscript i refers to the adjacent cells, and the subscript p denotes the feeding cell, i.e., the one yielding convective momentum, V is the cell volume, C the concentration in the cell, S the cell surface area where the particles settle from, Q the volume flow rate of fluid entering the cell, α an adjustable parameter that describes the turbulent diffusion, and v_t the particle settling velocity. In this model, the sedimentation and the diffusive flow were in the vertical direction, and the convective flow was radially oriented. These assumptions were justified on the basis of the radial discharge flow generated by the impellers. Brucato et al.[17] verified the model prediction with the experimental data of Fajner et al.[2] and Magelli et al.[1] It was shown that the axial concentration profile was very well predicted, however the validation was limited again to extremely low concentration suspensions.

The attractiveness of the network-of-zone method to compute solid–liquid mixing flows resides in its relative simplicity while being capable of capturing the main flow phenomena for a wide range of concentrations. The objective of the present work is to assess the real capability of this approach in the case of a complex mixing problem involving a coaxial mixer. Coaxial mixers are very popular for the preparation of pastes and slurries in the chemical, food, and coating industries. Another mixer setup is also tested, which consists of a single marine propeller rotating in a vessel (see description in the appropriate section below). This setup is used to validate the numerical model.

Description of the Coaxial Mixer Setup

The experimental setup is shown in Figure 11.1. It consists of the following items: (1) an AC drive with a nominal speed of 1760 rpm; (2) a gearbox with a reduction ratio 3.53:1 yielding a rotating speed of 500 rpm; (3) a torquemeter with a measurement range between 0 and 22.6 Nm (accuracy of ± 0.1% at full scale); and (4) a transparent vessel with a hemispherical bottom. The vessel diameter D_c and the height of the cylindrical section H_c are both equal to 40.64 cm, yielding a maximum volume of fluid of about 60 L. The impeller configuration is the following:

An anchor arm:

- Diameter $D_a = 36.83$ cm
- Width $W_a = 3.81$ cm
- Wall clearance $C_w = 1.9$ cm

Four rigid rods at 90° used for pigment wetting:[18]

- Length $D_t = 23.77$ cm
- Cross-section diameter $D_{st} = 0.95$ cm

A pitched blade turbine with two blades at 45°:

- Diameter $D_p = 20.2$ cm
- Width $W_p = 5$ cm
- Length $L_p = 7$ cm
- Bottom clearance $C_b = 20.32$ cm

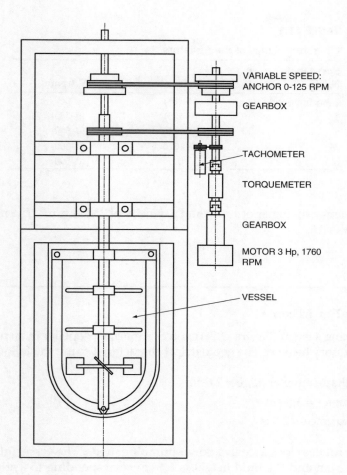

VARIABLE SPEED:
ANCHOR 0-125 RPM

GEARBOX

TACHOMETER

TORQUEMETER

GEARBOX

MOTOR 3 Hp, 1760 RPM

VESSEL

FIGURE 11.1
Coaxial mixer setup.

This configuration yields the following dimensionless ratios $C_w/D_c = 0.047$, $D_a/D_c = 0.906$, and $W_a/D_a = 0.103$ for the anchor, $D_t/D_c = 0.585$ for the rods; and $D_p/D_c = 0.5$ and $C_b/D_c = 0.5$ for the pitched blade turbine.

In this coaxial mixer, the primary role of the anchor is to clean up the wall from any accumulated solid lumps and reincorporate them back in the bulk. It also acts as a moving baffle, hampering the creation of a vortex at the liquid free surface. The purpose of the pitched blade turbine is to provide axial pumping so as to promote the resuspension of the solids, and radial dispersion to avoid solids reagglomeration. Finally, the aim of the wetting rods is to ease hydrophobic pigment incorporation by avoiding the creation of surface lumps.[19]

Two driving shafts are installed in this mixer: a fast rotating shaft drives the four rods and the turbine in a counterclockwise direction at speed N_c, whereas a slow rotating shaft entrains the scraping anchor in the clockwise direction at speed N_a. The operating range used in this work was as described in Table 11.1. In order to investigate the particle motion inside the coaxial mixer, a Newtonian solution of corn syrup with a viscosity of 1.05 Pa.s and a density of 1360 kg/m³ was used in conjunction with red Ballotini glass beads. The

TABLE 11.1

Operating Range of the Agitators

Speed Ratio N_c/N_a	N_a (rpm)	N_c (rpm)
0 (anchor only)	(0–125)	0
4	(0–125)	(0–500)
8	(0–62.5)	(0–500)
24	(0–20.8)	(0–500)
∞	0	(0–500)

beads had an average diameter of 1 mm and a density of 2500 kg/m³. The maximum packing factor (ϕ_m) was 0.6.

Marine Propeller Mixer

This simple mixing system (Figure 11.2) involves a marine propeller in an unbaffled vessel (actually a laboratory beaker). The geometrical characteristics are the following:

- Distance shaft-impeller edge = 2.6 cm
- Blade diameter = 1.8 cm
- Bottom clearance = 2.6 cm

The propeller rotates clockwise in a down-pumping mode. The vessel (glass beaker) has a diameter of 7.2 cm and the fluid height is 6.5 cm corresponding to a stirred volume of 264 cm³. The same fluid as the one considered in the coaxial mixer experiments was used. The rotating speed and the volume concentration of the particles were varied in order to investigate the resuspending mechanism in such a mixer. The operating conditions were

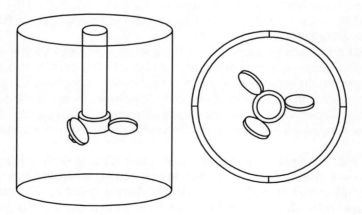

FIGURE 11.2
Marine propeller mixer.

as follows: $N = 173$ rpm and $\bar{\phi} = 2.8\%$; $N = 230$ rpm and $\bar{\phi} = 7.1\%$; and $N = 350$ rpm and $\bar{\phi} = 11.9\%$. For all the experiments, the particles were initially at rest on the tank bottom and the stirrer was suddenly set in motion.

Computational Model

Let us consider the incompressible flow of a suspension in a given domain Ω. The governing equations are:

$$\rho_m \left(\frac{\partial v}{\partial t} + v \cdot \text{grad } v \right) + \text{grad } p + \text{div } \tau = \rho_m g \quad \text{in } \Omega \tag{6}$$

$$\text{div } v = 0 \text{ in } \Omega \tag{7}$$

where v, p, g, and ρ are the velocity, pressure, gravity, and specific gravity, respectively. For a solid–liquid medium, the density ρ_m can be expressed as:

$$\rho_m = \rho_1(1 - \phi) + \rho_s \phi \tag{8}$$

where ρ_l and ρ_s represent the density of the liquid phase and the solid phase, respectively.

The stress tensor τ in Eq. (6) is related to the rate-of-strain tensor by a rheological equation of state such as:

$$\tau = -2\eta_s \dot{\gamma} \tag{9}$$

where η_s is a function of $|\dot{\gamma}|$ and ϕ,

$$\dot{\gamma} = \frac{1}{2}\left[\text{grad } v + (\text{grad } v)^T \right] \tag{10}$$

and

$$|\dot{\gamma}| = (\dot{\gamma} : \dot{\gamma})^{\frac{1}{2}} \tag{11}$$

The suspension viscosity η_s may or may not be a function of $|\dot{\gamma}|$ depending on the rheological behavior of the suspending medium. In a nondilute suspension, it is, however, always a function of the particle volume fraction ϕ. In this work, the Krieger–Dougherty model for a Newtonian suspension was used:

$$\eta_s(\phi) = \eta_l(1 - \phi / \phi_m)^{-1.82} \tag{12}$$

where η_l is the viscosity of the suspending liquid and ϕ_m the maximum packing factor.

For mathematical convenience, boundary conditions and initial conditions must be prescribed. For the simple marine propeller problem, a Lagrangian viewpoint was adopted. The frame of reference was attached to the propeller so that the propeller was fixed but the vessel was rotating. The boundary condition was then a zero velocity on the impeller, while the vessel wall rotated at $-\Omega_{impeller}$. The free surface was considered to be flat, therefore the normal velocity was zero and a shear-free condition was assumed. It should be noted that in the Lagrangian viewpoint, the frame of reference is in rotation. The fluid is therefore subjected to a constant acceleration and the momentum conservation equation [Eq. (6)] must be modified to account for centrifugal forces and Coriolis forces.[20] An advantage is, however, that the flow can be solved numerically at steady state provided the flow is fully periodic, which limits the computational efforts significantly.

In the case of the coaxial mixer, the rotation kinematics is much more complex since the two sets of agitators counter-rotate at different speeds. For the sake of simplicity, we decided to simulate the flow using the frame of reference of the anchor. In this Lagrangian viewpoint, the anchor is fixed but the vessel wall rotates at $-\Omega_{anchor}$ and the turbine rotates at $\Omega_{anchor} + \Omega_{turbine}$. In such a situation, contrary to the simple propeller problem, the resolution of the flow equations is time-dependent as the position of the central agitator changes with time.

The finite element method was used for the discretization of the flow equations. Considering the complex kinematics in the coaxial mixer and the associated change of topology at each time step, a new mesh should a priori be built for every topology considered in the time discretization. As a large number of time steps would be required to depict the agitator kinematics accurately, this approach would be a tremendous chore. To alleviate this difficulty, several alternatives have been proposed in the literature:

1. The description of the agitator by momentum sources or sinks inside the domain.[21] The major drawback of this method is the evaluation of a force equivalent to the representation of a body in rotation.

2. The arbitrary Euler–Lagrange method. It consists of moving the finite element mesh nodes with time.[22] This method works well as long as the mesh is not too distorted. In practice, remeshing is usually required after a few time steps.

3. Domain decomposition with sliding meshes.[23] This strategy is very popular in the finite volume literature and implemented in finite volume-based software (Fluent™, CFX™, and Star-CD™). The idea is to decompose the flow domain into several concentric cylindrical meshes and allow slipping of the meshes between the partitions. The continuity of the solution at the mesh interface is imposed by conservative interpolation. This method is powerful but limited to simple agitator configurations. It seems to work when very fine meshes are used although no error analysis has been published so far.

4. Mesh superimposition. This technique used in commercial finite element CFD software, like FIDAP™ and PolyFlow™, consists of generating a volume mesh without the moving impellers and a surface mesh of each impeller. At each time step, the surface mesh is projected in the volume mesh and a procedure has been developed to determine if the nodes of the volume mesh are located inside the surface mesh. When it is the case, the velocity of the impeller is imposed on these nodes. This technique is fairly simple to implement, however, it does not allow a precise representation of the impeller shape.

5. The virtual finite element method VFEM.[24] The method that was chosen for this work has been developed specifically to simulate flows around moving bodies in fixed enclosure. The principle of VFEM is the following: a volume mesh of the vessel with the anchor arm is first generated. Then, the surface of the moving bodies (in our case, the impeller and the rods) is meshed; the discretization nodes generated being stored as control nodes. Knowing the kinematics of the impellers, the velocity is imposed on these control nodes as constraints in the momentum equation and their position is updated with time according to the impeller rotation. A constrained optimization technique based on Lagrange multipliers and a penalty strategy is used to impose the velocity constraints. The VFEM approach was applied by Bertrand, Tanguy, and Thibault[24] to investigate the complex flow patterns in a planetary kneader and by Tanguy and Thibault[25] for the characterization of the hydrodynamics in a coaxial mixer provided by a helical ribbon and a Rushton turbine. Contrary to the mesh superimposition technique, the VFEM does not require that the control nodes coincide with the volume mesh nodes. They can be located inside finite elements as they are treated like external solicitations or optimization constraints. Figure 11.3 illustrates the VFEM concept in 2-D. The above method was implemented in our POLY3-D™ CFD finite element code using unstructured meshes made of tetrahedral elements. The reader is referred to Bertrand, Tanguy, and Thibault[24] for detailed information.

The modeling of solid–liquid mixing requires an additional equation to predict the dispersion of the solid phase in the vessel. As mentioned before, the network-of-zones approach was used in this work, which is based on unsteady mass balances on the solid phase carried out on a set of cells. In the literature, such mass balances are predominantly made on regular cells (finite volumes) in structured grids. The cells typically consist of quadrangles in 2-D and hexahedra in 3-D. In the present work, due to the use of unstructured grids, the mass balances were performed on the same elements as those used for the resolution of the flow equations, i.e., tetrahedral finite volumes (Figure 11.4).

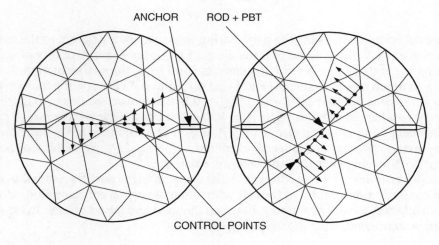

FIGURE 11.3
Virtual finite elements method concept in 2-D.

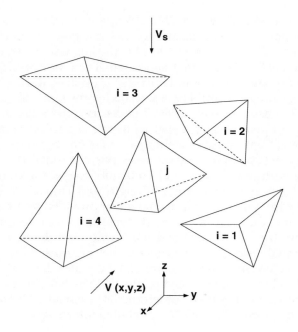

FIGURE 11.4
Tetrahedral finite volumes.

Let us consider the following mass balance on a tetrahedral element subjected to the velocity field v (x,y,z) of the suspension (as computed from the solution of the flow equations) and a sedimentation velocity, namely:

$$V_j \frac{d\phi_j}{dt} = \sum_i Q_{cj,i}\phi_{cj,i}^* + Q_{sj,i}\phi_{sj,i}^* \tag{13}$$

where the subscript i represents the neighboring finite elements adjacent to the four sides of finite element j, V_j the volume of finite element j, $Q_{cj,i}$ the convective flux of solid particles going through a common face to elements i and j, $Q_{sj,i}$ the sedimentation flux going through a common face to elements i and j, and $\phi_c j^*,i$ and $\phi_c j^*,i$ the volume fraction of solid particles defined as follows: $\phi_c j^*,i = \phi_i$, if $Q_{cj,i} > 0$ is the convective flux of particles leaving element i and entering element $j\phi_c$ $j^*_{,i} = \phi_j$, if $Q_{cj,i} < 0$ is the convective flux of particles leaving element j and entering element i $\phi_s j^*_{,i} = \phi_i$, if $Q_{sj,i} > 0$ is the sedimentation flux of particles leaving element i and entering element j $\phi_s j^*_{,i} = \phi_j$, if $Q_{sj,i} < 0$ is the sedimentation flux of particles leaving element j and entering element i.

This model assumes perfect mixing inside the element, as the particles enter the cell with different concentrations but leave the cell with a homogeneous concentration. Moreover, inertia is neglected, as well as slipping, between the solid and liquid phases. In Eq. (13), the fluxes can be expressed using the following equations:

$$Q_{cj,i} = \int_{S_{j,i}} \left(v_{cj,i}(x,y,z)\cdot\vec{n}_{j,i}\right)dS \tag{14}$$

$$Q_{sj,i} = \int_{S_{j,i}} \left(v_{sj,i} \cdot \vec{n}_{j,i} \right) dS \tag{15}$$

where $\vec{n}$ represents the unit vector of each face pointing to inward element j, S the surface area of the face, v_c the suspension velocity, and v_s the settling velocity determined using the classical sedimentation relations,[26] namely:

$$Ar = 24 Re_p \; (Ar < 4,8) \tag{16}$$

$$Ar = 24 \, Re_p + 3,6 \, Re_p \, 1.687 \; (4,8 < Ar < 10^5) \tag{17}$$

$$Ar = 4 / 9 Re_p^2 \left(Ar > 10^5 \right) \tag{18}$$

where Ar is the Archimedes number defined as:

$$Ar = \frac{4}{3} D_p^3 \left(\rho_s - \rho_l \right) \frac{\rho_l g}{\mu_l^2} \tag{19}$$

and Re_p is the particle Reynolds number defined as:

$$Re_p = \frac{D_p v_l \rho_l}{\mu_l} \tag{20}$$

In order to compute the unsteady term $d\phi/dt$, the following expression can be used:

$$\frac{d\phi}{dt} = \frac{a_t \phi^{t+1} + b_t \phi^t + c_t \phi^{t-1}}{\Delta t} \tag{21}$$

where $a_t = 1$, $b_t = -1$ and $c_t = 0$ at the first time step (Euler implicit scheme), and $a_t = 3/2$, $b_t = -2$ and $c_t = 1/2$ for the subsequent steps (second order Gear scheme). After substitution in Eq. (13), the following implicit numerical problem is obtained:

$$V_j \left[\frac{a_t \phi_j^{t+1} + b_t \phi^t + c_t \phi^{t-1}}{\Delta t} \right] = \sum_i Q_{cj,i} \phi_{cj,i}^{*t+1} + Q_{sj,i} \phi_{sj,i}^{*t+1} \tag{22}$$

This problem can then be resolved with the appropriate initial conditions of the mixing problem considered.

Several mixing cases were considered in this work:

1. Simple solid–liquid mixing experiments for the validation of the numerical model
2. Hydrodynamic studies in the coaxial mixer without solid particles
3. Coaxial solid–liquid mixing experiments

Validation of the Numerical Model

The propeller setup was used for this purpose. From a computational standpoint, a mesh of the vessel–propeller set was created containing 8746 elements yielding 54,333 velocity equations and 8746 concentration equations. The surface mesh of the propeller (Figure 11.5) comprised 964 control points. A maximum of three control points per element was used to avoid locking. Unsteady state flow simulations were performed with a 1-s time step and three coupling iterations between the Navier–Stokes equations and the solid transport equation were required per time step. Steady state was deemed obtained when the solids concentration coefficient of variation did not change.

Figure 11.6 illustrates the suspending mechanism versus time. At 350 rpm, all the particles are suspended and uniformly distributed. An attempt has been made to evaluate the minimum suspending speed N_{js} using the work of Armenante and Nagamine.[27] We found that $N_{js} = 855$ rpm for an average particle concentration of 11.9% v/v, which seems to contradict our visual observations. However, as mentioned by Ibrahim and Nienow,[28] published N_{js} correlations largely overestimate the suspending velocity when the suspending medium viscosity is larger than 0.1 Pa.s. Figure 11.6 also shows the time evolution of the computed volume fraction until stability is reached. As we do not know the experimental solids concentration distribution in the vessel, the numerical results can only be compared with the experimental results on a qualitative basis. We noted that the computation allows the prediction of the solid accumulation below the agitator. In agreement with the experiments, the network-of-zone model also predicts an increase in the resuspension mechanism when the rotation speed and/or the average concentration in the vessel increase (Table 11.2).

The accumulation of solids in the vessel bottom at equilibrium was also well captured. For instance at $N = 230$ rpm and a volume fraction of 7.1%, a solid layer with a volume

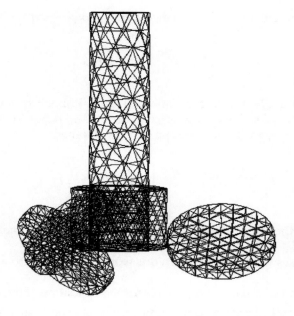

FIGURE 11.5
Mesh of the propeller.

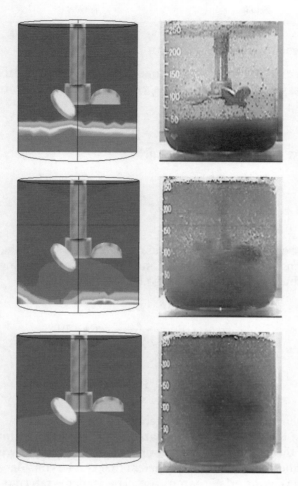

FIGURE 11.6
Suspending mechanism as a function of time.

fraction greater than 60% accumulates at the bottom and at the wall, which has been observed experimentally. At $N = 350$ rpm and a volume fraction of 11.9%, almost all the solid particles are suspended, except in a small region close to the bottom edge, again in agreement with the experiments. The performance of the computational model appears therefore satisfactory as fine hydrodynamic details can be predicted for intermediate concentration values. At very low concentration, as the solid layer thickness is reduced, a finer mesh would be required to enhance the model precision.

TABLE 11.2

Variation of the Average Concentration with the Rotation Speed

N (rpm)	Average Volume Fraction	Predicted Volume Fraction
173	2.8%	0.5%
230	7.1%	1.7%
350	11.9%	7.3%

Coaxial Mixer Results and Discussion

Considering the prediction of the power consumption, classically, the power drawn by the impeller is expressed with power curves, i.e., plots of the power number N_p versus the Reynolds number, Re, where:

$$N_p = \frac{P}{\rho N^3 D^5} \tag{23}$$

$$Re = \frac{\rho N D^2}{\mu} \tag{24}$$

P is the mixing power, N and D the rotation speed (in rps) and the impeller diameter respectively, and μ and ρ the fluid viscosity and density. The following relations apply for the laminar and turbulent regimes, respectively:

$$K_p = N_p \quad Re = \frac{P}{\mu N^2 D^3}, \quad Re < 10 \tag{25}$$

$$N_p = C \quad Re > 300 \tag{26}$$

where K_p and C depend only on the mixer geometry for a given fluid.

In a coaxial mixer, the selection of the characteristic speed and dimension that appear in the expression of these dimensionless numbers is ambiguous, as we have two different speeds (N_a and N_c) and three impeller diameters that can be considered. In this work, we used N_a and D_a as the characteristic parameters. The reader is referred to the discussion of this particular choice in Tanguy and Thibault[29] and Thibault and Tanguy.[30] The coaxial mixer power curve and the value of K_p have been obtained by numerical simulation, varying the impeller speed and the speed ratio. For each simulation, the velocity field was used to compute the power by a macroscopic energy balance, namely:

$$P = -\iiint_\Omega \tau : \nabla \upsilon \, d\Omega \tag{27}$$

Numerical simulations employed the virtual finite element method described above combined with a Lagrangian flow description (anchor frame of reference). The finite element mesh (generated by IDEAS™ software from SDRC) included 17,083 tetrahedral elements, yielding 106,295 velocities degrees of freedom. Twenty time steps per revolution (angular displacement of 18 degrees per time step) were used, the value of the time step depending on the revolution speed. Typically, the time step was in the range 0.03–0.135 s. For each time step, the control points located on the surface mesh of the moving impellers (Figure 11.7 had to be updated. The number of control points required is given in Table 11.3. The flow simulations carried out for several revolutions showed that the flow was periodic. Moreover, it was found that only one single revolution was enough to obtain

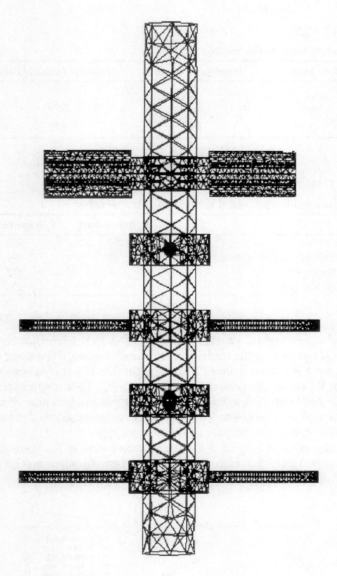

FIGURE 11.7
Surface mesh of the moving impellers.

a stable, converged solution at low Reynolds number. Finally each simulation required between 12 and 24 CPU hours on an IBM RISC6000/550 server.

Table 11.4 shows a comparison of the computed and experimental values of K_p for the anchor only. An excellent agreement is obtained. The comparison with literature data shows that the computed values are larger than the data published. The difference is believed to originate from the shape of the vessel bottom (hemispherical in the present work, flat in the literature results).

We show in Figure 11.8, a comparison of the numerical and experimental power curves for three different speed ratios ($R_N = N_c/N_a = 4$, 8 and 24). These results have been obtained with $\mu = 15$ Pa.s and $\rho = 1500$ kg/m^3. It can be seen that there is good agreement between

TABLE 11.3

Characteristics of the Surface Meshes

Surtface Mesh	Number of Elements	Number of Control Points
Shaft	400	208
Mixing rods	5022	2279
PBT	1968	884

TABLE 11.4

Numerical and Experimental Values of K_p for an Anchor

	K_p (numerical)	K_p (experimental)
This work	256	253
Tanguy, Thibault, and Brito de la Fuente[31]	206	199
Ho and Kwong[32]	—	215
Rieger and Novak[33]	—	206

the predicted and experimental values. The slope of –1 in the laminar region is captured by the simulation. The onset of the transition regime (decrease of the slope of the Np versus Re curve) occurs for a Re value below 10, and the threshold value is sensitive to the value of R_N. For a given Re value, the power increases with R_N. This result seems logical as the Reynolds number has been defined with the anchor parameters and when R_N increases, the central shaft speed rotates faster. This speed increase enhances the average shear rate in the tank, which entrains an increase in the power draw.

The values of the constant K_p of the coaxial mixer versus R_N were also established. Results are shown in Table 11.5. Here again, the agreement between the predictions and the experimental data is very good.

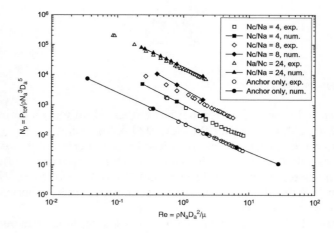

FIGURE 11.8
Numerical and experimental power curve for three different speed ratios.

TABLE 11.5

Numerical and Experimental Values of K_p for the Coaxial Mixer

Speed Ratio, R_N	K_p (numerical)	K_p (experimental)
0	256	253
4	1003	817
8	2651	2,284
24	17,411	16,486

It is usual in laminar mixing simulations to represent the flow using tracer trajectories. The computation of such flow trajectories in a coaxial mixer is more complex than in traditional stirred tank modelling due to the intrinsic unsteady nature of the problem (evolving topology, flow field known at a discrete number of time steps in a Lagrangian frame of reference). Since the flow solution is periodic, a node-by-node interpolation using a fast Fourier transform of the velocity field has been used, which allowed a time continuous

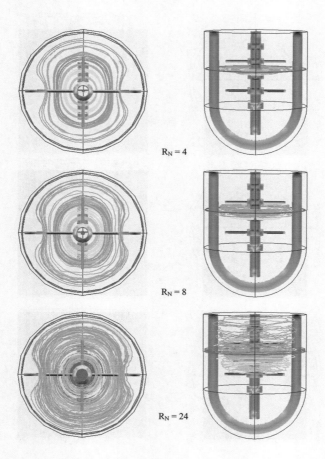

$R_N = 4$

$R_N = 8$

$R_N = 24$

FIGURE 11.9
Effect of speed ratio on the flow field (upper part of the vessel).

representation of the flow to be obtained. In other words, the velocity at node i was approximated with a Fourier series taking the following form:

$$v^i = a_0^i + a_n^i \cos nt + \sum_{k=1}^{n-1} \left(a_k^i \cos kt + b_k^i \sin kt \right) \tag{28}$$

where n is the number of harmonics and coefficients $a_k^i, k = 1, 2, ..., n-1$, and $b_k^i, k = 1, 2, ..., n-1$ are obtained from the whole set of velocity values at node i during an impeller revolution. In practice, 10 harmonics were employed according to the Shannon sampling theorem.[34]

We show in Figures 11.9 and 11.10 the effect of the speed ratio on the flow field for an anchor speed of 4.43 rpm. It can be seen that for the two injection points considered (upper part and lower part of the vessel), the radial dispersion increases significantly with the speed ratio. The axial dispersion is also enhanced but less dramatically. A speed ratio of

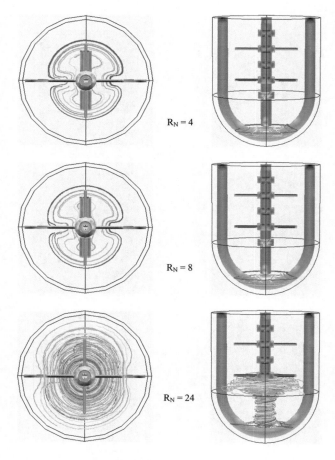

FIGURE 11.10
Effect of speed ratio on the flow field (lower part of the vessel).

four does not lead to good dispersion and therefore should yield longer mixing times. These results show the synergy between the anchor and the turbine, the axial pumping increasing with the speed ratio. To quantify this axial pumping, the axial velocity in the upward direction is plotted in Figure 11.11 versus the speed ratio. An additional advantage of the coaxial mixer can be seen as the axial pumping in the upper part of the vessel is enhanced when the speed ratio is augmented.

We now turn to the prediction of the suspension mechanism of the ballotini versus the speed ratio in the coaxial mixer. The average volume concentration is 1%, and the solids are initially at rest in the tank bottom. The first case investigated corresponds to the motion of the sole anchor arm at a speed of 40 rpm. Simulations are carried out in the Lagrangian frame of reference (fixed anchor, rotating vessel). Figure 11.12 shows the predicted and experimental solid volume fraction at equilibrium. The computation of the solid–liquid interface at the bottom is fairly well captured by the numerical model as in the previous case dealing with the propeller. The next case considered is the resuspension with the only central shaft in rotation at 160 rpm. This simulation has been carried out with the virtual finite element method described before. Steady state was achieved after 20 revolutions. In Figure 11.13, we compare the numerical and experimental distribution of particles in the tank. The agreement is again noteworthy, and the computation predicts that no particle has been resuspended. In fact, although the rotation is counterclockwise, particles have moved in the clockwise direction and accumulated behind the anchor arm. The Mann model is therefore capable of predicting this odd motion phenomenon at the bottom. Finally, we show in Figure 11.14 the resuspending of ballotini for a rotation speed of the central shaft

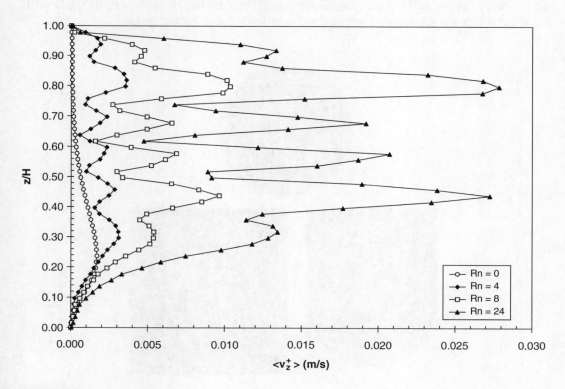

FIGURE 11.11
Axial velocity in the upward direction as a function of the speed ratio.

FIGURE 11.12
Predicted and experimental solid volume fraction at equilibrium.

of 160 rpm and a speed ratio R_N of 4 ($N_c/N_a = 4$) after 60 revolutions. This number of revolutions is the maximum that we were capable of computing in a reasonable timeframe (a few days of CPU time). Interesting information can already be obtained from this snapshot result. First, an overconcentration is noticeable in the anchor arm wake in agreement with the experiment. This overconcentration decreases with time as the tank bottom becomes leaner in particle, thereby decreasing the number of particles that can be entrained in the wake. As far as the solid–fluid interface is concerned, the interface rises in the bulk with time but it has not yet reached its eventual position after 60 revolutions.

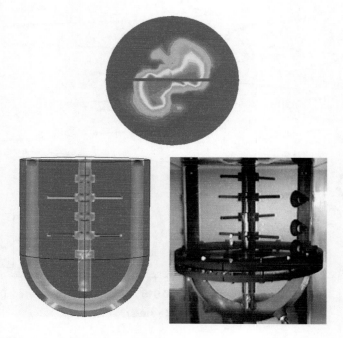

FIGURE 11.13
Numerical and experimental distribution of particles.

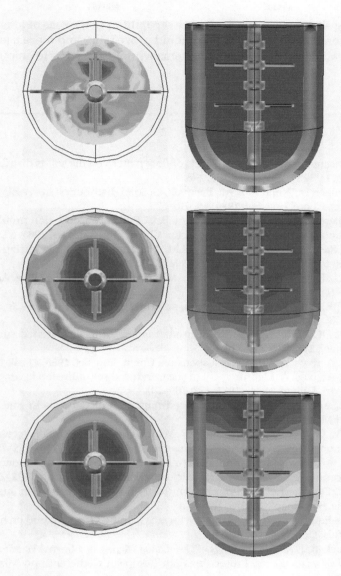

FIGURE 11.14
Resuspending of ballotini a central shaft speed of 160 r.p.m. and a speed ratio of 4.

Conclusions

Solid–liquid mixing processes can be simulated with good precision when sound CFD methods are used. The application of a combination of the virtual finite element method and the network-of-zone approach was used in this work to analyze the complex flow and suspension mechanisms in a coaxial mixer. Experiments carried on the laboratory scale confirmed the validity of the predictions.

Coaxial mixing shows strong performance capabilities in the case of tough mixing problems involving complex rheology, which should prove more and more useful in industry. Tools are now available to design these systems without resorting to empirical rules.

References

1. Bakker, A., Fasano, J.B., Myers, K.J. Effects of flow pattern on the solids distribution in a stirred tank. I. Chem. E. Symp. Ser. **1994**, *136*, 1–8.
2. Magelli, F., Fajner, D., Nonentini, M., Pasquali, G. Solid distribution in vessels stirred with multiple impellers. Chem. Eng. Sci. **1990**, *45*, 615–625.
3. Fajner, D., Magelli, F., Nocentini, M., Pasquali, G. Solids concentration profiles in a mechanically stirred and staged column slurry reactor. Chem. Eng. Res. Des. **1985**, *63*, 235–240.
4. Ferreira, P.J., Rasteiro, M.G., Figueiredo, M.M. A new approach to measuring solids concentration in mixing tanks. Adv. Powder Tech. **1994**, *5* (1), 15–24.
5. Rasteiro, M.G., Figueiredo, M.M., Freire, C. Modelling slurry mixing tanks. Adv. Powder Tech. **1994**, *5* (1), 1–14.
6. Richardson, J.F., Zaki, W.N. Sedimentation and fluidisation. Part I. Trans. Instn. Chem. Engrs. **1954**, *32*, 35–53.
7. Gadala-Maria, F., Acrivos, A. Shear-induced structure in a concentrated suspension of solid spheres. J. Rheol. **1980**, *24*, 799–814.
8. Leighton, D., Acrivos, A. Viscous resuspension. Chem. Eng. Sci. **1986**, *41* (6), 1377–1384.
9. Leighton, D., Acrivos, A. Measurement of shear-induced self-diffusion in concentrated suspensions of spheres. J. Fluid Mech. **1987**, *177*, 109–131.
10. Leighton, D., Acrivos, A. The shear-induced migration of particles in concentrated suspensions. J. Fluid Mech. **1987**, *181*, 415–439.
11. Altobelli, S.A., Givler, R.C., Fukushima, E. Velocity and concentration measurements of suspensions by nuclear magnetic resonance imaging. J. Rheol. **1991**, *35* (5), 721–735.
12. Phillips, R.J., Armstrong, R.C., Brown, R.A. A constitutive equation for concentrated suspensions that accounts for shear-induced particle migration. Phys. Fluids. **1992**, *A4* (1), 30–40.
13. Krieger, I.M., Dougherty, T.J. A mechanism for non-Newtonian flow in suspension of rigid spheres. Trans. Soc. Rheol. **1959**, *3*, 137–152.
14. Mann, R. Gas-liquid stirred vessel mixers: toward a unified theory based on network-of-zones. Chem. Eng. Res. Des. **1986**, *64*, 23–34.
15. Mann, R., Hackett, L.A. *Fundamentals of Gas–Liquid Mixing in a Stirred Vessel: An Analysis using Networks of Backmixed Zones*, Proceedings 6th European Conference on Mixing, Pavia, Italy, 1988; 321–328.
16. Brucato, A., Rizzuti, L. In *The Application of the Network-of-Zones Model to Solid-Liquid Suspensions*, Proceedings 6th European Conference on Mixing, Pavia, Italy, 1988; 273–280.
17. Brucato, A., Magelli, F., Nocentini, M., Rizzuti, L. An application of the network-of-zones model to solids suspension in multiple impeller mixers. Trans. Instn. Chem. Engrs. **1990**, *69*, Part A, 43–52.
18. Duquesnoy, J.A., Thibault, F., Tanguy, P.A. Dispersion of clay suspensions at high solids content. Private communication, 1995, MacMillan Bloedel, British Columbia.
19. Duquesnoy, J.A., Tanguy, P.A., Thibault, F., Leuliet, J.C. A new pigment disperser for high solids paper coating colors. Chem. Eng. Technol. **1997**, *20*, 424–428.
20. Tritton, D.J. *Physical Fluid Dynamics*; Clarendon Press: Oxford, **1988**; 325 pp.
21. Pelletier, D.H., Schetz, J.A. Finite element Navier–Stokes calculation of three-dimensional turbulent flow near a propeller. AIAA **1986**, *24*, 1409–1416.

22. Dermidzic, I., Peric, M. Finite volume method for the prediction of fluid flow in arbitrarily shaped domains with moving boundaries. Int. J. Numer. Methods. Fluids **1990**, *10*, 771–790.
23. Perng, C.Y., Murthy, J. A sliding-mesh technique for simulation of flow in mixing tranks. ASME 93-WA-HT-33 **1993**, Old SRN049.
24. Bertrand, F., Tanguy, P.A., Thibault, F. A three-dimensional fictitious domain method for incompressible flow problems. Int. J. Num. Meth. Fluids **1997**, *25*, 719–736.
25. Tanguy, P.A., Bertrand, F., Labrie, R., Brito-de la Fuente, E. Numerical modelling of the mixing of viscoplastic slurries in a twin-blade planetary mixer. Chem. Eng. Res. Des. **1997**, *74*, 499–504.
26. Coulson, J.M., Richardson, J.F. *Chemical Engineering*, 3rd Ed., Pergamon Press: New York, 1978.
27. Armenante, P.M., Nagamine, E.U. Effect of low off-bottom impeller clearance on the minimum agitation speed for complete suspension of solids in stirred tanks. Chem. Eng. Sci. **1998**, *53* (9), 1757–1775.
28. Ibrahim, S.B., Nienow, A.W. The effect of viscosity on mixing pattern and solid suspension in stirred vessels. I. Chem. E. Symp. Ser. **1994**, *136*, 25–32.
29. Tanguy, P.A., Thibault, F. Power consumption in the turbulent regime for a coaxial mixer. Can. J. Chem. Eng. **2002**, *80*, 601–603.
30. Thibault, F., Tanguy, P.A. Power draw characterization of coaxial mixer with Newtonian and non-Newtonian fluids. Chem. Eng. Sci. **2002**, *57*, 3861–3872.
31. Tanguy, P.A., Thibault, F., Brito de la Fuente, E. A new investigation of the Metzner–Otto concept for anchor mixing impellers. Can. J. Chem. Eng. **1996**, *74*, 222–228.
32. Ho, F., Kwong, A. A guide to designing special agitators. Chem. Eng. **1973**, July issue, 94–104.
33. Rieger, F., Novak, V. Power consumption of agitators in highly viscous non-Newtonian liquids. Trans. Instn. Chem. Engrs. **1973**, *51*, 105–111.
34. Ljung, L. *System Identification Theo for the User*; Prentice-Hall: Englewood, Cliffs, NJ, 1987.

12

Fluidization

A.-H. Park and L.-S. Fan

CONTENTS

Introduction 181
Particle and Regime Classification 182
   Classification of Fluidized Particles 182
   Fluidization Regime 183
      Minimum Fluidization and Particulate Fluidization 183
      Bubbling Fluidization 185
      Turbulent Fuidization 187
      Entrainment and Elutriation 187
      Slugging Beds 187
      Spouted Beds 188
      Fast Fluidization and Dilute Transport 188
Other Systems 189
   Downer 189
   Three-Phase Fluidized Bed 191
      Bed Contraction Phenomenon 191
      Moving Packed Bed Phenomenon 192
Computational Fluid Dynamics 193
Conclusions 194
Nomenclature 194
References 194

Introduction

Fluidization refers to the state of solid particles in a suspended condition owing to the flow of fluid, gas, and/or liquid. Contact schemes of fluidized bed systems can be classified on the basis of the states of solid motion. For a batch-solids system, the fluid at a low velocity merely percolates through the voids between packed particles, while the particles remain motionless. The solids in this case are in the fixed bed state. With an increase in the fluid velocity, particles move apart and become suspended; the bed then enters the fluidization state. The fluidization characteristics vary, depending on whether gas, liquid, or gas-liquid is the fluidizing medium.

Fluidized beds generally possess the following properties that promote their use in reactor applications:

1. Capability of continuous operation and transport of solids in and out of the bed.
2. High heat transfer rates from bed to surface and from gas to particle leading to temperature uniformity in the bed.
3. High mass transfer rates from gas to particle.
4. Applicability over a wide range of particle properties and high solids mixing rates.
5. Simplicity in geometric configuration and suitability for large-scale operation.

Fluidized beds have been used extensively for physical operations (e.g., adsorption and heat exchanger), chemical synthesis (e.g., acrylonitrile synthesis and maleic anhydride synthesis), metallurgical and mineral processes (e.g., roasting of sulfide ores, resid hydrotreating, and reduction of iron oxide), and other applications, such as coal combustion and microorganism cultivation.

In what follows, the fundamental properties of two-phase fluidization are described. Unless otherwise noted, the properties refer to gas–solid fluidization. This chapter's references suggest several books that can serve as general references for two-phase (gas–solid or liquid–solid) or three-phase (gas–liquid–solid) fluidization subjects.[1–9]

Particle and Regime Classification

Classification of Fluidized Particles

Particles can be classified into four groups (i.e., Groups A, B, C, and D) as shown in Figure 12.1, based on their fluidization behavior.[10] Group C comprises small particles ($d_p < 20$ μm) that are cohesive. Group A particles, with a typical size range of 30–100 μm, are readily fluidized. No maximum stable bubble size exists for Group B particles. Group D comprises coarse particles ($d_p > 1$ mm) that are commonly processed by spouting.

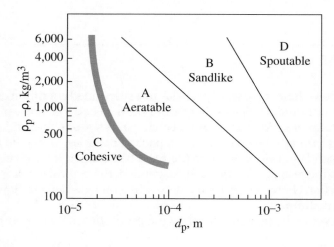

FIGURE 12.1
Geldart's classification of fluidized particles. (From Geldart.[10])

Fluidization Regime

Regime classification for dense- and lean-phase fluidization, in general, can be based on bubble or solid flow behavior. Dense-fluidization regimes include particulate fluidization, bubbling fluidization, and turbulent fluidization. In a broad sense, "dense-phase fluidization" also encompasses the slugging, spouting, and channeling conditions of operation. Lean-phase fluidization includes fast fluidization and dilute transport. The fundamental distinction between these regimes or conditions of operation is briefly described next. Figure 12.2 illustrates various flow regimes.

Minimum Fluidization and Particulate Fluidization

The state of fluidization begins at the point of minimum or incipient fluidization. At the minimum fluidization point, the pressure drop for a fixed bed and that for a fluidized bed are equivalent. This relationship is used as the basis for the formation of the predictive equation for the minimum fluidization velocity. The pressure drop in the fixed bed can be described by the Ergun equation. Under the minimum fluidization condition, the Ergun equation can be expressed as

$$\frac{\Delta p_b}{H_{mf}} = 150 \frac{\left(1 - \alpha_{mf}\right)}{\alpha_{mf}^3} \frac{\mu U_{mf}}{\varphi^2 d_p^2} + 1.75 \frac{\left(1 - \alpha_{mf}\right)}{\alpha_{mf}^3} \frac{\rho U_{mf}^2}{\varphi d_p} \tag{1}$$

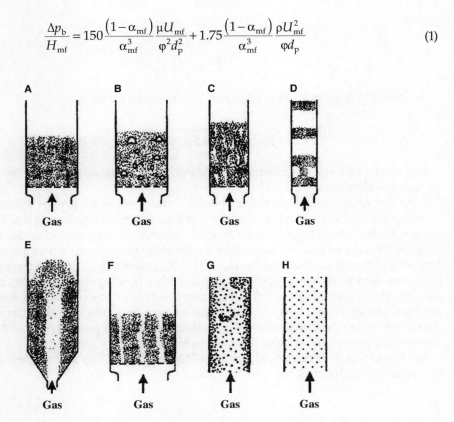

FIGURE 12.2
Various flow regimes or patterns in dense-phase fluidization: (A) particulate fluidization; (B) bubbling fluidization; (C) turbulent fluidization; (D) slugging; (E) spouting; (F) channeling; (G) fast fluidization. (From Fan and Zhu.[7])

In a fully fluidized bed, the pressure drop (cross-sectionally averaged) counterbalances the weight of the pseudocontinuum of the gas–solid mixture, which yields

$$-\frac{dp_d}{dH} = (\rho_p - \rho)(1 - \alpha)g \tag{2}$$

Under the minimum fluidization condition, Eq. (2) gives

$$\frac{\Delta p_b}{H_{mf}} = \frac{\Delta p_d}{H_{mf}} = (\rho_p - \rho)(1 - \alpha_{mf})g \tag{3}$$

Equating Eqs. (1) and (3) results in

$$Ar = 150\frac{(1 - \alpha_{mf})}{\alpha_{mf}^3 \varphi^2}Re_{pmf} + \frac{1.75}{\alpha_{mf}^3 \varphi}Re_{pmf}^2 \tag{4}$$

and

$$Re_{pmf} = \frac{\rho U_{mf} d_p}{\mu} \tag{5}$$

A semiempirical correlation based on Eq. (4) can be given by[11]

$$Re_{pmf} = \sqrt{(33.7)^2 + 0.0408 Ar} - 33.7 \tag{6}$$

The above analysis for minimum fluidization is applicable to both, the gas–solid and liquid–solid systems.

For a bed with Group A particles, bubbles do not form when the gas velocity reaches U_{mf}. The bed enters the particulate fluidization regime under this condition. This is also the regime under which liquid fluidization is operated. The operation under the particulate fluidization regime is characterized by a smooth bed expansion with an apparent uniform bed structure without bubbles for $U_{mf} < U < U_{mb}$, where U_{mb} is the superficial gas velocity at the minimum bubbling condition. In particulate fluidization, all the gas passes through the interstitial space between the fluidizing particles without forming bubbles. The bed appears grossly homogeneous. This regime exists only in a bed with Group A particles, under a narrow operating range of gas velocities. At high pressures or with gases of high density, the operating range of this regime expands.

For liquid fluidization, the Richardson–Zaki equation[12] as given below can be used to describe the bed expansion:

$$\varepsilon = \left(\frac{U_l}{U_i}\right)^{1/n} \tag{7}$$

where U_i is the extrapolated liquid velocity as the bed voidage approaches 1, and n is the Richardson–Zaki index (see Table 12.1 for estimation).

TABLE 12.1

Index for the Richardson–Zaki Equation

$n = 4.65 + 20d_p/D_c$	$\mathrm{Re_t} < 0.2$
$n = \left(4.4 + 18d_p/D_c\right)\mathrm{Re_t}^{-0.03}$	$0.2 < \mathrm{Re_t} < 1$
$n = \left(4.4 + 18d_p/D_c\right)\mathrm{Re_t}^{-0.1}$	$1 < \mathrm{Re_t} < 200$
$N = 4.4\mathrm{Re_t}^{-0.1}$	$200 < \mathrm{Re_t} < 500$
$N = 2.4$	$500 < \mathrm{Re_t}$

$\mathrm{Re_t} = U_t d_p \rho_l/\mu_l$. (From Fan and Tsuchiya.[13])

Bubbling Fluidization

Bubbles are formed as a result of the inherent instability of gas–solid systems. The instability of a gas–solid fluidized bed is characterized by fast growth in local voidage in response to a system perturbation. Because of the instability in the bed, the local voidage usually grows rapidly into a shape resembling a bubble.[13] Although it is not always true, the initiation of the instability is usually perceived to be the onset of bubbling, which marks the transition from particulate fluidization to bubbling fluidization. The theoretical expansion of the physical origin behind and prediction of the onset of the instability of gas–solid fluidized beds has been attempted (e.g., Anderson and Jackson,[14] Verloop and Heertjes,[15] and Rietema and Piepers[16]). The efforts have been focused on the primary forces behind the stability among the interparticle contact forces, particle–fluid interaction forces, and particle–particle interaction via particle velocity fluctuation.

Most bubbles in bubbling fluidized beds are of spherical cap or ellipsoidal cap shape. Configurations of two basic types of bubbles, fast bubble (clouded bubble) and slow bubble (cloudless bubble), are schematically depicted in Figure 12.3. The cloud is the region established by the gas, which circulates in a closed loop between the bubble and its surroundings. The cloud phase can be visualized with the aid of a color tracer gas bubble. The bubble wake shown in Figure 12.3 plays an important role in solid movement or -mixing in the bed and the freeboard. A bubble wake in a single-phase fluid is defined as the streamline-enclosed region beneath the bubble base. In a gas–solid fluidized bed, the emulsion phase can be treated as a pseudosingle-phase fluid. Hence, a bubble wake is defined as

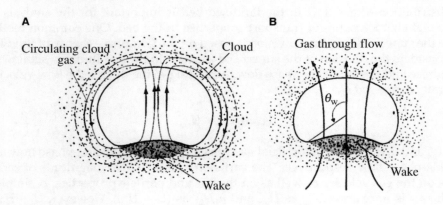

FIGURE 12.3
Bubble configurations and gas flow patterns around a bubble in gas–solid fluidized beds: (A) fast bubble (clouded bubble) $U_b > U_{mf}/\alpha_{mf}$; (B) slow bubble (cloudless bubble) $U_b < U_{mf}/\alpha_{mf}$. (From Fan and Zhu.[7])

the region enclosed by streamline of the pseudofluid behind the bubble base. In the bed, the wake rises with the bubble and thereby provides an essential means for global solid circulation and induced axial solids mixing.[17] In the freeboard, particles carried along the bubble wake are the primary origin of the particles there.

When gas enters the orifice of the distributor, it can initially form bubbles or jets. The formation of bubbles or jets depends on various parameters including types of particles, fluidization conditions around the orifice, orifice size, and the presence of internals in the bed. The initial bubble or jet is then transformed into a chain of bubbles. The jet is defined as an elongated void, which is appreciably larger than a bubble; it extends to some distance from the orifice into the inner bed. In general, bubbles tend to form in the presence of small particles, such as Group A particles; jets tend to form in the presence of large particles, such as Group D particles, when the emulsion phase is not sufficiently fluidized, or when internals are present and disrupt the solid flow to the orifice region.[18]

In bubbling fluidized beds under ambient pressure and low gas velocity conditions where the bubble size increases with the gas velocity, the bubble size may be estimated by various correlation equations, such as those developed by Mori and Wen,[19] Darton et al.[20] and Cai et al.[21] There is a similarity in the rise behavior of a single bubble in gas–solid and liquid media. The rise velocity of a single spherical cap bubble in an infinite liquid medium can be described by the Davies and Taylor equation.[22] Experimental results indicate that the Davies and Taylor equation is valid for large bubbles with bubble Reynolds numbers greater than 40.[23] Whereas for bubbles in fluidized beds, the bubble Reynolds numbers are typically of the order of 10 or less. By analogy, the rise velocity of an isolated single spherical cap bubble in an infinite gas–solid medium can be expressed in terms of the volume bubble diameter by[1]

$$U_{b\infty} = 0.71\sqrt{gd_{b\infty}} \tag{8}$$

In a free bubbling bed, the average bubble-rise velocity U_{bb} can be described by[1]

$$U_{bb} = U - U_{mf} + 0.71\sqrt{gd_{bb}} \tag{9}$$

The distribution of gas flow in the fluidized bed is important for the analysis of the fundamental characteristics of transport properties in the bed. One common method to estimate the superficial gas flow division is based on the two-phase theory of fluidization, which considers the division of the superficial gas flow in the bed into two subflows—the bubble-phase and the emulsion-phase flow. According to the theory, the flow velocity can be generally expressed as

$$U = \alpha_b U_{bb} + U_{em} \tag{10}$$

where U_{em} and $\alpha_b U_{bb}$ are the superficial gas velocities for the emulsion-phase flow and for the bubble-phase flow, respectively. The variation of the relative magnitude of each flow depends on the gas velocity as well as on the gas and particle properties. A simple two-phase theory is to express U_{em} as U_{mf} and $\alpha_b U_{bb}$ as $(U - U_{mf})$. However, $U - U_{mf}$ may overestimate the actual bubble phase flow in a fluidized bed, as a result of two prevailing effects of the gas flow pattern: 1) significant emulsion-phase flow or invisible gas flow through the bubble, and 2) larger interstitial gas flow in the emulsion phase than U_{mf}.[24]

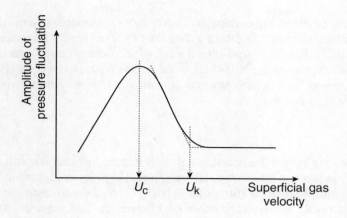

FIGURE 12.4
Variation of pressure fluctuation with the gas velocity for dense-phase fluidized bed with FCC particles. (From Clift.[23])

Turbulent Fuidization

The turbulent regime is often regarded as the transition regime from bubbling fluidization to lean-phase fluidization. At relatively low gas velocities, bubbles are present in the turbulent regime, while at relatively high gas velocities in the turbulent regime, the clear boundary of bubbles disappears and the nonuniformity of solid concentration distribution yields distinct gas voids, which become less distinguishable as the gas velocity further increases toward lean-phase fluidization. The onset velocity of the transition to the turbulent regime is commonly defined, based on the variation of the amplitude of the pressure fluctuation with the gas velocity, as shown in Figure 12.4, to be the gas velocity corresponding to the peak, U_c, whereas the leveling point, U_k, may be recognized as the onset of the turbulent regime proper.[25]

Entrainment and Elutriation

The freeboard region in a fluidized bed accommodates particles that are being entrained from the dense bed. Entrainment refers to the ejection of particles from the dense bed into the freeboard by the fluidizing gas. Elutriation refers to the separation of fine particles from a mixture of particles, which occurs at all heights of the freeboard, and their ultimate removal from the freeboard. The terms entrainment and elutriation are sometimes used interchangeably. The carryover rate relates to the quantities of the particles leaving the freeboard. Coarse particles with a particle terminal velocity that is higher than the gas velocity eventually return to the dense bed, while the fine particles eventually exit from the freeboard. The freeboard height required in design consideration is usually higher than the transport disengagement height, defined as a height beyond which the solids holdup, solids entrainment, or carryover rate remains nearly constant.

Slugging Beds

When bubbles grow to sizes comparable to the bed diameter, slugging occurs. Slugging is most frequently encountered in small-diameter beds with large bed heights, especially when large/heavy particles are employed. There are requirements of a minimum slugging velocity and a minimum bed height for a slug flow to take place. The slugs may appear in different forms including the round-nosed slug, which occurs in systems of fine particles, the wall slug (also known as the half slug), which takes place in the beds with rough walls, and the square-nosed slug, which appears in coarse particle systems where the particle bridging effect is significant.

Spouted Beds

In a spouted bed, gas enters the bed through a jet nozzle forming a spout. The surrounding annular region forms a downward-moving bed. Particles are entrained into the spout from the bottom and sidewall of the spout. A part of the gas seeps into the annular region through the spout wall, whereas the other part leaves the bed from the top of the spout. The particles carried into the spout disengage from the gas in a solid-disengagement fountain, just above the bed, and then return to the top of the annular region.[26] Group D particles are commonly used for the spouted bed operation.

Fast Fluidization and Dilute Transport

The fast fluidization regime is characterized by the clustering phenomenon with a core-annular heterogeneous flow structure, whereas the dilute transport regime is characterized by a homogeneous flow structure. Lean-phase fluidization that encompasses these regimes is carried out in a circulating fluidized bed (CFB), where solid particles circulate between the riser and the downer. The fast fluidization regime is the principal regime under which the CFB is operated. The operating variables for a CFB system include both the gas flow rate and the solid circulation rate, whereas in a dense-phase fluidized bed system, only the gas flow rate is the operating variable. The solid circulation is established by a high velocity of gas flow.

The fast fluidization regime is represented by a dense region at the bottom of the riser and a dilute region above it. The inter-relationship of the fast fluidization regime with other fluidization regimes in dense-phase fluidization and with the dilute transport regime is reflected in the variations of the pressure drop per unit length of the riser, Δ_p/Δ_z, gas velocity, and solid circulation rate as given in Figure 12.5.[27]

The transport velocity, U_{tr}, marks the lower limit of the gas velocity for fast fluidization operation. The transport velocity can be evaluated from the variations of the local pressure drop per unit length (Δ_p/Δ_z) with respect to the gas velocity and the solid circulation rate, J_p. An example of such a relationship is shown in Figure 12.6. It is seen in the figure that, along the curve AB, the solid circulation rates are lower than that of the saturation carrying capacity of the flow. Particles with low particle terminal velocities are carried over from the riser. With an increasing solid circulation rate, more particles accumulate at the bottom. At point B in the curve, the solids fed into the riser are balanced by the saturated carrying capacity. A slight increase in the solid circulation rate yields a sharp increase in the pressure drop. This behavior reflects the collapse of the solid particles into a dense-phase fluidized bed and is noted as choking. When the gas velocity is equal to or higher than the transport velocity (e.g., curve EF in Figure 12.6), there is no longer a sharp increase in the slope of the J_p vs. (Δ_p/Δ_z) relationship. Thus, U_{tr} is characterized as the lowest gas velocity at which the discontinuity in the curve of J_p versus (Δ_p/Δ_z) disappears.[25] U_{tr} varies with gas and solid properties as correlated by Bi and Fan.[28]

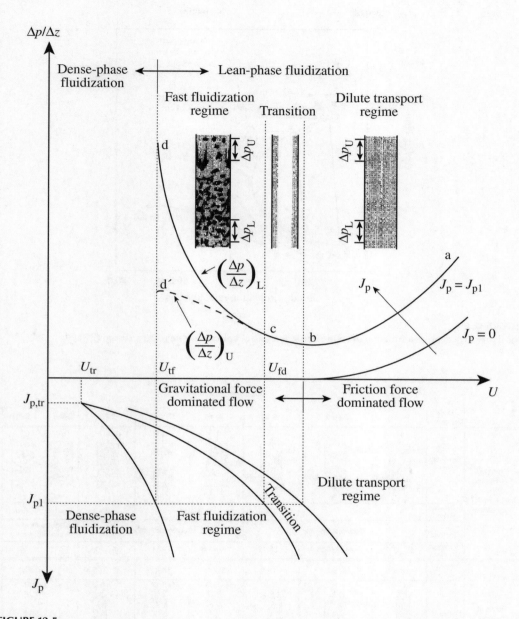

FIGURE 12.5
Variation of pressure drop per unit riser length with solid circulation rate and gas velocity for various fluidization regimes. (From Bai, Jin, and Yu.[27])

Other Systems

Downer

In contrast to the riser, the downer involves downward flow of gas and solid in the gravitational direction at a high velocity. The downer is characterized by the absence of the

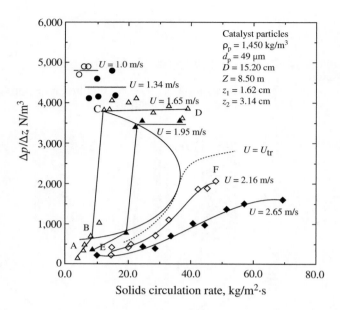

FIGURE 12.6
Local pressure drop as a function of solid circulation rate at various gas velocities. (From Clift.[23])

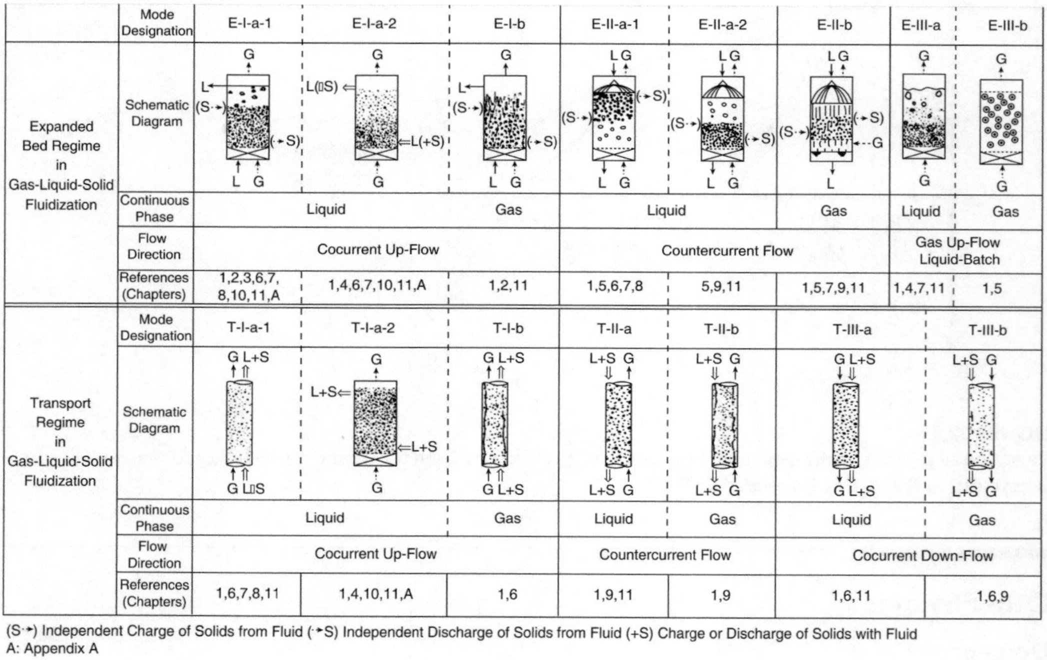

	Mode Designation	E-I-a-1	E-I-a-2	E-I-b	E-II-a-1	E-II-a-2	E-II-b	E-III-a	E-III-b
Expanded Bed Regime in Gas-Liquid-Solid Fluidization	Schematic Diagram								
	Continuous Phase	Liquid		Gas	Liquid		Gas	Liquid	Gas
	Flow Direction	Cocurrent Up-Flow			Countercurrent Flow			Gas Up-Flow Liquid-Batch	
	References (Chapters)	1,2,3,6,7, 8,10,11,A	1,4,6,7,10,11,A	1,2,11	1,5,6,7,8	5,9,11	1,5,7,9,11	1,4,7,11	1,5
Transport Regime in Gas-Liquid-Solid Fluidization	Mode Designation	T-I-a-1	T-I-a-2	T-I-b	T-II-a	T-II-b	T-III-a		T-III-b
	Schematic Diagram								
	Continuous Phase	Liquid		Gas	Liquid	Gas	Liquid		Gas
	Flow Direction	Cocurrent Up-Flow			Countercurrent Flow		Cocurrent Down-Flow		
	References (Chapters)	1,6,7,8,11	1,4,10,11,A	1,6	1,9,11	1,9	1,6,11		1,6,9

(S·→) Independent Charge of Solids from Fluid (·→S) Independent Discharge of Solids from Fluid (+S) Charge or Discharge of Solids with Fluid
A: Appendix A

FIGURE 12.7
Basic classification of gas–liquid–solid fluidization systems. (From Fan.[6])

minimum fluidization velocity, which allows a higher solid/gas loading ratio and a shorter flow development zone providing near plug flow conditions.[29] Thus, the downer has the benefits of uniform axial and radial gas–solid flow structures, thereby enhancing the reactant conversion for reactions in favor of the plug flow pattern.

Three-Phase Fluidized Bed

In gas–solid–liquid fluidization systems, the gas can be a continuous phase or discrete bubbles, while the liquid can be a continuous phase, a film or droplets, and the solids can be either in a continuous flow or a batch.[6] The operational modes of the three-phase fluidized beds can be classified by the directions of the gas and liquid flows: cocurrent upward, cocurrent downward, countercurrent, or crosscurrent as shown in Figure 12.7.[30]

The dispersed (or homogeneous) bubble and coalesced bubble (or churn-turbulent) flow regimes are the most common in operating the three-phase fluidized bed systems.[9] In the dispersed bubble flow regime, the bubbles with a fairly uniform size distribution rise without significant bubble coalescence. The vortical-spiral flow condition occurs in the coalesced bubble flow regime at lower gas velocities and marks a transition between the dispersed bubble regime and the churn-turbulent flow.[31] As the gas velocity further increases, the turbulent flow condition is developed, as large bubbles are generated by intensive bubble coalescence. The addition of fine particles leads to larger bubble sizes, and thus accelerates the transition.[32]

High-pressure and high-temperature operation of three-phase fluidized beds is commonly encountered in most industrial applications of commercial interest.[9] The flow characteristics of reactors at high pressure and temperature are distinctly different from those in ambient conditions. For example, elevated pressure leads to higher gas holdup and smaller bubble size in the system and, thus, dramatically affects the transport phenomena, including heat and mass transfer, and phase mixing. The effect of the operating pressure on the regime transition has been examined by many researchers in bubble columns,[32–40] in three-phase fluidized beds,[41] and in slurry bubble columns.[42]

The effects of pressure and temperature on fluid dynamics and transport properties are mainly due to variations in bubble characteristics, such as bubble size and bubble size distribution, and changes in the physical properties of fluid phases. The bubble size and distribution in the bed are closely associated with the initial bubble size, bubble coalescence rates, and bubble breakup rates. Under high-pressure conditions, bubble coalescence is suppressed and bubble breakup is enhanced, yielding smaller bubble sizes and narrower bubble size distributions. Thus, the flow regime transition is delayed in high-pressure bubble columns and slurry bubble columns.[42] Increasing temperature also delays the regime transition. In the coal liquefaction reactors, a gas holdup as high as 0.50 with the bubble size that may be as small as 0.7 mm was reported.[33]

For three-phase fluidization systems, two distinct phenomena pertaining to macroscopic hydrodynamic behavior—bed contraction and moving packed bed flow—are noted below.

Bed Contraction Phenomenon

Three-phase fluidized beds using small particles display unique bed expansion characteristics. Upon the initial introduction of the gas into the liquid–solid fluidized bed, contraction, instead of expansion, of the bed occurs.[43] An increasing gas flow rate causes further contraction up to a critical gas flow rate beyond which the bed expands.[44,45] A quantitative elucidation of the bed contraction phenomenon was reported by Stewart and Davidson[46] and El-Temtamy and Epstein.[47] Basically, bed contraction is caused by the behavior of the bubble wake, which entraps liquid and particles and, therefore, is associated with large bubble systems. The entrainment of the liquid and particles by the bubble wake reduces the effective amount of liquid in the bed used to fluidize the remaining particles. The bed contraction phenomenon has been extensively studied under ambient fluidization conditions and has also been observed at high pressure.[48]

On the basis of the generalized wake model of Bhatia and Epstein,[49] a criterion for the bed contraction was developed.[49,50] In the generalized wake model, the three-phase fluidized bed is assumed to consist of three regions, the gas bubble region, the wake region, and the liquid–solid fluidized region. Bed contraction will occur when the following criterion is satisfied, $\psi < 0$ where

$$\Psi = \left(\frac{n}{n-1}+k\right)\frac{U_1}{\varepsilon_1} + \frac{xk\left(U_g/\varepsilon_g\right)}{n-1}$$

$$-\left[(1+k)U_1 + \frac{k}{n-1}\left(\frac{U_g}{\varepsilon_g}-\frac{U_1}{\varepsilon_1}\right)\right] + xk\left(U_1-\frac{U_1}{\varepsilon_1}\right)\left(\frac{n}{n-1}\right) \tag{11}$$

Here, n is the Richardson–Zaki index, k is the ratio of wake size to bubble size, and x is the ratio of solid concentration in the wake region to that in the liquid–solid fluidized region.

Moving Packed Bed Phenomenon

The moving packed bed flow is characterized by the motion of solids in a piston flow in a three-phase fluidized bed. Moving packed bed flow, which usually occurs during the start-up, depends not only on the gas and liquid velocities, but also on how they are introduced into the bed. Moving packed bed flow is caused by the surface phenomena involving fine bubbles attached onto particles and subsequent formation of a fine bubble blanket under the packed solids; a liquid flow would move the entire bed upward. Thus, this phenomenon is associated with the small bubble system. The moving packed bed flow phenomenon in a three-phase fluidized bed is a known, anomalous event in the resid hydrotreating industry. This was observed in the 1960s, in the bench and pilot units, during the development and commercialization of the resid hydrotreating process.[51] The reactor was typically operated at pressures between 5.5 and 21 MPa and temperatures between 300°C and 425°C. In the early 1970s, moving packed bed flow was observed in a commercial three-phase fluidized bed reactor. The occurrence of a moving packed bed in a three-phase fluidized bed could simply be circumvented by utilizing a start-up procedure that involves degassing the bed first and then introducing liquid flow to expand the bed prior to commencing the gas flow. Commercial operators of three-phase fluidized bed reactors have long recognized and undertaken a proper start-up procedure of this nature since observing this anomalous event. As small bubbles can also be generated under ambient conditions using surfactants in an air–water system, the moving packed bed flow phenomenon was reported in open literature first by Saberian-Broudjenni et al.[52] and later by Bavarian and Fan[53,54] in small columns with small bubbles generated in the same manner.

Computational Fluid Dynamics

The discrete phases including particles, bubbles, and droplets in multiphase fluidization systems can be simulated numerically using the Eulerian continuum, and Lagrangian and direct simulation for solid particles, or the front tracking method for deformable surfaces, such as bubbles and droplets. In the continuum method, the individual phases are treated as pseudocontinuous fluids, each being governed by conservation laws expressed in terms of the volume per unit time or ensemble-averaged properties.[55–60] Numerical simulations using the continuum method have been extensively conducted for gas–solid,[61–64] gas–liquid,[65–68] and gas–liquid–solid flows.[68–71]

There are two main approaches for the numerical simulation of the gas–solid flow:[72] 1) Eulerian framework for the gas phase and Lagrangian framework for the dispersed phase (E–L) and 2) Eulerian framework for all phases (E–E). In the E–L approach, trajectories of dispersed phase particles are calculated by solving Newton's second law of motion for each dispersed particle, and the motion of the continuous phase (gas phase) is modeled using an Eulerian framework with the coupling of the particle–gas interaction force. This approach is also referred to as the distinct element method or discrete particle method when applied to a granular system.[73] The fluid forces acting upon particles would include the drag force, lift force, virtual mass force, and Basset history force.[74] Moreover, particle–wall and particle–particle collision models (such as hard sphere model, soft sphere model, or Monte Carlo techniques) are commonly employed for this approach. In the E–E approach, the particle cloud is treated as a continuum.[69] Local mean variables are utilized instead of point variables to describe the motions of both phases on the basis of the local average technique of Anderson and Jackson.[55] This approach is more suitable for modeling the dense multiphase system with a significant volume fraction for the dispersed phase (> 10%).

Because the effects of turbulence on particle motion are significant for most gas–solid fluidization systems, the numerical modeling of the carrier flow turbulence and motion of particles dispersed in a turbulent flow have been the subjects of extensive research. Direct numerical simulation (DNS) resolves the smallest scale structures (Komolgorov length scale) and can provide detailed point information for the flow field. No empirical closure models are needed in the DNS scheme. Large Eddy simulation (LES) involves both direct simulation and the Reynolds-averaging method. For the LES, large-scale motions are resolved rigorously, and the small-scale (sub-grid) motion is assumed to be homogeneous and independent of flow geometry.

On computation of gas–liquid bubble columns and gas–liquid-solid fluidized beds, numerical simulations provide a useful description of the hydrodynamics of bubble flows in liquids.[67,70,75–78] In the E–E method, both the continuous liquid phase and the dispersed bubble phase are treated as interpenetrating continua, occupying the same space based on the concept of volume averaging, with different velocities and volume fractions for each phases. In the E–L method,[75,77] on the other hand, liquid is treated as a continuous phase as described in the Eulerian mode, and bubbles are treated as a dispersed phase as tracked in the Lagrangian mode. Both the E–E and the E–L methods have proven to be more effective in modeling the gas–liquid flow in the homogeneous regime than in the heterogeneous regime.

More recently, the level-set technique has been developed for computing interfacial motion in two or three dimensions.[79] It is especially effective in simulating large topological changes such as bubble breaking and merging.

Conclusions

The fluidized bed systems have been utilized extensively in many physical, chemical, petrochemical, electrochemical, and biochemical processes. Successful applications of the fluidization systems lie in a comprehensive understanding of hydrodynamics, heat and mass transfer properties, and mixing. Various nonintrusive measurement techniques, such as electric capacitance tomography and radioactive particle tracking technique, are available to advance the fundamental understanding of the microscopic and macroscopic phenomena of fluidization. Till date, the design of the fluidized beds, however, still relies heavily on empirical relationships and engineering models. The computational fluid dynamics approach to fluidized bed simulation has gained considerable attention in recent years. However, many challenges remain, such as the formulation of closure relationships on turbulence. Fluidization of nanoparticles (< 100 nm) is an area of appreciable fundamental and applied interest for future research and development efforts.

Nomenclature

Ar	Archimedes number
d_{bb}	Averaged volume diameter of bubble in the bed
$d_{b\infty}$	Volume diameter of an isolated bubble
d_p	Particle diameter
g	Gravitational acceleration
H	Vertical distance from gas distributor
H_{mf}	Bed height at minimum fluidization
p_d	Dynamic pressure
Re_{pmf}	Particle Reynolds number at minimum fluidization velocity
U	Superficial gas velocity
U_{bb}	Average bubble-rise velocity in the whole bed
U_{mf}	Superficial gas velocity at the minimum fluidization
α	Bed voidage
α_b	Volume fraction of the bed occupied by bubbles
α_{mf}	Bed voidage at minimum fluidization
ϵ	Bed voidage
Δp_b	Pressure drop across the bed
μ	Dynamic viscosity
ρ	Density of fluid
ρ_p	Density of particle
Φ	Sphericity of particles

References

1. Davidson, J.F., Harrison, D. *Fluidized Particles*. Cambridge University Press: Cambridge, 1963.
2. Davidson, J.F., Clift, R., Harrison, D., Eds. *Fluidization*. 2nd Ed.; Academic Press: London, 1985.

3. Kunii, D., Levenspiel, O. *Fluidization Engineering*. 2nd Ed.; Butterworth-Heinemann: Boston, 1991.
4. Rietema, K. *The Dynamics of Fine Powders*. Elsevier Applied Science: London, 1991.
5. Gidaspow, D. *Multiphase Flow and Fluidization: Continuum and Kinetic Theory Descriptions*. Academic Press: New York, 1994.
6. Fan, L.S. *Gas–Liquid–Solid Fluidization Engineering*. Butterworths: Boston, 1989.
7. Fan, L.S., Zhu, C. *Principles of Gas–Solid Flows*. Cambridge University Press: Cambridge, 1998.
8. Jackson, R. *The Dynamics of Fluidized Particles*. Cambridge University Press: Cambridge, 2000.
9. Grace, J.R., Aridan, A.A., Knowlton, T.M., Eds., *Circulating Fluidized Beds*. Blackie Acasdemic and Professional: New York, 1997.
10. Geldart, D. Types of gas fluidization. Powder Tech. **1973**, *7*, 285.
11. Wen, C.Y., Yu, Y.H. Mechanics of fluidization. Chem. Eng. Prog. Symp. Ser. **1966**, *62* (62), 100.
12. Richardson, J.F., Zaki, W.N. Sedimentation and fluidization, part I. Trans. Inst. Chem. Eng. **1954**, *43*, 35.
13. Fan, L.S., Tsuchiya, K. *Bubble Wake Dynamics in Liquids and Liquid–Solid Suspensions*. Butterworths: Boston, 1990.
14. Anderson, T.B., Jackson, R.A. Fluid mechanical description of fluidized beds: stability of the state of uniform fluidization. I&EC Fundam. **1968**, *7*, 12.
15. Verloop, J., Heertjes, P.M. Shock waves as a criterion for the transition from homogeneous to heterogeneous fluidization. Chem. Eng. Sci. **1970**, *25*, 825.
16. Rietema, K., Piepers, H.W. The effect of interparticle forces on the stability of gas-fluidized beds: I. Experimental evidence. Chem. Eng. Sci. **1990**, *45*, 1627.
17. Clift, R., Grace, J.R., Weber, M.E. *Bubbles, Drops and Particles*. Academic Press: New York, 1978.
18. Massimilla, L. Gas jets in fluidized beds. In *Fluidization*, 2nd Ed.; Davidson, J.F., Clift, R., Harrison, D., Eds.; Academic Press: London, 1985.
19. Mori, S.; Wen, C.Y. Estimation of bubble diameter in gaseous fluidized beds. AIChE J. **1975**, *21*, 109.
20. Darton, R.C., La Nauze, R.D., Davidson, J.F., Harrison, D. Bubble growth due to coalescence in fluidized beds. Trans. Inst. Chem. Eng. **1977**, *55*, 274.
21. Cai, P., Schiavetti, M., DeMichele, G., Grazzini, G.C., Miccio, M. Quantitative estimation of bubble size in PFBC. Powder Tech. **1994**, *80*, 99.
22. Davies, L., Taylor, G.I. The mechanics of large bubbles rising through extended liquids and through liquids in tubes, Proc. R. Soc. Lond. **1950**, *A200*, 375.
23. Clift, R. Hydrodynamics of bubbling fluidized beds. In *Gas Fluidization Technology*. Geldart, D., Ed.; John Wiley & Sons: New York, 1986.
24. Clift, R., Grace, J.R. Continuous bubbling and slugging. In *Fluidization*, 2nd Ed.; Davidson, J.F., Clift, R., Harrison, D., Eds.; Academic Press: London, 1985.
25. Yerushalmi, J., Cankurt, N.T. Further studies of the regimes of fluidization. Powder Tech. **1979**, *24*, 187.
26. Mathur, K.B., Epstein, N. *Spouted Beds*. Academic Press: New York, 1974.
27. Bai, D., Jin, Y., Yu, Z. Flow regimes in circulating fluidized beds. Chem. Eng. Techno. **1993**, *16*, 307.
28. Bi, H.T., Fan, L.-S. Existence of turbulent regime in gas–solid fluidization. AIChE J. **1992**, *38*, 297.
29. Jin, Y., Zheng, Y., Wei, F. State-of-the-art review of downer reactors. In *Circulating Fluidized Bed Technology VII*, Proceedings of the 7th International Conference on Circulating Fluidized Beds, Niagara Falls, Ontario, Canada, May 5–8, 2002; Grace, J.R., Zhu, J.-X., de Lasa, H., Eds.; Canadian Society of Chemical Engineers: Ottawa, 2002; 40–60.
30. Fan, L.S. *Gas-Liquid-Solid Fluidization Engineering*. Butterworth: Boston, 1989.
31. Chen, R.C., Fan, L.S. Particle image velocimetry for characterizing the flow structure in three-dimensional gas–liquid–solid fluidized beds. Chem. Eng. Sci. **1992**, *47*(13,14), 3615–3622.
32. Clark, K.N. The effect of high pressure and temperature on phase distributions in a bubble column. Chem. Eng. Sci. **1990**, *45*, 2301.

33. Tamy, B., Chang, M., Coulaloglou, C., Ponzi, P. Hydrodynamic characteristics of three phase reactors. Chem. Eng. **1984**, *407*, 18.
34. Krishna, R., Wilkinson, P.M., Van Dierendonck, L.L. A model for gas holdup in bubble columns incorporating the influence of gas density on flow regime transitions. Chem. Eng. Sci. **1991**, *46*, 2491.
35. Krishna, R., Swart, J.W.A., Hennephof, D.E., Ellenberger, J., Hoefsloot, C.J. Influence of increased gas density on hydrodynamics of bubble-column reactors. AIChE J. **1994**, *40*, 112.
36. Wilkinson, P.M., Sper, A.P., Van Dierendonck, L.L. Design parameters estimation for scale-up of high-pressure bubble columns. AIChE J. **1992**, *38*, 544.
37. Hoefsloot, H.C.J., Krishna, R. Influence of gas density on the stability of homogeneous flow in bubble columns. Ind. Eng. Chem. Res. **1993**, *32*, 747.
38. Reilly, I.G., Scott, D.S., de Bruijn, T.J.W., MacIntyre, D. The role of gas phase momentum in determining gas holdup and hydrodynamic flow regimes in bubble column operations. Can. J. Chem. Eng. **1994**, *72*, 3.
39. Letzel, H.M., Schouten, J.C., Van den Bleek, C.M., Krishna, R. Influence of elevated pressure on the stability of bubbly flows. Chem. Eng. Sci. **1997**, *52*, 3733.
40. Lin, T.J., Tsuchiya, K., Fan, L.S. Bubble flow behavior at high pressure bubble column. Can. J. Chem. Eng. **1999**, *77*, 370–374.
41. Fan, L.-S., Yang, G. Gas–liquid–solid three-phase fluidization. In *Handbook of Fluidization and Fluid-Particle Systems*. Yang, W.-C., Ed., Marcel Dekker: New York, 2003; Chapter 27, 765–810.
42. Fan, L.S. Yang, G.Q. Lee, D.J. Tsuchiya, K. Luo, X. Some aspects of high pressure phenomena of bubbles in liquids and liquid-solid suspensions. Chem. Eng. Sci. **1999**, *54*, 4681.
43. Massimilla, L., Majuri, N., Signorini, P. Sull'assorbimento Di Gas in Sistema: Solido-Liquido Fluidizzato. La Ricerca Scientifica **1959**, *29*, 1934.
44. Turner, R. Fluidization. Soc. Chem. Ind. Lond. **1964**, 47.
45. Ostergaard, K. Fluidization. Soc. Chem. Ind. Lond. **1964**, 58.
46. Stewart, P.S.B., Davidson, J.F. Three-phase fluidization: water, particles and air. Chem. Eng. Sci. **1964**, *19*, 319.
47. El Temtamy, S.A., Epstein, N. Contraction or expansion of three-phase fluidized beds containing fine/light solids. Can. J. Chem. Eng. **1979**, *57*(4)520–522.
48. Jiang, P., Lin, T.J., Fan, L.-S. High temperature and high pressure three-phase fluidization—bed expansion phenomena. Powder Tech. **1997**, *90*, 103.
49. Bhatia, V.K., Epstein, N. Three-phase fluidization: generalized wake model. In *Fluidization and Its Applications*, Proceedings of the International Symposium, Toulose, October 1–5, 1973. Angelion, H., Coudere, J.P., Gibert, H., Laguerie, C., Eds., Cepadues-Editions: Toulose, 1974, 380–392.
50. Jean, R.H.; Fan, L.S. Bed contraction criterion for three phase fluidization. Can. J. Chem. Eng. **1987**, *65*, 351–352.
51. Fan, L.S. Moving packed bed phenomenon in three-phase fluidization. Powder Tech. **1999**, *103*, 300.
52. Saberian-Broudjenni, M., Wild, G., Charpentier, J.C., Fortin, Y., Euzen, J.P., Patoux, R. Contribution à L'ètude Hydrodynamique Des Rèacteurs à Lit Fludisè Gaz-Liquide-Solide. Entropie **1984**, *120*, 30.
53. Bavarian, F., Fan, L.-S. Mechanisms of hydraulic transport of a packed bed at the start-up of a three-phase fluidized bed. Chem. Eng. Sci. **1991**, *46*, 3081.
54. Bavarian, F., Fan, L.S. Hydraulic transport of a packed bed during the start-up of a three-phase fluidized bed with large gas holdups. Ind. Eng. Chem. Res. **1991**, *30*, 408.
55. Anderson, T.B., Jackson, R. A fluid mechanical description of fluidized beds. I&EC Fundam. **1967**, *6*, 527–539.
56. Ishii, M. *Thermo-Fluid Dynamic Theory of Two-phase Flows*. Eyrolles: Paris, 1975.
57. Drew, D.A. Averaged field equations for two phase media. Stud. Appl. Math. **1971**, *50*, 133.
58. Nigmatulin, R.I. Spatial averaging in the mechanics of heterogeneous and dispersed systems. Int. J. Multiphase Flow **1979**, *5*, 353.

59. Zhang, D.Z., Prosperetti, A. Averaged equations for inviscid disperse two-phase flow. J. Fluid Mech. **1994**, *267*, 185.
60. Jackson, R. Locally averaged equations of motion for a mixture of identical spherical particles and a Newtonian fluid. Chem. Eng. Sci. **1997**, *52*, 2457.
61. Sinclair, J.L., Jackson, R. Gas–particle flow in a vertical pipe with particle–particle interactions. AIChE J. **1989**, *35*, 1473.
62. Ding, J., Gidaspow, D. A bubbling fluidization modeling using kinetic theory of granular flow. AIChE J. **1990**, *36*, 523.
63. Pita, J.A., Sandaresan, S. Developing flow of a gas–particle mixture in a vertical riser. AIChE J. **1993**, *39*, 541.
64. Dasgupta, S., Jackson, R., Sundaresan, S. Turbulent gas–particle flow in vertical risers. AIChE J. **1994**, *40*, 215.
65. Torvik, R., Svendsen, H.F. Modeling of slurry reactors: a fundamental approach. Chem. Eng. Sci. **1990**, *45*, 2325.
66. Svendsen, H.F., Jakobsen, H.A., Torvik, R. Local flow structures in internal loop and bubble column reactors. Chem. Eng. Sci. **1992**, *47*, 3297.
67. Sokolichin, A., Eigenberger, G. Gas–liquid flow in bubble columns and loop reactor: part I. Detailed modeling and numerical simulation. Chem. Eng. Sci. **1994**, *49*, 5735.
68. Boisson, N., Malin, M.R. Numerical prediction of two-phase flow in bubble columns. Int. J. Numer. Methods Fluids **1996**, *23*, 1289.
69. Gidaspow, D., Bahary, M., Jayaswal, U.K. iN *Hydrodynamic Models for Gas–Liquid–Solid Fluidization: Numerical Methods in Multiphase Flows*. FED, ASME: New York, 1994; Vol. 185, 117–124.
70. Grevskott, S., Sannas, B.H., Dudukovic, M.P., Hjarbo, K.W., Svendsen, H.F. Liquid circulation, bubble size distributions, and solids movement in two- and three-phase bubble columns. Chem. Eng. Sci. **1996**, *51*, 1703.
71. Mitra-Majumdar, D., Farouk, B., Shah, Y.T. Hydrodynamic modeling of three-phase flows through a vertical column. Chem. Eng. Sci. **1997**, *52*, 4485.
72. Crowe, C., Sommerfeld, M., Tsuji, Y. *Multiphase Flows with Droplets and Particles*. CRC Press LLC: Boca Raton, 1998.
73. Xu, B.H., Yu, A.B. Numerical simulation of the gas–solid flow in a fluidized bed by combining discrete particle method with computational fluid dynamics. Chem. Eng. Sci. **1997**, *52* (16), 2785.
74. Ranade, V.V. *Computational Flow Modeling for Chemical Reactor Engineering*. Process Systems Engineering; Academic Press: San Diego, 2002; Vol. 5.
75. Lapin, A., Lübbert, A. Numerical simulation of the dynamics of two-phase gas–liquid flows in bubble columns. Chem. Eng. Sci. **1994**, *49*, 3661–3674.
76. Sokolichin, A., Eigenberger, G. Applicability of the standard k-e turbulence model to the dynamic simulation of bubble column: part 1. Detailed numerical simulation. Chem. Eng. Sci. **1999**, *54*, 2273–2284.
77. Delnoij, E., Kuipers, J.A.M., Van Swaaij, W.P.M. Computational fluid dynamics applied to gas–liquid contactors. Chem. Eng. Sci. **1997**, *52*, 3623–3638.
78. Mudde, R.F., Simonin, O. Two- and three-dimensional simulations of bubble plume using a two-fluid model. Chem. Eng. Sci. **1999**, *54*, 5061–5069.
79. Chen, C., Fan, L.-S. Discrete simulation of gas–liquid bubble columns and gas–liquid–solid fluidized beds. AIChE J. **2004**, *50* (2), 288.

13

Fluidized Bed Reactor

John R. Grace, Jamal Chaouki, and Todd Pugsley

CONTENTS

Introduction 199
Key Features of Gas-Solid-Fluidized Bed Reactors..200
Fluidized Bed Catalytic Reactor Processes..203
    Fluid Catalytic Cracking 203
    Partial Oxidation Reactions 203
    Fischer-Tropsch Synthesis...203
    Oxychlorination of Ethylene...204
    Propylene Ammoxidation..204
    Polymerization...204
    Catalytic Reforming...205
Gas-Solid Reactions 205
    Combustion..207
    Gasification..208
    Pyrolysis 209
    Calcination 210
    Roasting...210
    Other Processes...211
        Ultrapure Silicon and Iron Ore Reduction...211
        Chlorination and Fluorination of Metal Oxides 211
        Nano- and Ultrafine Particles...211
Liquid-Solid Reactions in Fluidized Beds 211
Gas-Liquid-Solid (Three-Phase) Chemical Processes in Fluidized Beds 212
    Biochemical Processes 212
    Hydrocarbon Processing..213
Fluidized Bed Reactor Modeling 213
Conclusions..214
References..214

Introduction

This chapter covers key features of fluidized bed reactors for catalytic and noncatalytic reactions involving solid particles and one or more fluids, most commonly a gas, and also liquids and gas-liquid mixtures. Fluidized beds find wide application in several industries—chemical processes, petrochemicals, polymers, mineral processing, pharmaceuticals, and food processing—because of their unique features that are advantageous in a number of applications. Enough background is provided for the reader to understand the major advantages and disadvantages of fluidized bed reactors, the most important design challenges and considerations, and the principal applications. Emphasis is placed on gas-solid-fluidized beds, but liquid-solid and gas-liquid-solid (three-phase) fluid bed reactors are also treated in brief. For more extensive coverage, the reader should consult standard reference works on fluidization.[1–6]

Key Features of Gas-Solid-Fluidized Bed Reactors

The advantages of gas-solid-fluidized bed reactors relative to packed bed reactors are:

- Greatly improved bed-to-wall and bed-to-immersed-surface heat transfer
- Reduced axial and lateral temperature gradients, minimizing the probability of hot spots, catalyst sintering, and unwanted side reactions
- Ability to add or remove particles continuously or intermittently, without shutting down the process
- Reduced pressure drops (The pressure drop across the bed, once fluidized, essentially remains equal to only that required to support the weight of the bed.)
- Smaller catalyst particles, leading to improved catalyst effectiveness factors
- Ability to introduce (usually as a spray) modest quantities of liquid reactants that vaporize before reacting or yield solid products upon reaction inside the bed

Fluidized beds, however, have some important disadvantages relative to packed (fixed or moving) bed reactors:

- Substantial axial gas mixing, causing much larger deviations from plug flow than for packed bed reactors, thereby adversely affecting conversions and selectivities
- For reactions where the particles themselves react, substantial particle mixing, greatly broadening solid residence time distributions relative to moving beds
- Particle attrition because of particles colliding with each other and with fixed surfaces
- Wear on immersed tubes and other interior surfaces because of the particle impingement
- Entrainment of particles, causing loss of catalyst and/or solid product, contributing to air pollution and requiring gas-solid separation equipment
- Increased risk because of complex hydrodynamics and difficulties in characterizing and predicting reactor performance

Gas-solid contacting and axial dispersion of both gas and particles depend on the particle properties, operating conditions, column geometry, and scale, all of which affect the motion of gas and solids, commonly referred to as hydrodynamics. For extensive coverage of hydrodynamics[2–6] may be consulted. Here we summarize the key features that affect the performance of gas-solid-fluidized beds as chemical reactors.

Overall Configuration: Common configurations for fluidized bed reactors are shown in Figure 13.1. Bubbling beds and turbulent beds [Figure 13.1(A)] usually operate at excess superficial gas velocities ($U - U_{mf}$) of <0.25 m/s and between ~0.3 and 1.2 m/s, respectively, whereas circulating beds (Figure 13.1B) generally feature superficial gas velocities of 2–12 m/s.

Particle Properties: Particles should be free-flowing (not sticky and with rounded shapes), ideally within either group A or group B of the powder classification scheme of Geldart.[7] Catalyst particles primarily belong to group A (mean diameter 50–100 μm), whereas particles that react in the bed fall into group B or D (mean diameter 200–2000 μm). Particle size distributions should be reasonably broad, e.g., covering a range from 10 to 200 μm. Particle densities should, however, be as uniform as possible. The particles must be able to withstand frequent collisions without breaking. Ideally, they should also not be subject to major electrostatic charging effects.

Operating Pressure and Temperature: Fluidized beds generally operate more smoothly with increasing absolute pressure, so that elevated pressures do not present a problem from a fluidization point of view. Many fluidized bed reactors also operate at high temperatures.

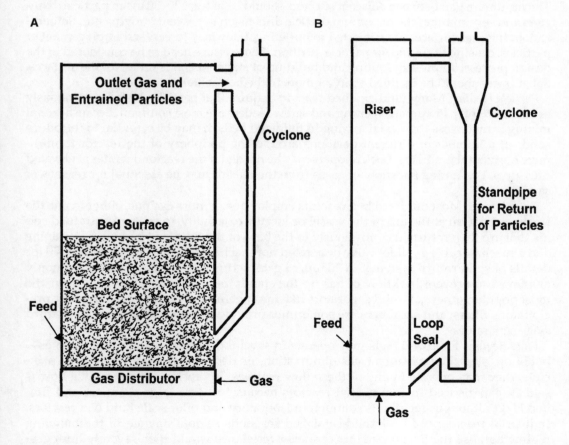

FIGURE 13.1
Typical equipment configurations for: (A) dense (bubbling, slugging or turbulent) fluidized bed reactors; (B) circulating fluidized bed reactors. Heat transfer surfaces or baffles that might be present in (A) or (B) are not shown.

The effects of temperature differ from system to system, probably because temperature affects particle properties (e.g., surface hardness and stickiness) in addition to causing well-characterized changes in gas properties. The effects of pressure and temperature have been reviewed by Yates.[8]

Gas Distributor: The gas distributor plate at the bottom of the reactor must introduce gas uniformly, prevent hole-plugging or weeping of solids, promote good gas-solid contacting, and support the weight of the bed when it is defluidized. During steady state operation, it is usually important to ensure that the pressure drop across the distributor is at least 30% of that across the bed itself. Gas introduction points should not be more than 0.3 m apart. Many different distributor geometries are found in industry, including tuyères, bubble caps, pipe grids with orifices oriented obliquely downward, and multiorifice plates (see Karri and Werther[9] for more details).

Internals: Fluidized bed reactors commonly contain heat transfer surfaces, and baffles are sometimes added to improve gas-solid contacting and/or to reduce axial dispersion. Surfaces should, whenever possible, be horizontal or vertical, not inclined at other angles. The minimum gap between adjacent surfaces should be at least 20–30 mean particle diameters and several times the maximum particle dimension to prevent bridging and defluidization. Internal surfaces are subjected to buffeting, buoyancy forces, and impingement of particles, causing wear and/or particle attrition. These issues need to be considered in the design process. To prevent fouling and build-up of stagnant solids on top of the surfaces, flat surfaces should be inclined at 60° or more to the horizontal.

Particle Feeding: Many fluidized bed reactors require that particles be fed continuously or intermittently. Pneumatic feeding and screw feeders are most common. Because lateral mixing is much less than axial mixing in fluidized beds, it may be essential to introduce solids at a number of different positions around the periphery of the reactor to minimize horizontal gradients. Lock hoppers may be needed if the reactor operates at elevated pressures. Preventing backflow of gases from the reactor may be essential for reasons of safety.

Entrainment: Most fluidized bed reactors employ one or more cyclone, either inside the freeboard region at the top of the vessel or located externally, to capture entrained solids that are then returned continuously to the base of the fluidized bed via a standpipe and a mechanical (e.g., slide) valve or aerated nonmechanical valve (see Knowlton[10] for details of solid return systems). A "flapper" gate, acting as a check valve, is commonly employed to prevent backflow of gas up the standpipe. While cyclones are by far the most popular, other gas-solid separators like impingement separators, electrostatic precipitators, filters, and scrubbers are sometimes provided, especially as second- or third stage separators.

Flow Regime: Fluidized beds may operate in several different distinct flow regimes—bubbling, slug flow, turbulent, fast-fluidization, or dense suspension upflow. In principle, they may operate in any of these flow regimes. In practice, however, slug flow is seldom experienced in commercial reactors because of their large diameter and limited H/D, although slugging is common in laboratory and pilot-scale fluid bed reactors. Industrial reactors used for solid-catalyzed reactions seldom operate in the bubbling regime because the interphase mass transfer resistance would then severely limit gas-solid contacting. However, bubbling is often relevant for the larger particles found in gas-solid reactions. Catalytic reactors most often operate in the turbulent or dense suspension upflow regime, because of favorable gas-solid contacting combined with limited gas axial mixing, whereas the turbulent and the fast-fluidization regimes are common for gas-solid reactions.

Fluidized Bed Catalytic Reactor Processes

The advantages of fluidized bed reactor technology, discussed earlier, have led to several commercial fluidized bed catalytic processes. The key ones are surveyed here.

Fluid Catalytic Cracking

Fluid catalytic cracking (FCC) represents the major industrial application of fluidized bed technology. As of 1997, there were over 350 units operating worldwide, processing more than 16 million barrels of feed per day.[11] For a summary of typical FCC process conditions, see Avidan.[11]

The process is operated as a circulating fluidized bed, with fine catalyst particles continuously circulated between the riser, where the cracking reactions occur and catalyst is deactivated because of coke deposition, and the regenerator, where the coke on the catalyst particles is combusted. Matsen[12] reported the slip factor in FCC risers to be approximately 2, which is consistent with radial segregation of the catalyst, leading to radial variations in gasoline yield.[13,14] Data[15] indicate that the catalyst particles travel upward at the wall of the riser because of the high gas velocity and solid flux. This results in reduced gas backmixing, thereby minimizing overcracking of gasoline.

In early FCC process designs, the regenerator was operated as a bubbling or a turbulent fluidized bed combustor. Mobil and UOP developed riser regenerators in the 1970s.[11] The riser regenerator is advantageous in operations in which there is no external CO recovery boiler. With today's trend of processing heavier and more variable feedstocks in FCC units, the preference is to operate in a partial combustion mode in the regenerator, in which case riser regenerators are of little benefit. Most modern FCC systems use a turbulent fluidized bed regenerator.

Partial Oxidation Reactions

This class of reactions, carried out in fluidized beds, involves parallel and series reactions, with reaction intermediates being the desired products. Industrial examples include partial oxidation of *n*-butane to maleic anhydride and *o*-xylene to phthalic anhydride. The vigorous solid mixing of fluidized beds is valuable for these reactions because they are highly exothermic. However, gas backmixing must be minimized to avoid extended gas residence times that lead to the formation of products of total combustion (i.e., CO_2 and H_2O). For this reason, fluidized bed catalytic partial oxidation reactors are operated in the higher velocity regimes of turbulent and fast-fluidization.

Most fluidized bed partial oxidation processes are operated in the turbulent flow regime of fluidization. However, DuPont operated a circulating fluidized bed catalytic reactor process for maleic anhydride production in Spain,[16] featuring regeneration of the catalyst (by oxidation) on the downcomer side of the circulating system.

Fischer-Tropsch Synthesis

Fischer-Tropsch reactions convert synthesis gas into hydrocarbons. The fluidized bed process for this reaction, known as the Synthol process, was developed by Sasol of South Africa to produce synthetic gasoline from coal. Duvenhage and Shingles[17] describe the development of the Sasol reactor technology from pilot studies in the 1940s to today's commercial operation. Sasol 1A and 1B circulating fluidized bed reactors started up in 1955 and were fraught with problems: the planned run length of 340 days was in actuality more like 40 days. Many of the problems were resolved over 20 years,[17] and two additional trains of larger circulating fluidized bed reactors were commissioned uneventfully in the early 1980s. In the late 1980s, Sasol began pilot studies that led to the installation of two large turbulent fluidized bed reactors in 1995 and 1999. Among the advantages of this design are lower installed and operating costs associated with smaller vessels, reduced bed pressure drops, and lower catalyst usage.[17] The Sasol reactors rely on immersed heat exchangers to remove the heat of reaction. Plugging and cutting of the tube sheet coolers in the risers of the Sasol 1A and 1B reactors proved to be a problem.[17] The tube sheets were eventually replaced with cooling coils because of the tendency for tube sheet coolers to plug with solids.[18]

Oxychlorination of Ethylene

Vinyl chloride monomer, the basic building block of polyvinylchloride (PVC), is commercially manufactured by dehydrochlorination of 1,2-dichloroethane. The modern process for producing 1,2-dichloroethane involves oxychlorination of ethylene in a fluidized bed catalytic reactor:

$$C_2H_4 + 2HCl + \frac{1}{2}O_2 \rightarrow Cl_2C_2H_4 + H_2O \tag{1}$$

This is a highly exothermic reaction that takes place in the gas phase in the presence of a $CuCl_2$ catalyst. Small amounts of alkali and rare-earth metals in the catalyst inhibit by-product formation. The reactor is operated at 220–240°C and 2–4 bar. Immersed heat exchangers again remove the excess exothermic heat of reaction.

Propylene Ammoxidation

Ammoxidation refers to the catalytic oxidation of a feedstock with ammonia. When propylene is the feedstock, acrylonitrile is produced. Most of the world's acrylonitrile is based on the Sohio (now BP) process in which stoichiometric amounts of propylene and ammonia are reacted with a slight excess of air in a fluidized bed operated in the turbulent fluidization flow regime. The reactor temperature and pressure are 450°C and 1.5 bar, respectively. The reaction usually takes place over a bismuth molybdenum oxide catalyst; other metal oxides (iron-antimony, uranium-antimony, and tellurium-molybdenum) have also been used. The catalyst contains iron compounds that increase selectivity to acrylonitrile. The exothermic heat of reaction is removed by immersed heat exchangers in which boiler feed water is circulated and high-pressure steam is produced.

Polymerization

The gas-phase polymerization of ethylene to produce linear low-density polyethylene (LLDPE) is often carried out in fluidized bed reactors. The Unipol process technology is licensed worldwide by Union Carbide. It can also be used to make high-density polyethylene (HDPE). In the early years, ethylene was fed to the reactor with a 1-butene comonomer. The process was later modified to introduce higher α-olefins such as 1-hexene, yielding stronger LLDPE. The LLDPE resin is produced with Ziegler catalysts in large fluidized bed reactors (~5 m tall and 3–4 m in diameter) operated at pressures of 15-25 bar and temperatures of 70–90°C. Unipol reactors employ flared sections at the top to promote gas-solid disengagement. The monomer/comonomer mixture, together with nitrogen and hydrogen, is fed to the reactor in which the solid resin is vigorously fluidized. The gaseous feed reacts and condenses on the solid product. Particles continue to grow; large resin particles sink to the bottom of the reactor and are continuously withdrawn. The heat of polymerization is removed by immersed heat transfer surfaces. Per pass conversion is only ~2%. The unconverted gaseous feed exiting the top of the reactor is compressed, cooled, and recycled. In the condensing mode of operation, the recycled gas mixture is cooled below its dew point so that a portion of the recycled olefins enters the reactor in the liquid phase. The liquid droplets rapidly evaporate, helping to remove the heat of polymerization. This strategy increases the rate of production from fluidized bed polymerization reactors.

Catalytic Reforming

Catalytic reforming is the key refinery process for converting low-octane paraffinic hydrocarbons in the gasoline range (i.e., naphtha) to high-octane naphthenes and aromatics. The noble metal catalysts and promoters (e.g., platinum and rhenium) deactivate because of coke deposition and must be regenerated. The first generation catalytic reforming technology developed in the 1950s employed fixed bed reactors operated at pressures of 30 bar or more. The rate of coke formation is reduced at high pressure, facilitating longer run times before the reactors are shut down for regeneration (after ~6 months). However, reformate yield and octane number are adversely affected by high pressure. Continuous catalytic reforming (CCR) technology, featuring four reactors in series, has been developed to overcome this problem. The reforming catalyst forms a slowly moving bed that descends through each reactor. In the UOP stacked design, the catalyst is transferred between reactors because of gravity. In the IFP design, the reactors are side by side, and the catalyst leaving the bottom of one reactor enters a lift-pot from which it is transported to the top of the next reactor. From the lift-pot at the base of the fourth reactor, the deactivated catalyst is sent back to the regenerator, located upstream of the first reactor. Modern CCRs operate at ~5 bar, substantially increasing the reformate yield and octane number. The CCR process is similar to the FCC process in that the catalyst is continuously circulated. However, the solid circulation flux is much lower in the CCR process. Furthermore, the regenerators of FCC units provide combustion environments to remove coke from the catalyst. Combustion also occurs in the CCR regenerator, but oxygenation and chlorination take place to restore the acidity of the bifunctional catalyst.

Gas-Solid Reactions

This section concerns noncatalytic gas-solid reactions:

$$A \text{ (gas)} + B \text{ (solid)} \rightarrow \text{products (gas and/or solid)} \tag{2}$$

Table 13.1 describes the nature of gases and solids depending on the reaction. These reactions include combustion, gasification, pyrolysis, calcination, and roasting. Some

TABLE 13.1

Main Characteristics of Fluidized Bed Combustion, Gasification, Pyrolysis, Calcination, and Roasting

Reaction Type	A	B	Main Products	Other Products	Temperature Range (°C)	Solid Residence Time (sec)
Combustion	Air	Coal, MSW, or biomass	CO_2, H_2O, ash	SO_x, NO_x	850–930	3–500
Gasification	Air, oxygen, or stream	Biomass, MSW, or coal	CO, H_2, CH_4, char	Tar, SO_x, NO_x	650–850	1–600
Pyrolysis or fast pyrolysis	Recycle gases with or without air	Biomass, hydrocarbons	Liquid fuel (bio-oil), H_2O, char, coke	Lower molecular weight hydrocarbons	450–550	0.5–500
Calcination	Air	Limestone, dolomite, alumina, phosphates, etc.	Oxides	SO_x, NO_x	600–800	1–60
Roasting	Air or oxygen-enriched air	Sulfide ores	Oxides	SO_2	600–1000	30–120

specific information is also provided for these types of reaction. Many noncatalytic gas-solid reactions use coal, biomass, and municipal solid wastes as feedstock. Table 13.2 summarizes their chemical compositions. These compositions differ in many ways, including the organic/inorganic content and chemical and physical properties.[19] The most important noncatalytic gas-solid reactions using fluidized bed technology are discussed here.

TABLE 13.2

Physical, Chemical, and Fuel Properties of Biomass, Coal, and Municipal Solid Waste

Property	Biomass	Coal	MSW
Fuel density (kg/m³)	~500–800	~1300–1500	~700–800
Mean particle size (mm)	~3	~0.1–3	~3–5
C content (wt.% of dry fuel)	42–54	65–85	25–30
O content (wt.% of dry fuel)	35–45	2–15	20–25
S content (wt.% of dry fuel)	Max. 0.5	0.5–7.5	Max. 0.25
SiO_2 content (wt.% of ash)	23–49	40–60	10–40
K_2O content (wt.% of ash)	4–48	2–6	1.5–9
Al_2O_3 content (wt.% of ash)	2.4–9.5	15–25	2–10
Fe_2O_3 content (wt.% of ash)	1.5–8.5	8–18	2–6
Ignition temperature (°C)	418–426	490–595	400–450
Friability	Low	High	Medium
Dry heating value (MJ/kg)	14–21	23–28	10–13[a]

[a] Dry heating value could be as low as 3 MJ / kg for developing countries.

Combustion

Fluidized bed steam generators burning solid fuels have operated in the energy industry since 1980. This technology penetrated the energy market surprisingly quickly. Outstanding features of fluidized bed steam generators are high fuel flexibility, integrated emission control, flexibility in meeting the operating requirements of various industries, and a proven record of reliability. Fuel flexibility has become an important criterion in selecting boiler technology for new utility power plants: the type and the grade of fuels also affect the solid handling provisions, energy recovery, emissions control, and solid waste handling.[20] For example, the 200 MW$_e$ Tonghae thermal power plant CFB boiler (2 units) began commercial operation in 1998–1999 by firing low-quality Korean anthracite for electrical power generation.

Although combustion efficiency is a function of many variables (the most important being combustion temperature, excess air, residence time, feed size, cyclone separation efficiency, and mixing of gas and solids), carbon burnout efficiency is high (usually 99% or more) in fluidized bed combustion (FBC) systems. Many comprehensive FBC simulation models are available in the literature.[21] One of them has the advantage of being integrated with Aspen Plus, an advanced computer based simulation package for process engineering that is able to simulate a variety of processes ranging from single unit operations to complex multiunit processes.[21]

Several clean coal municipal waste or biomass technologies [bubbling fluidized bed (BFB) and circulating fluidized bed (CFB)] at atmospheric or at higher pressure, typically 6–15 bar) have been developed that reduce pollutant emissions.[22] Limestone is employed as a sorbent to capture sulfur in the bed, where it undergoes calcination followed by sulfation. The Ca to S molar feed ratio must typically be 2 to 3 for 90–95% retention of sulfur (SO$_2$ emission of ~200 mg/MJ). Unfortunately, the limestone utilization is normally low, typically 30–40%. NO$_x$ emissions are low for FBC boilers relative to those of conventional pulverized coal boilers because of lower operating temperatures (typically 830–880°C). Compared with BFB combustors (135–1800 mg/MJ), CFB combustors tend to have lower NO$_x$ levels (135–180 mg/MJ) because of air staging. Due to the catalytic effect of SO$_2$ and limestone, N$_2$O emissions show a maximum for a bed temperature of ~800°C. Ultralow emissions of NO$_x$ (45 mg/MJ) can be obtained by combining FBC, and selective noncatalytic reduction with ammonia or selective catalytic reduction (over Pt, Au or V$_2$O$_5$) is commonly used. However, achieving ultralow NO$_x$ emissions may compromise sulfur oxide reduction and/or carbon utilization. Compared with BFB combustors, CFBs also produce less CO (~40 mg/MJ) because of the more intensive gas turbulence and better mixing, although lateral solid mixing remains an issue. Before discharge to the atmosphere, the flue gas is passed through a baghouse to remove most of the entrained dust. Particulate emissions from FB combustors are, on average, ~10 mg/MJ. Cofiring biomass and coal, compared with firing coal alone, helps reduce both NO$_x$ and SO$_2$ and also fuel costs, minimize waste, and reduce soil and water pollution, depending upon the chemical composition of the biomass.[22]

Frequent operational problems have been encountered in biomass combustors. One common issue is agglomeration of bed material. Biomass ash is often relatively rich in alkali and alkaline-earth metals, causing it to melt at relatively low temperatures. Fouling of the heat exchangers may cause additional problems.

Several countries have introduced stringent emission limits (0.1 ng-TE/Nm³) for chlorinated dioxins and furans emitted from combustion sources, in particular solid waste incinerators, because of concerns over their adverse health effects. Technologies for reducing their formation and emission in incineration processes have been studied extensively and can be applied in modern incineration plants.[23] Activated carbon injection and fabric filtration are currently practiced in many installations. However, to minimize capital cost, a more fundamental approach is needed to control and limit formation of these pollutants in incineration processes, e.g., involving the postcombustion zone, the combustion chamber, and waste feeding.[23]

Gasification

Gasification is the conversion of coal, municipal solid waste (MSW), or biomass to a gaseous fuel by heating in the presence of oxygen and/or steam. The resulting fuel gas is more versatile than the original solid reactant. For example, it could be used to power gas engines and gas turbines, or as a chemical feedstock to produce liquid fuels. Air-blown processes produce low-calorific-value gases with a typical higher heating value (HHV) of 4–7 MJ/m^3, while oxygen- and steam-blown processes result in gases with an HHV of 10–18 MJ/m^3.[24] The reactions in the gasifier include partial oxidation and complete oxidation of the carbon content, steam-carbon reaction, Boudouard reaction, shift reaction, and methane formation.[25] A number of reactor configurations have been used including BFBs and CFBs.

Lurgi's atmospheric CFB technology, originally developed for alumina calcination, was later adapted to coal combustion. It has since been applied to biomass gasification. Recently, a consortium of Japanese utilities announced the construction of a commercial-scale 250 MW (1700 t/d) integrated gasification combined cycle (IGCC) plant at Nakoso, Japan, based on two stages. In the first (combustion) stage, the high temperature simplifies separation of liquid slag from the gas. In the second (reduction), only coal is introduced, in the absence of additional oxidant. The temperature drop over the reductor stage is 700°C, with a reactor outlet temperature of ~1000°C.[25] Many fluidized bed gasification processes have been and are being developed, but none incorporates a heat carrier in such a way that the tars are combusted and char reacts with the gasifying agent to produce synthesis or fuel gas. Such a system could produce pure syngas, with the gas free of tars and carbon conversion virtually complete.[26]

The major operational difficulty is gas cleaning, which is still the major bottleneck, limiting the use of biomass for electricity generation. The gas formed contains a number of impurities: particulates (ash and char), tar, and nitrogen, sulfur, and alkali compounds. The tar, consisting of high molecular weight compounds that typically condense at temperatures below 450°C, can cause blockages and corrosion, while also reducing overall efficiency. Moreover, impurities (such as methane) can affect the end usage of syngas.

As a first approximation, a nonstoichiometric equilibrium model based on direct minimization of Gibbs free energy may be applied to predict gasifier performance, although experimental data deviate somewhat from model predictions.[24] Extensive research is being carried out to develop a process to produce a low-tar or tar-free gas. To reach these goals, two strategies are applied: improvement of the gasification technologies (updraft gasifier and cocurrent moving bed gasifier with throat[26]) and development of tar-removal processes. Since the mid-1980s, interest has grown with respect to catalytic biomass gasification.[27] The catalysts may be divided into two groups depending on the position of the catalyst reactor relative to the gasifier. In the first, the catalyst is added directly to the biomass prior to gasification. These catalysts consist of cheap disposable materials, e.g., dolomite (MgCO$_3$ · CaCO$_3$), calcite (CaO), or magnesite (MgO), which reduce tar content.

The second group of catalysts is placed in a secondary reactor downstream from the gasifier. Commercially available and relatively inexpensive, nickel reforming catalyst is highly effective at removing hydrocarbons and adjusting the gas composition to syngas quality. It is most active and has a longer lifetime when operated at 780°C in a fluidized bed.[27]

Pyrolysis

Pyrolysis is a moderate temperature (~500°C) process in which biomass, coal, or MSW is decomposed in the absence of an oxidizing agent at very high heating rates and vapor residence times typically less than 1 sec.[28] The product is collected as a liquid (bio-oil) and a solid char after rapid cooling. For fast pyrolysis, BFBs and CFBs have been used extensively. They are designed and operated to maximize the liquid fraction at up to 75 wt.% on a dry feed basis. The bio-oil can be stored and transported. Its typical higher heating value is ~18 MJ/kg, and its chemical composition differs significantly from that of petroleum-derived oils. Several fast pyrolysis technologies have reached near-commercial status. Six CFB plants have been constructed by Ensyn Technologies, the largest with nominal capacities of 50 t/day. DynaMotive demonstrated a BFB process at 10 t/day of biomass and is scaling up the plant to 110 t/day.[28] Bio-oil has been successfully burned in boilers and furnaces, although some problems have been reported with char levels which can block filters and atomizers. Other problems have also been encountered—the alkaline nature of the ash in turbines, the high viscosity of the bio-oil product (40–100 cP at 40°C), water content with possible phase separation, and corrosion from its low pH (pH 2–3). Comparison of three routes from biomass to electricity—combustion, gasification, and fast pyrolysis—based on cost and performance indicates that fast pyrolysis is expensive up to 5 MW_e. However, competition will be fierce as electricity production costs vary only slightly among these options.[29]

Upgrading bio-oil (oxygen content typically 35–40 wt.%) to quality transport liquid fuel requires full deoxygenation, e.g., by hydrotreating or catalytic vapor cracking over zeolite catalysts. Neither of these processes appears to be economically viable. The hydrotreating process is realized at high pressure and high temperature with sulfided CoMo or NiMo catalysts supported on alumina.

Very few chemicals produced from the fractionation of the bio-oil are already available. The most promising are glycoaldehyde as a meat-browning agent and levoglucosan, which has potential in the synthesis of antibiotics and flavor compounds. Some phenols were even found in coal pyrolysis, which may produce high value specialty chemicals. Specialty chemicals generally require further work to develop reliable low-cost separation procedures.

Fluid coking is a thermal cracking process developed by Exxon (now ExxonMobil) in which heavy residual hydrocarbons, heavy oils, or bitumen derived from oil sands are upgraded in fluidized bed reactors at high temperatures and moderate pressures. The hydrocarbon to be upgraded is fed as a liquid. The lower-molecular-weight product is withdrawn at the top, while heavy "coke" coats the particles. The coke particles are then extracted near the bottom, after being passed through a stripper in which steam removes liquid hydrocarbon from the surfaces of the particles. The coke particles are circulated to a combustion chamber where the exothermic heat of reaction heats the coke particles, some of which are then recirculated to the fluid coker to provide the heat needed by the endothermic thermal pyrolysis reactions.

Calcination

An early fluid bed calciner for limestone and dolomite, built by England Lime Company in 1949, was 4 m in diameter, 14 m tall, and operated at 1000°C. As the reactions are highly endothermic and both the gas and solid exit at high temperature, several attempts were made to save energy, including multistaging and combining a calciner with suspension preheating.[4]

In the late 1960s, Lurgi adopted the CFB for calcination of alumina trihydrate. Design studies have shown that CFBs can be applied for capacities as high as ~1000 t/day. A high degree of automation, economy of space, and a simplified process flowsheet are important design criteria.

Apatitic rock phosphates (a poor-quality phosphate) are commonly calcined in fluidized beds (with one, two, or even three stages) to remove impurities (carbonates, water, and organic matter) and to increase phosphorus concentration. The hydrocarbon content (up to 5%) of this phosphate provides most of the heat needed for calcination. Increasing the process temperature improves the upgrading of the ore, but decreases the agronomic effectiveness in direct application to soils. This operation has a maximum temperature of ~900°C.[30]

Roasting

Roasting is often used in the metallurgical industry to convert mineral sulfides into oxides or other compounds such as sulfates suitable for further treatment and recovery of metals. Oxides and sulfates are readily dissolved in leach solutions, while sulfides only dissolve with great difficulty. In these pyrometallurgical processes, sulfide ores as fine particles (100–500 µm) are normally roasted to oxides in fluidized beds at 600–1000°C. The residence time of the solids is around a minute, and short circuiting of solids must be avoided. Fluidized bed roasters are widely used, especially for zinc sulfide ores, which account for >80% of all zinc production. As the reactions are exothermic, the temperature is kept constant by direct (water addition) or indirect (immersed cooling tubes) heat exchangers. In the late 1950s, Dorr-Oliver designed the first fluidized bed roaster producing SO_2 from pyrite, zinc blend, and sulfide ores. The roaster was 5.5 m in diameter and 7.6 m high; it operated at 700°C and superficial gas velocities of 0.45–0.5 m/s.

Pyrometallurgical processes operating at high temperatures are usually associated with expensive emissions control, as dioxins, chloride compounds, and mercury can be generated. Competitive technologies include hydrometallurgical operations. In such processes, pressure reactors (autoclaves) are gaining acceptance for the leaching of ores and concentrates, and in the recovery of metals from leach solutions. To continue to be attractive, pyrometallurgy faces two major challenges: reducing environmental emissions and increasing the capacity of existing plants. Although many plants attempt to increase the roasting capacity by applying an oxygen-enriched roasting condition, optimization of the concentrate flow and the process hydrodynamics is an alternative way to improve capacity, versatility, and stability. Improvements in mass/heat transfer in conventional fluidized beds have led to numerous innovative reactor configurations. One example is the toroidal fluidized bed developed by Torftech for sulfide ore roasting. This roaster has a fixed annular gas distributor consisting of angled blades, which force the particles into a toroidal motion that enhances heat/mass transfer, even at low gas velocities.[31]

Other Processes

Ultrapure Silicon and Iron Ore Reduction

Silicon of exceptional purity for semiconductor and photovoltaic industries is obtained from metallurgical-grade silicon or liquid silicon tetrachloride. At least four of six steps involve fluidized beds. In Osaka Titanium's process, step 1 (obtaining $SiHCl_3$ from Si and HCl) takes place at 300°C, whereas in the Union Carbide process, $SiHCl_3$ is obtained by hydrogenating gaseous $SiCl_4$ over $CuCl_2$ catalyst at 500°C. Other noncatalytic gas-solid hydrogenation reactions are also carried out in fluidized beds. For example, Bethelem Steel developed the H-Iron process. Alan Wood Steel, Kawasaki Iron and Steel, and others have developed processes that produce iron by reducing fine iron oxides in a single stage at ~700°C. Other processes for iron ore reduction with hydrogen include the Stelling process to form cementite and the Armco or Exxon process with multistage reactors.[4]

Chlorination and Fluorination of Metal Oxides

These reactions are in some cases the only way to obtain highly purified metals such as titanium and U-235. The chloride process, in which $TiCl_4$ is produced from titaniferous feedstocks, chlorinated in the presence of solid carbon at 950–1150°C in a BFB, is of significant importance in producing titanium metal and TiO_2 pigments. A recent techno-economic study identified the potential of a CFB reactor for the chlorination of fine slag, which can be obtained at lower cost than standard chloride-grade slag.[32]

Nano- and Ultrafine Particles

Several fluid bed processes are under development for production and encapsulation of nanoparticles, for example, WC-Co composites, bioceramics (i.e., calcium phosphate hydroxyapapite), carbon encapsulation of iron magnetic nanoparticles, and carbon nanotubes. These nano- or ultrafine powders have broad industrial and pharmaceutical applications. Production processes usually include solution preparation (sol-gel), drying, calcination, and sintering. The last three steps may be realized in a fluidized bed, but fluidization of nano- and ultrafine powders is very difficult because of strong interparticle forces.

Liquid-Solid Reactions in Fluidized Beds

Liquid-fluidized beds predate gas-fluidized beds, but they have considerably fewer applications because of a smaller number of advantages. Most applications are physical,[33] with bioreactors being the sole significant reactor application. Much of the recent attention has focused on aerobic wastewater treatment and fermentation processes, e.g., with methane as the organic substrate (see Fan[34] and Grady, Daigger, and Lim[35] for more details). In these processes, microbial cells are attached to the surface of inert particles (e.g., sand or activated carbon) as a biofilm, or trapped within the pores or interior of particles, causing the particle size and/or density to vary with time. Loaded particles therefore have different fluidization characteristics, and this difference can be used to separate product from fresh particles to recover the biomass.

Advantages of liquid fluidization include lack of plugging, reduced pressure drops, good heat and mass transfer, favorable surface area, and ability to readily separate excess biomass. The fluidization may be upward, or downward (referred to as "inverse fluidization") when the particles have a lower density than the liquid. Tapering may be used to reduce entrainment. Various configurations, including circulating beds,[36] are possible.

Gas-Liquid-Solid (Three-Phase) Chemical Processes in Fluidized Beds

In gas-liquid-solid (three-phase) fluidized beds, solid particles are simultaneously contacted with both gas and liquid. The gas and liquid may flow cocurrently upward, or the liquid may descend, while the gas rises. The liquid usually forms the continuous phase in which the solid particles and gas bubbles are dispersed. The bubbles are larger when the particles are smaller, and bed contraction can occur when gas is introduced into a liquid-fluidized bed of fine particles. Higher pressures lead to smaller bubbles and increased gas hold-ups.

Advantages of three-phase fluidized beds over trickle beds and other fixed bed systems are temperature uniformity, high heat transfer, ability to add and remove catalyst particles continuously, and limited mass transfer resistances (both external to the particles and bubbles, because of turbulence and limited bubble size, and inside the particles owing to relatively small particle diameters). Disadvantages include substantial axial dispersion (of gas, liquid, and particles), causing substantial deviations from plug flow, and lack of predictability because of the complex hydrodynamics. There are two major applications of gas-liquid-solid-fluidized beds: biochemical processes and hydrocarbon processing.

Biochemical Processes

Three-phase fluidized beds can be used as bioreactors for aerobic biochemical processes, including both fermentation processes and wastewater treatment. The gas phase is air, required for biological growth, while the solid particles provide immobilized surfaces on or in which cell growth can occur. The aqueous liquid phase provides the culture medium needed for the growth and maintenance of the cells. Air may be introduced separately from the liquid, or be premixed with the aqueous medium. The liquid medium may exhibit non-Newtonian rheology. A disadvantage of three-phase fluidization for some processes, e.g., those involving mammalian cells, is that fluidization results in significant shear stresses, which may disrupt or damage the cells. Extensive background information, literature references, and a summary of recent advances in understanding three-phase hydrodynamics are provided by Fan[34,37] and Wright and Raper.[38] Various geometries are possible, for example, tapered columns, draft tubes, and external circulation.[34]

Wastewater treatment is relatively simple in that the sole purpose is the degradation of all the organic species present in the liquid to remove both biological oxygen demand (BOD) and chemical oxygen demand (COD). Three-phase fluidization is also of interest for bioremediation of contaminated soils. Production of alcohols by fermentation (e.g., ethanol from glucose) has been practiced commercially. Other fermentation processes have been examined for production of enzymes, acetic acid, stem cells, monoclonal antibodies, antibiotics, and other pharmaceutical products.

Hydrocarbon Processing

A number of energy-related processes have been developed employing three-phase fluidization. Early examples were in coal liquefaction and hydrogenation. In the 1960s and 1970s, there was significant activity to develop three-phase fluidization processes, including the H-Oil, H-Coal, and LC-Fining processes. The latter remains in commercial operation, and practical aspects have been described in the literature.[34,39] The reactor in this case is called an "ebullated bed." Hydrogen gas is bubbled through a heavy hydrocarbon liquid (e.g., a residue, heavy oil, or bitumen) at elevated temperature (>400°C) and pressure (>10 MPa) in the presence of solid catalyst particles. Complex reactions including thermal cracking, hydrocracking, and hydrodesulfurization occur, leading to an upgraded product with a higher H/C ratio. Fresh catalyst is added periodically to maintain the overall activity level. Efforts to resolve practical problems in a large industrial unit have been described by McKnight et al.[39] A more recent development is the liquid-phase methanol synthesis process where syngas is reacted to produce methanol in a catalyst-inert oil slurry.

Fluidized Bed Reactor Modeling

Considerable effort has been expended to devise models for fluidized bed reactors. These models differ greatly in complexity. It is important to adopt a model with the optimum degree of sophistication, complex enough to capture the mechanistic elements affecting the principal aspects of concern, but not so complex that they contain features whose inclusion is not justified in view of the uses to which the model will be put.

Models that are entirely empirical are seldom of much value for fluidized bed reactors, given the hydrodynamic complexity and the large number of variables. The simplest mechanistic models treat the reactor as simple single-phase vessels (well-mixed or plug flow) and have little relation to reality. Two-phase models were developed beginning in the 1950s and 1960s, with bubbling beds primarily in mind. In such models, reviewed in a number of studies (e.g., in Yates,[1] Kunii and Levenspiel,[4] and Grace[40]), the bed is divided into two parallel one-dimensional paths, one of high voidage (e.g., to represent the bubble phase) and the other of much lower voidage (representing the dense phase). Mole balances are then written for each phase, incorporating interphase mass transfer between the phases, as well as terms to account for the reaction(s). The predicted performance is then dependent on the interphase transfer, as well as the fraction of solids, total volume of each phase, gas flow assigned to each phase, and chemical kinetics. Such models are widely used and have been quite successful for steady state fluidized bed gas-phase solid-catalyzed reactions.

The two-phase models have been extended to gas-solid reactions (e.g., in Yates,[1] Kunii and Levenspiel,[4] and Grace[40]), requiring that the mole balances also account for reaction of the solid phase and that the stoichiometry of the reaction be satisfied. Population balances may then be added to account for changes in particle size distribution. Single-particle reaction models (e.g., shrinking core model) may also be needed to account for changes in particle properties resulting from the reaction(s). Two-phase models have also been extended to other flow regimes—slugging beds,[41] turbulent beds,[42] and fast fluidization.[5] They also form the basis of a more generic probabilistic approach,[43] which covers the bubbling, turbulent, and fast-fluidization flow regimes.

In recent years, there has been considerable effort to develop computational fluid dynamic (CFD) models to predict the hydrodynamics and performance of fluidized beds. While this approach will no doubt yield valuable tools in the future, CFD models are not yet at the point where they can be used with confidence for design and scale-up of fluidized bed processes.

Conclusions

Few fluidized bed reactors in practice operate in the bubbling regime.[18] The trend is toward higher gas velocities. Numerous potential applications of circulating fluidized bed technology as a catalytic reactor have been listed.[44] Improved understanding of the flow inside fluidized beds is reducing the risk associated with scale-up and leading to new processes.

A promising area involves integration of product separation with reaction by using fluidized bed membrane reactors, where semipermeable membranes, immersed in the bed, allow selective withdrawal of products like hydrogen, or distributed introduction of reactants like oxygen. Continuous product removal can overcome reaction equilibrium limitations of industrially important reactions such as steam methane reforming.[45] Controlled addition of oxygen through membranes can also enhance selectivity to desired products and reduce the risk of explosions in catalytic partial oxidation reactions.[46]

Other areas for future development captured in this chapter are nanoparticle processes, the development and validation of CFD, and other advanced reactor models for fluidized beds. In addition, incremental improvements in fluidized bed processes will be made, helping to ensure that fluidized beds continue to gain ground as reactors in many industries.

References

1. Yates, J.G. *Fundamentals of Fluidized-Bed Chemical Processes*. Butterworths: London, 1983.
2. Davidson, J.F., Clift, R., Harrison, D. *Fluidization*; 2nd Ed.; Academic Press: London, 1985.
3. Geldart, D. Ed.; *Gas Fluidization Technology*. Wiley and Sons: Chichester, U.K., 1986.
4. Kunii, D., Levenspiel, O. *Fluidization Engineering*. 2nd Ed.; Butterworth-Heinemann: Boston, 1991.
5. Grace, J.R., Avidan, A.A., Knowlton, T.M. *Circulating Fluidized Beds*. Chapman and Hall: London, 1997.
6. Yang, W.C., Ed.; *Handbook of Fluidization and Fluid-Particle Systems*. Marcel Dekker: New York, 2003.
7. Geldart, D. Types of gas fluidization. Powder Technol. **1973**, 7, 285–292.
8. Yates, J.G. Effect of temperature and pressure. In *Handbook of Fluidization and Fluid-Particle Systems*. Yang, W.C., Ed.; Marcel Dekker: New York, 2003; 129–154.
9. Karri, S.B.R., Werther, J. Gas distributor and plenum design in fluidized beds. In *Handbook of Fluidization and Fluid-Particle Systems*. Yang, W.C., Ed.; Marcel Dekker: New York, 2003; 155–170.
10. Knowlton, T.M. Standpipes and non-mechanical valves. In *Handbook of Fluidization and Fluid-Particle Systems*. Yang, W.C., Ed.; Marcel Dekker: New York, 2003; 599–617.
11. Avidan, A.A. Fluid catalytic cracking. In *Circulating Fluidized Beds*. Grace, J.R., Avidan, A.A., Knowlton, T.M., Eds.; Chapman and Hall: London, U.K., 1997; 466–488.

12. Matsen, J.M. Some characteristics of large solids circulation systems. In *Fluidization Technology*. Keairns, D.L., Ed.; Hemisphere: Washington, 1976; Vol. 2, 135–149.

13. Fligner, M., Schipper, P.H., Sapre, A.V., Krambeck, F.J. Two-phase cluster model in riser reactors impact of radial density distribution on yields. Chem. Eng. Sci. **1994**, *49*, 5813–5818.

14. Derouin, C., Nevicato, D., Forissier, M., Wild, G., Bernard, J. Hydrodynamics of riser units and their impact on FCC operation. Ind. Eng. Chem. Res. **1997**, *36*, 4504–4515.

15. Karri, S.B.R., Knowlton, T.M. A comparison of annulus solids flow direction and radial solids mass flux profiles at low and high mass fluxes in a riser. In *Circulating Fluidized Bed Technology VI*. Werther, J., Ed.; Dechema: Frankfurt, Germany, 1999; 71–76.

16. Contractor, R.M. DuPont's CFB technology for maleic anhydride. Chem. Eng. Sci. 1999, *54*, 5627–5632.

17. Duvenhage, D.J., Shingles, T. Synthol reactor technology development. Catal. Today **2002**, *71*, 301–305.

18. Bolthrunis, C.O., Silverman, R.W., Ferrari, D.C. Rocky road to commercialization: break-throughs and challenges in the commercialization of fluidized bed reactors. In *Fluidization XI: Present and Future for Fluidization Engineering*. Arena, U., Chirone, R., Miccio, M., Salatino, P., Eds.; Engineering Foundation: New York, 2004; 547–554.

19. Demirbas, A. Combustion characteristics of different biomass fuels. Prog. Energy Combustion Sci. **2004**, *30*, 219–230.

20. Sondreal, E.A., Benson, S.A., Hurley, J.P., Mann, M.D., Pavlish, J.H., Swanson, M.L., Weber, G.F., Zygarlicke, C.J. Review of advances in combustion technology and biomass cofiring. Fuel Proces. Technol. **2001**, *71*, 7–38.

21. Sotudeh-Gharebaagh, R., Legros, R., Chaouki, J., Paris, J. Simulation of circulating fluidized bed reactors using ASPEN PLUS. Fuel **1998**, *77* (4), 327–337.

22. Leckner, B., Lyngfelt, A. Optimization of emissions from fluidized bed combustion of coal, biofuel and waste. Int. J. Energy Res. **2002**, *26*, 1191–1202.

23. Buekens, A., Huang, H. Comparative evaluation of techniques for controlling the formation and emission of chlorinated dioxins/furans in municipal waste incineration. J. Hazardous Mater. **1998**, *62*, 1–33.

24. Li, X.T., Grace, J.R., Lim, C.J., Watkinson, A.P., Chen, H.P., Kim, J.R. Biomass gasification in a circulating fluidized bed. Biomass Bioenergy **2004**, *26*, 171–193.

25. McKendry, P. Energy production from biomass (part 3): gasification technologies. Bioresource Technol. **2002**, *83*, 55–63.

26. Higman, C., Van der Burgt, M. *Gasification*; Gulf Professional Publishing: Amsterdam, 2003.

27. Sutton, D., Kelleher, B., Ross, J.R.H. Review of literature on catalysts for biomass gasification. Fuel Proces. Technol. **2001**, *73*, 155–173.

28. Czernik, S., Bridgwater, A.V. Overview of applications of biomass fast pyrolysis oil. Energy Fuel **2004**, *18*, 590–598.

29. Bridgwater, A.V., Toft, A.J., Brammer, J.G. A techno-economic comparison of power production by biomass fast pyrolysis with gasification and combustion. Renew. Sustain. Energy Rev. **2002**, *6*, 181–248.

30. Lim, H.H., Gilkes, R.J. Beneficiation of apatite rock phosphates by calcination: effects on chemical properties and fertilizer effectiveness. Aust. J. Soil Res. **2001**, *39* (2), 397–402.

31. Shu, J., Lakhmanan, V.I., Convey, J. Sintering and ferrite formation during high temperature roasting of sulfide concentrates. Can. Metall. Quart. **1999**, *38* (4), 215–225.

32. Luckos, A., den Hoed, P. The carbochlorination of titaniferous feedstocks in a fluidized bed. In *Fluidization XI*; Arena, U., Chirone, R., Salatino, P., Eds.; Engineering Conferences International: New York, 2004.

33. Epstein, N. Applications of liquid-solid fluidization. Int. J. Chem. Reactor Eng. **2003**, *1*(Review) 1.

34. Fan, L.-S. *Gas-Liquid-Solid Fluidization Engineering*. Butterworths: Boston, 1989.

35. Grady, C.P.L., Daigger, G.T., Lim, H.C. *Biological Wastewater Treatment*. Marcel Dekker: New York, 1999.

36. Zhu, J.-X., Zheng, Y., Karamanev, D.G., Bassi, A.S. (Gas-)liquid-solid circulating fluidized beds and their potential applications to bioreactor engineering. Can. J. Chem. Eng. **2000**, *78*, 82–94.
37. Fan, L.-S. Advances in gas-liquid-solid fluidization. In *Fluidization XI*; Arena, U., Chirone, R., Miccio, M., Salatino, P., Eds.; Engineering Conferences International: New York, 2004; 1–20.
38. Wright, P.C., Raper, J.A. A review of some parameters involved ion fluidized bed bioreactors. Chem. Eng. Technol. **1996**, *19*, 50–64.
39. McKnight, C.A., Hackman, L.P., Grace, J.R., Macchi, A., Kiel, D., Tyler, J. Fluid dynamic studies in support of an industrial fluidized bed hydroprocessor. Can. J. Chem. Eng. **2003**, *81*, 338–350.
40. Grace, J.R. Fluid beds as chemical reactors. In *Gas Fluidization Technology*. Geldart, D., Ed.; Wiley: Chichester, U.K., 1986; 285–339.
41. Hovmand, S., Freedman, W., Davidson, J.F. Chemical reaction in a pilot scale fluidized bed. Trans. Inst. Chem. Eng. **1971**, *49*, 149–162.
42. Chaouki, J., Gonzalez, A., Guy, C., Klvana, D. Two-phase model for a catalytic turbulent fluidized bed reactor: application to ethylene synthesis. Chem. Eng. Sci. **1999**, *54*, 2039–2045.
43. Abba, I.A., Grace, J.R., Bi, H.T., Thompson, M.L. Spanning the flow regimes: generic fluidized-bed reactor model. AIChE J. **2003**, *49*, 1838–1848.
44. Berruti, F., Chaouki, J., Godfroy, L., Pugsley, T.S., Patience, G.S. Hydrodynamics of circulating fluidized bed risers: a review. Can. J. Chem. Eng. **1995**, *73*, 579–602.
45. Adris, A.M., Grace, J.R. Characteristics of fluidized-bed membrane reactors: scale-up and practical issues. Ind. Eng. Chem. Res. **1997**, *36*, 4549–4556.
46. Ramos, R., Pina, M.P., Menendez, M., Santamaria, J., Patience, G.S. Oxidative dehydrogenation of propane to propene: simulation of a commercial inert membrane reactor immersed in a fluidized bed. Can. J. Chem. Eng. **2002**, *79*, 902–912.

Section IV

Polymer and Supercritical Fluid Applications

14

Powder Coating Application Processes

Harry J. Lader

CONTENTS

Introduction 219
Background 220
    Advantages 220
    Disadvantages 220
    Powder Coating Markets 221
Overview of the Powder Coating Application Process 221
Powder Charging and Application 222
    Corona Charging 222
    Back-Ionization 224
    The Faraday Cage Effect 225
    Internal Charging 226
    Tribocharging 226
    Nozzles 228
    Bells and Discs 229
    Electrostatic Fluidized Bed Technology 230
Powder Recycling 230
    Particle Size 230
    Powder Recovery 231
Meeting New Challenges 232
    New Equipment Technology 232
    New Powder Technology 232
Conclusions 233
Acknowledgment 233
References 233

Introduction

Powder coating originated in the 1950s as a method to coat parts by immersing preheated parts in a fluidized bed. This method was used primarily for corrosion resistance or electrical insulation since only thick films were possible. These powder coatings were dry-blended mixtures of pigmented epoxy resins, flow agents, and curing agents. Over time, electrostatic fluidized beds were developed that provided better control over film thickness and are now used mostly for coating small parts. Powder coatings were compounded using extruders in the 1960s. This development meant that powder coatings could now be applied as an electrostatic spray, instead of using fluidized beds. The resin of choice was epoxy due to its good storage stability, good flow properties, and reasonable appearance. Epoxy powders had the ability to retain an electrostatic charge that allowed the powder to be attracted to and adhere to a grounded part during the application process. The charged powder remained on the part until the powder was cured. The ability to formulate a powder coating as compounded particles spurred tremendous growth for the powder coating industry.

While compounded epoxy powders stimulated the development of the powder coating industry, epoxies were prone to chalking and gloss reduction because of poor stability to ultraviolet (UV) radiation for exterior applications. This property deficiency encouraged research on other resin systems and eventually led to a range of powder coatings that now include epoxy/polyester hybrids, TGIC (triglycidyl isocyanurate)–polyesters, polyurethanes, acrylics, in addition to epoxies. These powders are known as "thermosetting powders" as they crosslink on heating and comprise a majority of the powder coatings in use today. On the other hand, thermoplastic powders, which can also be used in electrostatic powder coating applications, do not crosslink on heating and therefore can be remelted. Examples of these types of powders include polyolefins, polyvinyl chloride, polyamides, and polyvinylidene fluoride.

In this chapter, we review basic charging concepts, corona charging, internal charging, tribocharging, and electrostatic fluidized beds. Along with these methods, some of the properties of powders needed for their application are also discussed.

Background

Advantages

The major advantage of powder coatings is that there are no solvents with which to contend; hence, compliance with environmental regulations is much easier. The problems and costs associated with them, such as toxicity, mixing, thinning, and labor, are eliminated while reducing time and material variability. Another major benefit is the high material utilization through recycling. While a primer is generally not required, the part must be properly cleaned to ensure a good quality coating.

After powder has been applied, a convection oven, infrared heating, or a combination of infrared followed by UV curing is used to cure the part. In some cases, induction heating of metals parts is used. Powder coated parts are fully cured when they leave the oven; hence the total processing time tends to be shorter as compared with wet coatings. This results in a reduction of rejects due to damage that can occur after painting. Powder coating allows the parts to be racked closer together on a conveyor allowing more parts to be processed in a shorter time. Powder coated parts are cured at elevated temperatures, usually between 160°C and 220°C (320–430°F), which generally results in a tougher, more chip-resistant coating.

Disadvantages

While there are many advantages to using powder coatings, there are also some disadvantages. Some of the challenges include the ability to quickly change colors, especially to take advantage of the benefits of recycling; color matching and uniformity; particle size variation during recycling that could lead to color variation, especially with regard to metallic powders; high cure temperatures; film thickness nonuniformities, especially at edges, corners, or in recessed areas; thin films; and good surface appearance with high transfer efficiency. Advances in equipment technology are addressing some of these challenges, while others are being improved through material technology.

Powder Coating Markets

Powder coatings are used in a wide range of markets. Epoxy powders are typically used for functional applications such as steel pipelines and rebar that is used in concrete construction. These functional coatings typically have film thicknesses of over 250 μ (10 mil) and are usually applied onto preheated substrates. Many decorative coatings also serve a protective function, but have to be visually appealing as well. These coatings are typically applied between 25 and 100 μ (1–4 mil). The choice of powder depends on the environment in which the part is to be used. Today, polyester/epoxy hybrids comprise almost half of all the powders used worldwide. Their improved weatherability over epoxies and lower cost have allowed them to capture the largest percentage of the powder market. Polyesters have captured almost 40% of the world market, while epoxies have diminished to only about 10%. Polyurethane and acrylic powders are in low single digits.[1] Typical applications for powder coatings include metal furniture and fixtures, machinery, major and small appliances, automotive, electrical components and equipment, building and architectural, farm, and garden equipment.

The types of powder coatings available are gradually evolving with the advent of new powder chemistries that allow for numerous formulation variations. For example, powder coatings have been developed to replace porcelain in laundry applications, heat-resistant polymers for range door trim, and flexible powder coatings for refrigerator doors. This wide latitude in formulation capabilities means that there may be more than one resin chemistry used in a particular powder formulation.

Overview of the Powder Coating Application Process

Powder charging can use either a corona gun or a tribocharging gun. Other types of application devices are discussed later. Corona charging uses a gun applicator with an electrode at the tip, usually at a negative high voltage, while a tribogun does not need an external power supply as the powder is charged by friction. A fluidized bed or a box feeder is used to feed the powder to the gun. The powder in the fluidized hopper is mixed with dry compressed air through a porous fluidizing plate so that the powder acts like a liquid. A pump transports the powder through a delivery tube into a corona- or tribogun applicator. The charged powder is sprayed through the gun onto a part. Nozzles are used to shape the spray pattern to the part so that most of the powder attaches itself to the grounded part. The booth contains the powder. This powder can be recovered and transported back to the feed hopper, if necessary. Virgin powder is added to the system to maintain powder levels. (See Liberto[2] for an excellent practical guide on powder coating.)

Recently, the adoption of lean manufacturing practices, which focuses on meeting customers' needs to quickly deliver any quantity of any parts in specified colors, has resulted in many color changes daily. This need has encouraged equipment manufacturers to provide the ability to change powder color quickly. In addition, equipment improvements have increased transfer efficiency and have reduced the need to recover powder. By increasing first pass transfer efficiency, spray-to-waste is gradually becoming accepted.

Powder Charging and Application

Corona Charging

A basic knowledge of corona charging is important for understanding how powders charge, the limitations of charging efficiency, and how this impacts transfer efficiency and surface appearance. A corona discharge occurs when a high voltage is applied to an electrode in the presence of a ground. When the electric field at the electrode is very high and exceeds the breakdown of air, negative ions, present naturally in air, are split into both an electron and a neutral atom when they enter the high field region. The electron is driven away from the high electric field and gains enough energy to ionize an air molecule, creating a cascading effect. At the same time, naturally occurring positive ions are attracted to the negative point electrode, where they attain sufficient energy from the high electric field and release secondary electrons that add to the discharge. This process is self-sustaining above the threshold voltage. Light emitted during ionization creates a glow around the sharp electrode causing a "corona."[3]

Numerous negative ions are created when the electrode goes into corona. These ions will follow the field lines toward the grounded part. Figure 14.1 is a schematic of a corona gun spraying powder and shows the electric field lines between the electrode and a flat part. The powder dispersed from the nozzle will enter the region between the electrode and the grounded part. Most of the air-diffused powder particles will attract ions and become charged. The electric field lines will accelerate the charged particles toward the grounded part and the free ions will also continue along the same path.

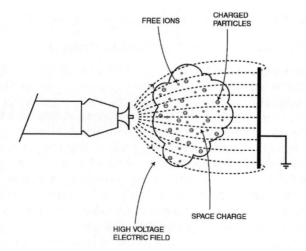

FIGURE 14.1
Schematic of a corona gun with an electrode at high negative voltage with charged powder and ions in the presence of a grounded flat part.

Figure 14.1 also shows that the electric field lines concentrate on the sharp edges of the part. Some of the electric field lines actually wrap around the back of the target. In addition to the electric field lines, aerodynamics will also contribute to this "wrap" effect. Increased film thickness at the edge is usually referred to as the "picture frame" effect. The charged powder and free ions between the gun and ground comprise a charged cloud known as the "space charge." This space charge creates its own electric field lines to the ground.

Figure 14.2 is a schematic representation of the electric field lines from the gun to a part that has a protrusion at its center. In this case, the electric field lines will concentrate at the protrusion. The charged powder and free ions will build up faster here and contribute to back-ionization. Because the powder particles are strong dielectrics, their charge will be retained for many hours, depending on a number of factors such as powder resistivity, as well as the ambient relative humidity.

The maximum charge that can be acquired by a spherical, nonconducting particle in a uniform electric field was determined by Pauthenier[4] and is shown in Eq. (1):

$$q_{max} = 4\pi\varepsilon_0 a^2 Ep, \quad \text{where } p = \frac{3\varepsilon_r}{\left(\varepsilon_r + 2\right)} \tag{1}$$

In Eq. (1), ε_0 is the permittivity of free space (8.8×10^{-12} F/m), E is the maximum electric field before breakdown of air occurs (3×10^6 V/m), a is the particle radius in m, and ε_r is the relative dielectric constant of a powder particle. This equation shows that the maximum charge that can be acquired is proportional to the electric field and the square of the particle radius. It is also weakly dependent on the relative dielectric constant of the particle. The charge-to-mass ratio takes into account the mass of the particle and is determined by dividing Eq. (1) by the mass of a spherical particle, $(4/3)\pi a^3\rho$, where ρ is the density of the powder (kg/m^3), as shown in Eq. (2):

$$\frac{q_{max}}{m} = \frac{3\varepsilon_0 Ep}{a\rho} \tag{2}$$

Eq. (2) shows that the maximum charge-to-mass ratio is inversely proportional to the size of the particle. This equation shows that the maximum charge-to-mass ratio of a 100 μ particle will be 10 times lower than a 10 μ particle. It is also important to note that because the mass of a spherical particle is proportional to the third power of a particle's radius, the mass of one 100 μ spherical particle is the same as that of a thousand 10 μ particles.

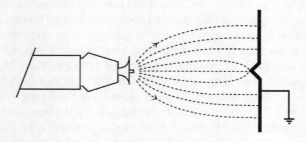

FIGURE 14.2
Schematic of a corona gun with an electrode at high voltage in the presence of a part containing a protrusion at its center.

The actual charge-to-mass ratio of a charged particle will depend on the mechanism of charging used, such as the electrical characteristics of the charging gun, as well as the surface chemistry and electrical resistivity of the particle, particle shape, ambient relative humidity, and the charging time. Powder particles are not perfectly spherical. Rather, they have a gravel-like shape and, as they have a high electrical resistivity, they will not be uniformly charged. Also, the particles spend only a finite amount of time in the charging zone. Therefore, the theoretical maximum charge-to-mass ratio cannot be achieved. Under ideal conditions, smaller particles can have a larger charge-to-mass ratio. However, when they are mixed with larger particles, the field lines will interact more strongly with the larger particles, making it more difficult for the smaller particles to become charged.

Aerodynamics and gravity also come into play. Smaller particles will tend to follow the airflow lines across the part while the larger particles, having a greater inertia, will generally move in the direction toward the part. Gravity will tend to act more strongly on larger particles, pulling them downward. While some of the coarser particles will bounce or fall off the target and be pulled downward by gravity, the finer particles will tend to overspray the target.

Eq. (3) shows how long it will take a particle to reach its maximum charge.

$$q(t) = \frac{q_{max}}{1 + \tau/t} \tag{3}$$

Here, τ is the time for a particle to reach half of its maximum charge and was calculated to be about 0.1 sec for a typical corona discharge.[5] After 1 sec, 90% of the maximum theoretical charge will be realized. Particles having a lower velocity will spend more time in the charging zone and will have a higher charge-to-mass ratio. Reducing the powder feed rate and its velocity will allow the powder cloud to be more diffuse and spend more time in the charging zone, improving powder charging that will result in more powder being deposited onto the target.

Corona charging is very widely used due to its ability to charge a wide range of powders at fairly high throughput. This charging method is relatively insensitive to ambient humidity, plus the equipment is robust for a wide variety of manufacturing processes.

Back-Ionization

Increasing the voltage on the electrode tends to increase transfer efficiency, up to a point. As voltage is increased, back-ionization also increases until powder deposition is reduced. This is due to the fact that less than 1% of the total number of ions generated contributes to charging the powder and more than 99% of the ions are deposited on the surface of the part and any other nearby grounds.[6] Ions are much smaller and have a much faster mobility than powder particles. As powder is sprayed onto the part, the particles adhere where little or no powder has previously been deposited. Back-ionization begins with as little as a single monolayer of powder.[7] As powder is deposited onto a surface, electrical discharges in the layer can be observed by using an image intensifier. Air that is trapped in the powder layer is highly stressed, which causes it to break down into both positive and negative ions at these discharge points. The positive ions move away from the part and "back" toward the negative electrode, reducing the efficiency of powder deposition onto the part by neutralizing the oncoming negatively charged powder. The effects of back-ionization can usually be seen in the uncured powder as "starring" or "orange peel" in the finished coating. In severe cases, back-ionization can cause craters and pinholes down to the substrate.

Equipment manufacturers have found ways to deal with this problem. One method simply uses a grounded metal rod or ring, known as an "ion collector," positioned behind the electrode but at a distance closer to the electrode than the distance between the electrode and the part. The idea is to create a stronger electric field between the electrode and the ion collector than between the electrode and grounded part, so that fewer ions will be attracted to the grounded part.[8] Internal charging is another way to reduce the effects of back-ionization. In this case, the ground electrode is located inside the powder applicator, instead of outside, which neutralizes a significant portion of free ions during the charging process.

A recent study was done to examine the effectiveness of ion collectors in reducing ion current and its effect on surface appearance.[9] A smooth ring and a ring with 11 sharp points for concentrating the electric field were investigated. The authors found that both the charge-to-mass ratio and transfer efficiency correlate directly with an increase in ion current on the target. That is, a reduction in ion current responsible for back-ionization occurs at the expense of transfer efficiency. Interestingly, they also found an improvement in surface appearance with the smooth ring that collected fewer ions. Therefore, care needs to be taken when using ion collectors to reduce back-ionization while optimizing transfer efficiency and surface appearance.

Another technique to reduce back-ionization is to control the gun current. Controlling gun current helps reduce excessive free ions by automatically reducing the electrode voltage when the gun-to-part distance decreases. This method reduces back-ionization and the Faraday Cage effect (see below).

Free ions also affect the ability to apply a second layer of powder on top of a coating that has already been cured. This application is known as "recoating." When charged powder and ions are applied to this surface, the charge builds rapidly as charge cannot bleed off. Therefore, minimizing the amount of free ions makes recoating easier to achieve.

The Faraday Cage Effect

As discussed earlier, electric field lines concentrate on sharp points or edges. Figure 14.3 shows what happens to the electric field lines when the part has a recessed area or a channel. Here, the electric field lines concentrate on the edge of the corner. This means that the powder and ions will deposit more rapidly on these areas and will tend to go into

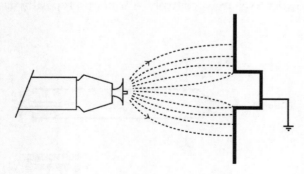

FIGURE 14.3
Schematic of a corona gun with an electrode at high voltage in the presence of a part containing a recess.

back-ionization. Furthermore, as there are very few field lines inside the channel, the powder has difficulty penetrating there. Back-ionization at the edge of the recess will also neutralize incoming charged particles and prevent them from depositing inside the corners. The use of ion traps, controlling gun current, or an internal charging gun can be used to overcome the Faraday Cage effect.

In addition to the Faraday Cage effect, aerodynamic effects do not favor spraying inside a corner. Airflow considerations based on part geometry must also be taken into account to deliver the highly charged powder inside a corner or a channel, while not using so much air that the powder is prevented from depositing.

Internal Charging

Corona guns have the disadvantage of the Faraday Cage effect from strong electric fields and back-ionization due to excess ions. As seen later, tribocharging largely overcomes the problem of the Faraday Cage effect and back-ionization, but is sensitive to the chemistry of the powder and the extreme ambient relative humidity conditions. Internal charging guns can minimize these disadvantages by putting both the high-voltage electrode and the ground inside the gun.[8] Figure 14.4 is a schematic of an internal charging gun. As both the charging electrode and the ground are inside the gun, few electric field lines will escape outside the gun minimizing back-ionization resulting in a smoother finish, similar to a tribogun. Furthermore, the powder does not have to be specially formulated as it does for the tribogun. External electrodes are used to improve transfer efficiency. Powder flow rates are similar to a conventional corona gun.[10] A patent on an internal charging gun[11] refers to a chamber used to slow down and swirl the powder to increase the time of the particles in the charging zone.

Tribocharging

This method charges powder as a result of frictional contact with the inside surfaces of a tribogun. No source of high voltage is required. Transfer of charge occurs when two materials that differ in their ability to accept electrons are brought into close contact with one another. Materials that have the ability to accept electrons are called electronegative, while those that can donate electrons are called electropositive. The internal surfaces of a tribocharging gun are usually made of a material that is very electronegative, such as polytetrafluoroethylene or Teflon®.[12] This material accepts electrons from powder particles, causing the powder to charge positively. A tribogun typically uses additional air to

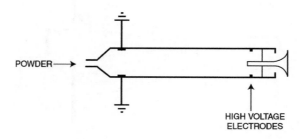

FIGURE 14.4
Schematic of an internal charging gun.

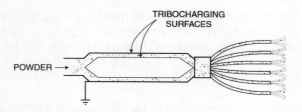

FIGURE 14.5
Schematic of a tribocharging gun.

help charge the powder. The path inside the gun is designed to ensure multiple contacts between the powder and the internal walls.

Figure 14.5 shows a schematic of a tribocharging gun. As negative charge builds up inside the gun, a means is needed to bleed off this excess charge. This is accomplished by grounding the gun, sometimes through a microammeter to monitor or record the current. The current-to-ground is an indication of the amount of charging that occurs inside the gun. If the powder flow rate is constant, the charge-to-mass ratio can be determined by dividing the current (in µA) by the powder flow rate (in g/sec) to obtain µC/g.

Because triboguns have no charging electrode, the Faraday Cage effect is greatly reduced. This allows the powder to penetrate recesses, corners, and deep channels. In addition, as triboguns do not generate free ions, as corona guns do, they can coat parts that are not highly conductive more easily than a corona gun. This means that triboguns are better at recoating parts, as they do not build up charge as fast as corona guns. The resulting surface appearance is usually smoother and has considerably less back-ionization at a given film thickness. Triboguns, therefore, can build thicker films than corona guns before back-ionization becomes a problem.

The amount of charge on powder is determined by many factors including the composition of the powder, particle size, velocity of the powder on contact, and ambient relative humidity. Some powder formulations will tribocharge better than the others depending on their ability of being charged positive. Most polyamide and epoxy powders will tribocharge well. Some epoxy/polyesters, polyurethanes, and TGIC–polyester powders can also be tribocharged. Powder formulators can adjust the tribocharging properties of the powder through selection of the resin and the use of additives.

Another factor that controls charging efficiency is particle size distribution, which is discussed in more detail later. Larger particles will impact the tribosurface with greater energy, resulting in better charging. If the level of fine particles is too high, they will tend to build up in the reclaim system as they would not charge as effectively as larger particles. Typically, tribocharging powders have a median particle size between 30 and 45 µ, slightly greater than corona or internal charging guns. The level of fine particles (under 10 µ) should be minimized to enhance the charging process and also to minimize problems associated with fluidization.

Recently, a new tribocharging method has been developed that charges the powder negatively[13] instead of positively, as conventional tribocharging. This method relies on air jets to force the powder particles to contact the internal walls. In this case, the tribocharging surface gives up an electron to the powder. This method has the same advantages as conventional tribocharging. In addition, this charging gun can be used together with a conventional corona gun for touch-up of difficult-to-coat areas, because the charged powder has the same negative polarity as a corona gun. However, it also suffers from the same disadvantages as conventional tribocharging in that the charging is sensitive to the type of powders that can be used. As with conventional tribocharging, most powder suppliers have learned how to optimize the powder formulation for negative tribocharging.

Nozzles

After the powder becomes charged, whether by corona, internal charging, or tribocharging, the mixture of air and powder particles is shaped into a spray pattern. The air velocity and the nozzle influence the spray pattern. Because corona charging uses a high voltage electrode, an electric field between the gun and the grounded part causes the charged powder and ions to follow the field lines to the part. If the part geometry is complex, electric field lines will not penetrate inside recesses and corners making it difficult for powder to penetrate these areas due to the Faraday Cage effect described earlier.

Flat spray (also called fan spray) nozzles can partially overcome this effect by forcing powder into recesses. However, if the velocity is too high, it can also blow powder off the part. Figure 14.6 shows a perspective of a flat spray nozzle. The electrode is slightly recessed inside the nozzle and the powder emerging from the slot somewhat shields the electrode from the grounded workpiece. These factors combine to reduce the electric field between the gun and the part. This nozzle generates a directed fan spray pattern that has a pattern width of approximately 15 cm (6 in.) to more than 30 cm (12 in.). The flat spray nozzle is ideal for large flat areas. This nozzle is not recommended for wire goods because of its high powder velocity.

Conical nozzles use deflectors to generate a cone-shaped pattern as shown in the schematic in Figure 14.7. Pattern widths can vary from as small as about 2.5 cm (1 in.) to approximately 60 cm (2 ft) in diameter, depending on the size and the radius of the deflector. The size of the pattern also depends on the powder flow rate. Conical nozzles generate a "softer" spray pattern because the deflector creates a partial vacuum that "pulls" the inner part of the cone pattern toward itself. This reduces the forward velocity of the powder.

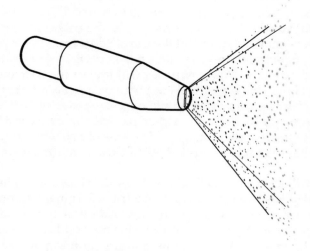

FIGURE 14.6
Schematic of a flat spray (fan spray) nozzle.

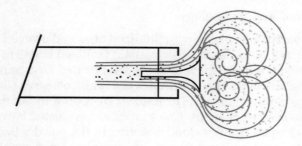

FIGURE 14.7
Schematic of a conical nozzle.

Conical nozzles are usually used in manual coating operations, where they are used for large flat parts and complex shapes.

Electrostatic forces dominate when the charged powder comes within about 10 mm (<0.5 in.) of the grounded part. As the conical nozzle has a lower forward velocity, the powder cloud tends to hover in front of the target, allowing the charged powder particles to be attracted to the grounded surface. In addition, the powder from a conical nozzle has a longer residence time in the charging zone allowing it to pick up more charge. In contrast, the flat spray nozzle is more directional, allowing aerodynamic forces to dominate making it easier for the powder to penetrate recessed areas. For both nozzles, the best transfer efficiency is achieved by good dispersion of the powder in air, reducing the velocity and the powder flow rate as low as possible for the specific application. Some powder guns swirl the air to control the velocity of powder, which improves particle charging.

Nozzles used for tribocharging can be very flexible, as tribocharging does not depend on a high voltage electrode at the front of the gun. The powder exiting the gun can be split into different streams and shaped in a multitude of ways allowing the powder to conform to the part geometry. The velocity of the powder from each "finger" of the nozzle is relatively low, allowing the gun to be closer to the part resulting in uniform powder deposition. Parts that can be coated using this technique are, for example, radiators, transformers, bicycles frames, or wire goods.

Bells and Discs

Powder can also be applied by bells or discs, similar to that used in the liquid paint industry. A powder bell uses a rotating turbine that rotates an enclosed bell head. The powder is delivered to the head where centrifugal forces disperse and atomize the powder. The particles are charged by corona either from the high voltage at the edge of the bell or from the external electrodes. Shaping air around the head controls the spray pattern and the forward velocity of the particles. The powder bell delivers high powder output uniformly over a wide flat area, making it ideal for the automotive industry.

Discs use large conical-type deflectors usually mounted in the vertical downward position so that powder is deflected in a horizontal position in a 360°radius. An omega (Ω) loop conveyor system normally surrounds the powder disc that is usually mounted on an oscillator or reciprocator for vertical up-and-down motion. The powder can be charged by a corona edge or by tribocharging guns. Typical applications for the powder disc are vertical extrusions or wire goods.

Electrostatic Fluidized Bed Technology

Powder application technology began with fluidized beds and evolved to electrostatic fluidized beds that are still in use today. Electrostatic fluidized beds are a method of powder application that does not use guns to apply the powder. Dry, compressed air passes through the charging media that has electrodes at a high voltage as shown in Figure 14.8. This causes the air to become ionized. The ionized air is used to fluidize the powder and causes the powder to become charged. An electric field is created between the grounded part above and the charged powder cloud resulting in the powder being attracted to the part. As the grounded part becomes coated, additional charged powder will have difficulty in adhering to the coated surface due to back-ionization discussed earlier. The film thickness is controlled by voltage and exposure time.

Powder Recycling

Particle Size

Any discussion of recycling powder must include information about particle size. As powders are similar to gravel-shaped particles, their size is not easily defined. There are many ways to define particle size, as well as different techniques that can give different results. An excellent review of this subject is presented by Rawle.[14] Low angle laser light scattering or commonly referred to as "light scattering" or "laser diffraction" is used extensively in the powder coating industry. Assumptions used in the older instruments that use the

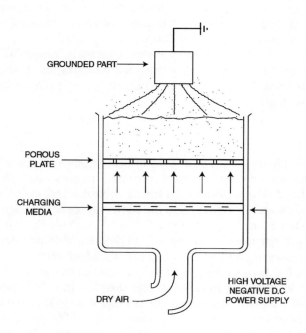

FIGURE 14.8
Schematic of an electrostatic fluidized bed.

Fraunhofer approximation can make it difficult to compare the results between different instruments; the algorithms used to calculate particle size can vary depending on the type and model instrument that is used. Recently, particle size instruments have been developed that are based on Mie scattering theory, which gives more accurate results, as long as the refractive index of the powder and medium are known.

Laser diffraction methods generate a volume mean distribution. For laser light scattering techniques, the median particle size, commonly referred to as $D(v, 0.5)$, is defined as the value that divides half of the volume distribution, i.e., 50% of the distribution by volume is above this value and 50% is below this value. It is also common to define the volume percentage that is either above or below certain particle sizes. For example, particles below 10 µ agglomerate due to van der Waals forces causing fluidization and pumping problems. Therefore, it would be important to know the volume percentage of particles less than 10 µ. Coarse particles can also be a problem, especially when trying to obtain film uniformity; so it would be prudent to know the volume percentage of particles greater than, for example, 100 µ. Knowledge of these percentages helps to define the particle size distribution used.

Powder Recovery

Figure 14.9 shows a simplified schematic of how powder is recycled in a system. After the powder is charged, the mixture of air and powder particles is sprayed onto a part. The powder that does not adhere to the part is contained in the spray booth and can be recycled. The oversprayed powder and the air mixture are drawn into the collector, where either a cartridge filter system or a cyclone (not shown) is used to separate the powder from the air. The collected powder is then transferred to the feed hopper where it is sieved (not shown) and remixed with virgin powder and pumped to the guns. Sieving the powder removes foreign particulates and conditions the powder prior to spraying. The particle size distribution in a cartridge filter system usually shifts toward finer particles. In a cyclone system the fines are scalped off, shifting the particle size distribution to the coarser side.

A model was developed that accurately predicts how particle size changes in a recovery system.[15] The basis of the model is that each particle size in a distribution has a certain

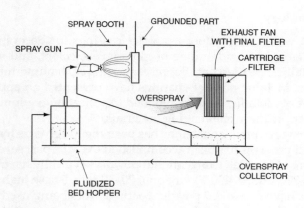

FIGURE 14.9
Schematic of a powder recycling system.

probability of depositing onto the part. Probability factors for each particle size can be determined by comparing the particle size distribution of virgin powder to the powder that has been oversprayed one time. Factors calculated for one cycle are used to calculate the particle size distribution of subsequent cycles. In addition to these factors, a transfer efficiency value is needed to determine how much virgin powder needs to be added to maintain a constant level in the system. The particle size results calculated using this model were in good agreement with those measured from a production run of primer surfacer for exterior vehicle application. The model also showed that the median particle size in that system achieved steady state after only four complete cycles.

Meeting New Challenges

New Equipment Technology

Recent developments in equipment technology are addressing some of the challenges of powder coatings such as fast color change and improved transfer efficiency. New powder feed technology has been developed[16,17] that allows greater powder loading at slower powder velocity to the spray guns using much less compressed air than a standard venturi pump. The advantages of this new technology are reduced powder consumption for better transfer efficiency using less energy, gentler transport of powder with less wear on the powder and the equipment, constant high delivery of powder over longer distances, and faster color change.

Fast color change developments have recently been made that utilize a two-canister system to apply color-specific powders for automotive applications.[18] While one canister is used for powder coating, the other one is getting ready for the next job. This allows a continuous operation with 8 sec claimed between jobs. Another recent development[19] consists of a color-changing manifold that uses automated valves to control color change. Powders from the different feed hoppers are attached to the manifold. The powder of choice is fed into the manifold that consists of a central channel. Air is used to purge the hose and block between colors, which allows for a color change of usually less than 8 sec.[10]

New Powder Technology

Much development work has produced powder coatings for decorative niche markets. They are now available in a wide range of gloss levels, colors, and specialized surface effects such as metallics, pearlescent, fluorescent, antique, hammertone, textured, wrinkled, and clearcoats. Metallic powder coatings have presented an application challenge, because some types of metallics are dry blended. Application equipment manufacturers have learnt how to apply these challenging materials.[20]

A major disadvantage of powder coatings has been their high cure temperature requirement. A great deal of progress has been accomplished over the past several years to reduce the cure temperature of powder. Conventional powders typically have cure temperatures between 160°C and 220°C (320–430°F) for about 20–30 min. These high cure temperatures and long cure times have prevented powder coatings from being used for plastics, wood, and medium-density fiberboard products, or metals with heat-sensitive parts such as low melting solder joints. Recent developments in curing technology have produced powders that can be melted at much lower temperatures of 100–140°C (212–285°F) for powder flow-out followed by curing with UV light.[21] Recent advances also include antimicrobial and high-temperature resistant coatings, improvements in weatherability, resistance to loss of gloss, improved mar- and scratch-resistance, and superdurable coatings. Nanotechnology will gradually play an increasing role as new powder coating compositions are developed having improved properties.[22]

Conclusions

An overview of powder coating application technology that includes the basics of powder charging has been described. From early developments, powder coating and powder application equipment technology have also been examined. Application methods have made great strides from its early beginnings using fluidized bed technology to provide thick coatings for corrosion protection and insulation. Today, the emphasis is on faster color change, higher transfer efficiency, and in some cases spray-to-waste. Powder coatings have evolved into the decorative markets that include specialty surface textures and appearances, antimicrobial powders, scratch-resistant, heat-resistant and low-cure temperature coatings. Application equipment has kept pace with new powder introductions. Improvements in powder coatings and application technology have steadily eliminated many of the drawbacks of using powder coatings. It is clear that equipment technology, as well as powder coating technology will continue to evolve in new directions creating new opportunities.

Acknowledgment

My sincere appreciation to Leah Lader for beautifully creating all the line drawings for the figures.

References

1. Harris, S., Ed., *Focus on Powder Coatings, Chemquest Powder Market Survey*. Elsevier Advanced Technology; 2004; Sept, 6–7.
2. Liberto, N.P., Ed., *Powder Coating—The Complete Finisher's Handbook*, 2nd Ed.; The Powder Coating Institute: Alexandria, VA, 1999.
3. Cross, J.A. Electrification of solids and liquids. In *Electrostatics—Principles, Problems, and Applications*. Adam Hilger by IOP Publishing Ltd.: Bristol, England, 1987; 47.
4. Cross, J.A. Electrification of solids and liquids. In *Electrostatics—Principles, Problems, and Applications*. Adam Hilger by IOP Publishing Ltd.: Bristol, England, 1987; 51–52.
5. Cross, J.A. Electrification of solids and liquids. In *Electrostatics—Principles, Problems, and Applications*. Adam Hilger by IOP Publishing Ltd.: Bristol, England, 1987; 52.

6. Hughes, J.F. Electrostatic effects. In *Electrostatic Particle Charging—Industrial and Health Care Applications.* John Wiley & Sons, Inc.: New York, 1997; 30.
7. Hughes, J.F., Ting, Y.C. Surface disruption phenomena in electrostatically deposited powder layers. In *Proceedings of the 12th Annual Meeting of the IEEE.* Industrial Applications Society: Los Angeles, CA, 1977; IEEE; 35-G, 906–909.
8. Campbell, D.H. Powder charging techniques. In *Symposium on Powder Coatings.* Birmingham, England, Apr 4–5, 1995; Paint Research Association, 1995.
9. Biris, A.S., Mazumder, M.K., Sims, R.A., Yurteri, C.U., Farmer, S., Snodgrass, J. The effect of ring electrodes attachment to a corona gun on control of free ion concentration and back corona for improving powder paint appearance. IEEE Trans. Ind. Appl. **2003**, *39* (6)1614–1621.
10. Milojevic, K. Multi-color powder application with internal charging guns: spray pattern and thickness uniformity control. In *21st International Conference on Automobile Body Finishing SURCAR*, Cannes, June 26–27, 2003; 165–174.
11. Börner, G., Nienburg, H.-C., Wittmann, J., Böhme, H. Powder-Spraying Appliance. US Patent 6,254,684, July 3, 2001.
12. Teflon® is a registered trademark of DuPont.
13. Rehman, W.R., Lader, H.J., Messerly, J.W. Unipolarity Powder Coating Systems Including Improved Tribocharging and Corona Guns. US Patent 6,645,300, November 11, 2003.
14. Rawle, A. The importance of particle sizing to the coatings industry. Part 1: particle size measurement. Adv. Colour Sci. Technol. **2002**, *5* (1), 1–12.
15. Lader, H.J. Particle size modeling of powder paint in a recovery system. In *Powder Coating '94 Proceedings;* Powder Coating Institute, 1994; 254–267.
16. Moser, J. Soft spray: a new powder coating technology. JOT/Oberflaeche **2003**, *43* (10), 54.
17. Nordson signs licensing agreement with Borger. Finishing 2004, February.
18. GM honors top innovators of 2003. Powder Coating **2004**, *15* (6), 10.
19. Attinoto, R.A., Ciarelli, G.J., Koster, M.L., Milojevic, D.K., Rennie, C.M. Powder Paint Color Changer. US Patent 6,589,342, July 8, 2003.
20. Binder, J., Lader, H., Schroeder, J. Making the world a brighter place. Product Finishing 2003, (April), 64–68.
21. Udding-Louwrier, S., Witte, F.M., Sjoerd de Jong, E. Radiation curable powder coating system. Product Finishing (April), **1997**, 26–30; Buysens, K., Tielemans, M., Randoux, T. Radiation-curable coatings: a variety of technologies for a variety of applications. Surf. Coat. Int.—Part A **2003**, *2003* (05) 179–186; Christianson, R. Powder-coated wood OEM has bold ambitions. Wood and Wood Products 2004, July 1.
22. Anderson, L.G., Barkac, K.A., DeSaw, S.A., Hartman, M.E., Hayes, D.E., Hockswender, T.R., Kuster, K.L., Nakajima, M., Olson, K.G., Sadvary, R.J., Simpson, D.A., Tyebjee, S., Truman, F.W. Coating Compositions Having Improved Scratch Resistance, Coated Substrates and Methods Related Thereto. US Patent 6,803,408, October 12, 2004.

15

Supercritical Carbon Dioxide Processing of Polymer–Clay Nanocomposites

Matthew J. Factor and Sunggyu Lee

CONTENTS

Polymer–Clay Nanocomposite Background 235
   Clay 236
      Clay Configurations 237
   Polymer–Clay Compatibility 237
Polymer–Clay Nanocomposite Processing with Supercritical Carbon Dioxide (ScCO$_2$) 238
   ScCO$_2$ Background 238
   Effect of ScCO$_2$ on Polymers 238
      Polymer Solubility in ScCO$_2$ 238
      ScCO$_2$ as a Solvent 240
      Plasticization by ScCO$_2$ 240
      Diffusion in ScCO$_2$ 240
      Applications 241
   Effect of ScCO$_2$ on Clay 241
   Effect of ScCO$_2$ on Polymer–Clay Nanocomposites 242
      Clay Intercalation in ScCO$_2$ 242
      Clay Infusion/Impregnation in ScCO$_2$ 242
Methods of Polymer–Clay Nanocomposite Creation with ScCO$_2$ 243
   In Situ Polymerization 243
   In Situ Polymerization with ScCO$_2$ 244
   Melt-Mixing 244
   Melt-Mixing with ScCO$_2$ 244
      Melt-Mixing with ScCO$_2$: Foaming 245
   Solvent Casting 246
   Vessel (Batch) Infusion/Dispersion in ScCO$_2$ 246
      Vessel (Batch) Infusion/Dispersion in ScCO$_2$: Foaming 247
References 247

Polymer–Clay Nanocomposite Background

Polymer–clay nanocomposites (PCNs) were first developed by Toyota Central Research and Development Laboratories in 1986. The nanocomposite consisted of nylon-6 and a modified montmorillonite clay intercalated with amino acids. The addition of clay exhibited a significant effect on the mechanical, thermal, and gas-barrier properties of the nylon.[1,2] Due to the substantial enhancement of the polymer properties by adding such a common material, PCNs have become of great interest to researchers.

Clay

Clay is abundant, inexpensive, and environmentally benign. In nature, clays typically exist as agglomerations of stacked platelets that exhibit thixotropic characteristics, manifesting a reduction in viscosity under applied stress. The thixotropic nature of clays prompted their original use in 1940s as additives to help modify the flow characteristics of products such as paints, oils, greases, cosmetics, and printing inks.[3]

Many types of clay exist, but sodium montmorillonite (MMT) is the most popularly used for polymer–clay composites mainly due to its abundance.[3,4] An individual MMT clay platelet can have an aspect ratio of 1500:1 (commercial clay typically has a moderate aspect ratio from 10 to 300), allowing a high surface area in contact with the polymer matrix. This permits greater property enhancement with lower clay loadings and increased property transfer from the clay to polymer matrix. Dispersion of clay as individual platelets throughout the polymer is difficult to achieve due to strong van der Waals forces holding platelets together in conjunction with the incompatibility of the hydrophilic clay with the organophilic (hydrophobic) polymer matrix, giving way to clay agglomeration. To weaken the forces holding platelets together and reduce agglomeration, clay is often converted into an organoclay by replacing the sodium cations (Na^+) in the interlamellar gallery (region between clay platelets) with organic cations, typically alkylammonium salts.[3,4] The organic cations weaken van der Waals forces by pushing the platelets further apart and reduce agglomeration by making the clay more organophilic, increasing compatibility with the polymer. As illustrated in Figure 15.1, expansion of interlamellar galleries is

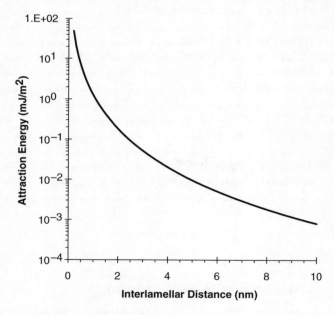

FIGURE 15.1
Energy of attraction for van der Waals forces as a function of the interlamellar distance between two MMT clay platelets. (Source: From equation in Ngo.[5])

an effective route toward exfoliation since the attraction energy of van der Waals forces decreases as the distance between platelets increases. The MMT interlayer spacing (~1 nm) can be detected by X-ray diffraction (XRD) analysis as the d_{001}-spacing.

Clay Configurations

Clay can exhibit three primary configurations within a polymer matrix: conventional, intercalated, and exfoliated. The most easily achieved configuration is referred to as a conventional configuration and is the typical result of dispersing clay throughout a molten polymer without any compatibilizing compounds. A conventional composite contains stacks of clay that are agglomerated in a polymer matrix, interacting with the matrix on a micrometer scale (size of agglomerated clay particles). In contrast, nanocomposites contain clay that is intercalated or exfoliated which interact with the polymer matrix via their platelets with a dimension on the nanometer scale (thickness of clay platelet). The interaction of clay with a polymer on the nanometer scale permits lower clay loadings to achieve the same or better property enhancements as compared to micrometer-sized agglomerated clay. Clay is intercalated when polymer chains are present between individual clay platelets, still retaining their natural stacked configuration. In this configuration, the clay stack as a whole imparts its strength on the polymer matrix via the intercalated and surrounding polymer chains. When the clay stacks are completely separated into individual platelets, an exfoliated structure is attained. Due to their high aspect ratio, each clay platelet has a large surface area capable of allowing greater stress transfer from the host polymer to the filler clay when clay is exfoliated, enhancing the nanocomposite's properties. Nanocomposites containing exfoliated clays can also exhibit different properties depending on whether the clay platelets are in a random or aligned orientation. Figure 15.2 displays the clay configurations within a polymer matrix. Additionally, uniform platelet distribution is optimal to ensuring the property enhancement afforded by the clay is evenly spread across the polymer, so the entire polymer is used to absorb external stress and it is not focused in one area, creating a weak point for initial structure failure. Lower clay loadings afforded by well-dispersed and delaminated clays permit the same processing methods to be used for PCNs as would be used for the pure polymer.

Polymer–Clay Compatibility

Compatibility between the polymer matrix and clay filler plays a significant role in the properties of the resultant composite due to its ability to affect the clay's configuration.

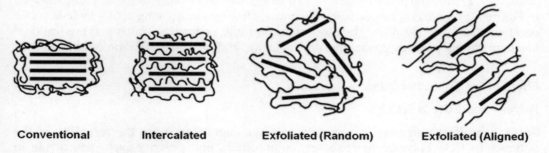

| Conventional | Intercalated | Exfoliated (Random) | Exfoliated (Aligned) |

FIGURE 15.2
Clay configurations within a polymer matrix.

Clay that is incompatible with the polymer matrix will favor agglomeration, resulting in a conventional clay structure.[6] In addition to increasing interlamellar gallery distance, use of clay modifications have been motivated by the intent of increasing the material compatibility between clay and polymer for use in PCNs. Typically, clays are modified by organic cations (normally alkylammonium salts) exchanged with metal cations (Na^+) intercalated between clay platelets and are commonly referred to as organoclays.[3,4] By compatibilizing the ions between clay platelets with the polymer matrix, the polymer chains are more likely to enter the interlamellar galleries of the clay to create an intercalated structure. In many instances of PCN creation, intercalation or partial intercalation/exfoliation is the optimal configuration obtained with the use of organoclays. The absence of complete exfoliation could be due to the still large difference between the hydrophobicity of polymers and the hydrophilicity of the organoclay. To reconcile this difference, further compatibilization can be employed in addition to the modifications offered by the organoclay. Grafting maleic anhydride (MA) to non-polar polymers (e.g., polyolefins) adds a hydrophilic (polar) component to a hydrophobic polymer that is capable of interacting with the oxygen groups on the surface of the hydrophilic clay platelet. The hydrophilic nature of MA aids the polymer in entering the clay gallery, while the hydrophobic polymer pushes the platelets apart into an exfoliated structure. The hydrophobic nature of the polymer is important to exfoliating the clay since a polymer that is hydrophilic will be attracted to the clay platelets, holding them together as it enters the clay gallery to produce an intercalated morphology.[7,8] However, it is desirable to minimize the use of compatibilizers as they can lead to a prohibitive increase in cost.

Polymer–Clay Nanocomposite Processing with Supercritical Carbon Dioxide ($ScCO_2$)

$ScCO_2$ Background

A novel technique to create improved PCNs involves incorporating CO_2 at supercritical conditions into current processing methods. $scCO_2$ possesses properties of both gas and liquid with high diffusion rates and dissolving characteristics, respectively. Altering these density-driven properties can be accomplished by adjusting temperature and pressure, making it a tunable fluid. In addition, CO_2 is inexpensive, non-toxic, non-combustible, and chemically stable. CO_2 reaches its critical point at 31.06°C and 7.38 MPa, exhibiting a critical density of about 0.466 g/cm^3. The phase diagram and pressure–density isotherms for CO_2 are displayed in Figures 15.3 and 15.4, respectively. Current applications employing $scCO_2$ include coffee decaffeination, dry cleaning, chemical extraction and separation, fluoropolymerization,[9] low-temperature polymer processing, and biological and pharmaceutical processing.

Effect of $ScCO_2$ on Polymers

Polymer Solubility in $ScCO_2$

$ScCO_2$ can affect polymers in a variety of ways depending upon the solubility of the polymer in CO_2. Polymer morphology, composition, and polarity each play a role in determining polymer solubility in CO_2. Considering morphology, CO_2 is absorbed only by the amorphous regions of the polymer,[11–13] thus CO_2 can more substantially modify

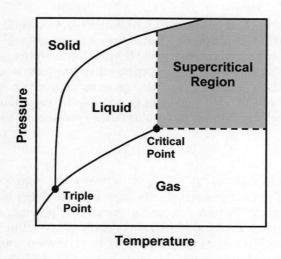

FIGURE 15.3
Carbon dioxide phase diagram.

amorphous polymers in comparison to crystalline polymers. Shieh et al.[12,13] observed that CO_2 absorption for select amorphous polymers led to a maximum weight increase more than four times greater than for select crystalline polymers. The elemental composition of a polymer influences its solubility as fluorine [e.g., polyvinylidenefluoride (PVDF)]- and siloxane [e.g., polydimethylsiloxane (PDMS)]-containing polymers can absorb more CO_2 due to their CO_2-philicity.[12–16] To take advantage of this, fluorination of polymers has been undertaken to make them soluble in CO_2 at low pressures.[17–20] Polymer polarity also contributes to its dissolution in CO_2 since higher CO_2 dissolution is possible with compounds containing groups exhibiting polarity[13,14,21] including –CN, –COOR, Ph–O–, and S=O.[13] Polar groups of the polymer interact favorably with the alternating polarity of

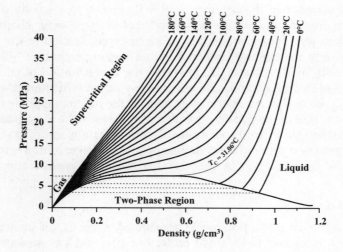

FIGURE 15.4
Carbon dioxide density–pressure isotherms. (From Peng–Robinson equation of state with Mathias-type volume shift in Boyle and Carroll.[10])

the considerable quadrupole moment of CO_2.[17] Furthermore, position of the polar groups influences the amount of CO_2 absorption possible as a result of steric hindrance.[13,18] Shieh et al. observed that poly(methyl methacrylate) (PMMA), containing ester groups in its side chain, exhibited a higher degree of CO_2 absorption than poly(ethylene terephthalate) modified with glycol (PETG), containing ester groups in its main chain.[13] In addition, higher equilibrium sorption of CO_2 into polymers is possible by employing higher pressures, lower temperatures,[11,22,23] and lower molecular weight polymers.[18,23]

ScCO₂ as a Solvent

Due to its properties of dissolution, $scCO_2$ can possess the high solvating ability of an organic liquid solvent in the presence of polymers. CO_2 solubility can be greater than 30 wt.% in certain polymers,[24] which can lead to significant polymer swelling, depending on a variety of factors such as chemical makeup, degree of crystallinity, molecular weight, sample dimensions, etc.[12] Occurring simultaneously to create a single phase, as the CO_2 dissolves in the polymer, the polymer dissolves in the CO_2. CO_2 dissolution in the polymer reduces the polymer's viscosity, permitting processing at moderate temperatures. The small molecules of $scCO_2$ afford a low surface tension and allow it to penetrate smaller spaces than the larger molecules of liquid solvents. Furthermore, separation of $scCO_2$ from processed materials is easily achieved by depressurization, facilitating recycling. Making use of the attributes of $scCO_2$ as a solvent, fluoroalkyl-substituted polythiophene-semiconducting polymers were produced in $scCO_2$ and were observed to exhibit comparable properties to those made in the organic solvent chloroform.[15]

Plasticization by ScCO₂

By dissolving in a polymer, $scCO_2$ is capable of reversibly plasticizing the polymer at a low temperature, i.e., avoiding thermal degradation.[21,26] The phenomenon of plasticization occurs when the glass transition temperature of the polymer is reduced below the processing temperature.[24] In addition, plasticization is accompanied by a reduction in the melting temperature and viscosity of the polymer.[12,13,21,26] Reduction in viscosity primarily occurs by increasing the free volume of the polymer,[21] boosting polymer chain mobility.[11,12,21] The added freedom of polymer chains can promote crystallization of the polymer,[11,12] as evidenced by a rise in the melting temperature and melting enthalpy.[12] Plasticization can also reduce viscosity by dissolved CO_2, lowering the concentration of entwined polymer chains.[21] In terms of thermodynamics, an effective plasticizer will interact with the polymer via intermolecular forces that are on the same scale as the polymer–polymer intermolecular forces, where, as the size of the plasticizer molecule decreases, the plasticizer efficiency will increase.[26] Due to their ability to absorb more CO_2, plasticization occurs to a higher degree in amorphous polymers as compared to crystalline polymers. Polymer impregnation and foaming are some applications that benefit from the phenomenon of plasticization.

Diffusion in ScCO₂

The rate of CO_2 diffusion into a polymer is accelerated when CO_2 is pressurized. The diffusion rate of CO_2 is related to its small molecular size and low surface tension, facilitating its sorption into the polymer to dictate polymer free volume.[11,22] Studies have shown that polymer sorption/desorption of CO_2 follows Fickian diffusion kinetics.[11,22–25] Consequently, CO_2 diffusivity is related to the initial linear slope of a plot of the mass uptake of dissolved CO_2 in the polymer per total CO_2 mass uptake possible [$M(t)/M_\infty$] versus the square root of desorption time divided by the initial polymer thickness ($t^{1/2}/L$). Fick's second law of diffusion (one-dimensional) for changes in concentration with respect to time is written as:

$$\frac{\partial C}{\partial t} = D\frac{\partial^2 C}{\partial x^2}$$ (1)

Using Fick's second law, diffusion for a slab (e.g., film) of thickness L can be estimated as:

$$\frac{M(t)}{M_\infty} = 1 - \frac{8}{\pi^2}\sum_{n=0}^{\infty}\frac{1}{(2n+1)^2}\exp\left(\frac{-D(2n+1)^2\pi^2 t}{L^2}\right)$$ (2)

where $M(t)$ is the mass of the substance diffusing into the polymer at time, t, M_∞ is the mass at equilibrium sorption (after infinite time), and D is the diffusivity.[11,22,23,25] Eq. (2) assumes the polymer thickness remains constant during CO_2 sorption. Although swelling of the polymer can occur from CO_2 sorption, studies have circumvented the issue by measuring polymer swelling to validate the assumption that it is small enough to be negligible[11] or have adjusted treatment conditions to ensure it is small.[22] The rate of diffusion increases with increasing pressure and temperature,[11] relying on solvent density and polymer plasticization that increase respectively with increasing pressure and temperature.[11] High diffusion rates in conjunction with its ability for polymer swelling enable compressed CO_2 to deposit additives within a polymer matrix. Berens et al.[24] reported that the sorption rate of dimethyl phthalate (DMP) into poly(vinyl chloride) (PVC) in pressurized CO_2 could be six orders of magnitude higher than in PVC without CO_2 treatment.

Applications

Applications of scCO₂ use with polymers depend on the degree to which a polymer is affected by scCO₂. Polymers significantly affected by scCO₂ can experience extensive polymer swelling and CO_2 dissolution, which can be beneficial for processes involving extraction,[12,27,28] impregnation,[23,24,28] foaming,[12,13,21,26,29–31] blending,[21,26] polymerization,[15,21,26,32] fiber production,[28] and particle production.[21,26] Polymers that undergo little to no change in scCO₂ have uses in applications involving the polymer surface, such as cleaning,[12] modifying,[12,13,15,21,26] and coating.[12]

Effect of ScCO₂ on Clay

Clay can also benefit from scCO₂ processing as the distance between platelets can be expanded in this environment, resulting in intercalation or delamination. The mechanism proceeds by CO_2 molecules, reduced in size from pressurization, entering the interlamellar galleries of the clay. Once an adequate soak period has elapsed to allow sufficient time for CO_2 to enter the clay galleries and swell the organic cations, the system is catastrophically depressurized. Quickly expanding, the CO_2 molecules separate the clay platelets to permit intercalation or platelet delamination.[16,33,34] As mentioned above, expansion of interlamellar galleries can prompt delamination since van der Waals forces holding platelets together progressively diminish as platelets are separated (Figure 15.1).

Organoclays Cloisite 10A (C10A),[33] Cloisite 93A (C93A),[16] and I.30P[16] have undergone varying degrees of platelet delamination due to catastrophic depressurization in $scCO_2$. Horsch et al.[16] attributed the delamination of C93A and I.30P to the CO_2-philicity of the clay modification. Cloisite 93A, exhibiting high CO_2-philicity, and I.30P, exhibiting moderate CO_2-philicity, experienced respective high and moderate degrees of delamination. As postulated by Manitiu et al.[33] delamination occurs predominantly to the exterior clay layers of the clay tactoid, whereas interior layers only undergo a disruption of their parallel registry (reason for loss of the coherent XRD diffraction peak) which they term "disordered intercalated." They noted that although some parallel registry was regained after the clay was incorporated into the polymer (as evidenced by the reemergence of the XRD diffraction peak), significant rheological improvement (hence dispersion) was observed as compared to samples created without clay pretreatment in $scCO_2$. In addition, $scCO_2$ can promote further exchange of organic cations, which previously were only bound between clay layers by van der Waals interactions, with Na^+ by increasing their mobility via plasticization. Mobilization of the organic cations can give rise to an increase in interlayer spacing due to realignment of organic cations within clay galleries.[35]

Effect of $ScCO_2$ on Polymer–Clay Nanocomposites

Clay Intercalation in $ScCO_2$

Polymer–clay nanocomposites can be improved by intercalating clay with polymer chains via $scCO_2$ processing. The proposed mechanism occurs by the compressed CO_2 particles intercalating between the clay platelets, swelling the organic cations and pushing the layers further apart to allow the polymer chains to enter.[36] This process typically employs a soaking period in which the polymer–clay mixture is given sufficient time for the $scCO_2$ to dissolve the polymer and aid in intercalating the clay. During decompression, the expanding CO_2 molecules have been theorized to push clay layers even further apart to allow more chains to intercalate and possibly cause delamination.[37,38] Upon returning to a subcritical state, the CO_2 diffuses out of the polymer, and the chains remain intercalated between clay layers.

Clay Infusion/Impregnation in $ScCO_2$

The implementation of $scCO_2$-assisted processing for the purpose of infusing (impregnating) and dispersing additives within a polymer matrix has applications in PCN synthesis. Infusion and dispersion are mass transfer processes that benefit from the dissolving and diffusing abilities of $scCO_2$. Dissolving of CO_2 in certain polymers causes swelling and density reduction due to the dilution and dissolution of polymer chains. Plasticization from dissolving increases the mobility of the polymer chains and thus additives within the matrix. The fast diffusion rate of $scCO_2$ lends to quickly impregnating and dispersing additives throughout the polymer. Upon depressurization, the CO_2 quickly diffuses out while the additive is left behind to diffuse out at a much slower rate consistent with ambient conditions.[23,24]

$ScCO_2$ processing has realized the impregnation[39] and improved dispersion[16] of an organoclay and natural clay, respectively. Other cases benefitting from $scCO_2$-assisted infusion and dispersion involve time-release devices and temperature-sensitive particles. The low critical temperature of CO_2 allows temperature-sensitive particles, i.e., biocompatible particles and organically modified clays, to be processed without degradation.

Processing conditions for clay infusion/impregnation in scCO₂. The processing conditions of the $scCO_2$ environment play a large role for infusing clay into a polymer matrix. Slight changes in temperature and pressure of CO_2 in the supercritical region can have a large effect on the fluid density, as shown in Figure 15.4, which dictates the solvating characteristics of the fluid.[27] High temperatures and low pressures facilitate infusion of clay into the polymer matrix, with the contrary being true for extraction processes.[27] Higher temperatures increase plasticization and the rate of diffusion,[11] but care should be taken to remain below the degradation temperatures of the processing materials (e.g., the organic modification of organoclays begins to degrade around 180–200°C).[3,40] It is also worth mentioning that at temperatures well below the degradation and melting points for polymers, certain polymers can deform due to dissolving and reforming in CO_2.[12,13] More extreme processing conditions, like higher temperatures, pressures, and decompression rates, increase the chance of deformation for certain polymers undergoing $scCO_2$ treatment.[12,13] As the pressure increases up to the solubility pressure of the polymer, the polymer viscosity decreases as a result of an increasing amount of dissolved CO_2 plasticizing the polymer. Above the solubility pressure, dissolved CO_2 content remains constant and the polymer viscosity increases due to the hydraulic pressure effect,[41] curbing diffusion of CO_2 molecules and clay into the polymer. In addition, low pressures decrease the density of the fluid, creating a better environment for infusion. Sufficient processing time is also a determinant for clay infusion as CO_2 and clay particles need time to diffuse into the polymer, and the polymer needs time to dissolve, swell, and plasticize from the CO_2. The duration in which the components are within the $scCO_2$ is frequently referred to as the "soak time"; when it is too short, clay infusion into the polymer can be insufficient.

Methods of Polymer–Clay Nanocomposite Creation with ScCO₂

ScCO₂ can be co-opted by PCN production methods to improve processing and nanocomposite properties. Typical techniques to create PCNs consist of in situ polymerization, melt-mixing, and solvent casting.

In Situ Polymerization

In situ polymerization reacts to monomer and initiator in the presence of clay to synthesize PCNs. To begin the process, a solution of dissolved monomer in organic solvent is added to a clay-organic solvent mixture. The dissolved monomer spontaneously intercalates between the layers of the clay due to its favorable interaction with the intercalated organic cations.[2] The monomer is thermally polymerized upon addition of the initiator, ideally exfoliating the clay platelets as a result of the propagating polymer chain pushing them apart.[2,31] In situ polymerization is also conducted in high shear mixers by thermally reacting initiator, monomer and clay.[31] To purify the resulting nanocomposite, an organic solvent effectively removes residual monomer, initiator and oligomers via Soxhlet extraction.[31,42]

A Toyota research group reported the first successful PCN, which they created by using ε-caprolactam, a monomer for nylon-6, to intercalate MMT clay modified with 12-amino-lauric acid during in situ polymerization of nylon-6.[1,2] They noted that interlayer distances between platelets remained unchanged when samples were polymerized without heating due to the lack of heat-induced cationic catalytic ring-opening reactions of ε-caprolactam.[2]

The resulting nanocomposite displayed considerable enhancement in tensile, flexural, heat deflection, and gas-barrier properties as compared to the pure polymer.

In Situ Polymerization with ScCO$_2$

In situ polymerization is an effective route to creating exfoliated PCNs. However, this method typically employs large amounts of expensive and toxic organic solvents, deterring from use on the industrial scale. The in situ polymerization with scCO$_2$ process is similar to conventional in situ polymerization, except that the organic solvent is replaced by scCO$_2$ and the reaction is undertaken in a pressurized vessel. Functioning as the solvent, scCO$_2$ dissolves the monomer to facilitate its transport between the clay platelets, where it is allowed to polymerize and exfoliate the clay. This process can typically occur as a dispersion polymerization, owing to the insolubility of the product (polymer) and the solubility of the reactants (initiator and monomer) in the reaction medium (scCO$_2$).[43] To help manage the reaction, stabilizing molecules are employed containing a segment capable of physically adsorbing or chemically bonding to the polymerizing molecule and a segment that sterically stabilizes the propagating polymerization reaction by interacting with the CO$_2$ medium via CO$_2$-philic groups (such as fluorine or siloxane in the cases with PVDF and PDMS).[19,32,43] Furthermore, scCO$_2$ has been noted as a means to extract residual monomer, initiator and oligomers (owing to its ability to readily dissolve low molecular weight compounds).[42,44]

Using this method, Zhao et al. (2005) were successful in creating partially exfoliated/intercalated nanocomposites by polymerizing methyl methacrylate (MMA) in the presence of hydrocarbon surfactant-modified and fluorinated surfactant-modified clays, resulting in nanocomposites with enhanced properties. The clays with CO$_2$-philic fluorinated surfactant exhibited higher yields (85%) than the hydrocarbon surfactant-modified clays (38%).[32] In a later work, Zhao et al. (2006) created completely exfoliated and uniformly dispersed nanocomposites with polydimethylsiloxane (PDMS)-modified clay and PMMA.[45]

Melt-Mixing

Processing PCNs by melt-mixing, employing devices such as batch mixers or extruders, results in nanocomposites with various clay configurations. Melt-mixing is accomplished by heating a polymer until it is molten, while ideally remaining below its degradation temperature; some degradation is unavoidable with certain polymers such as polypropylene (PP) and polytetrafluoroethylene (PTFE).[21] As the polymer is heated, mixing applies shear forces that serve to break up the clay agglomerations into smaller tactoids.

Primarily, intercalated clays are the optimum result of melt-mixing.[6,30,31] Further compatibilization, in addition to the modifications offered by the organoclay, like the implementation of MA-grafted copolymers, has realized exfoliated clay structures in melt-mixing processes. The use of MA has allowed exfoliated clay structures in linear low-density polyethylene (LLDPE), styrene–ethylene/butylene–styrene block copolymer (SEBS), and PP to be attained with devices such as twin-screw extruders and roller mixers.[8,46,47]

Melt-Mixing with ScCO$_2$

Injection of scCO$_2$ into melt-mixing devices serves to enhance dispersion and increase delamination of clay within a PCN due to the dissolving, plasticizing, and diffusing characteristics of scCO$_2$. ScCO$_2$ dissolves in the polymer to reduce viscosity and create a single phase, which helps to increase additive dispersion. In addition, the polymer is plasticized by $scCO_2$, facilitating intercalation of the polymer chains between clay platelets.[29,38] The enhanced transport offered by the diffusive properties of $scCO_2$ help to quickly move the CO_2 molecules and polymer chains between clay platelets before exiting continuous melt-mixing devices like extruders. Upon extrusion through the die, it has been suggested that as the $scCO_2$ molecules become subcritical, they can further expand the interlayer distance between clay platelets.[38,48] Using a twin-screw extruder with $scCO_2$ injection capability, one study noted that there exists an optimum amount of dissolved CO_2 in the polymer associated with the maximum achievable clay intercalation/exfoliation. Too low of dissolved CO_2 content can hinder diffusion of polymer chains between clay platelets, while too high of dissolved CO_2 content is unfavorable to delaminating tactoids into individual clay platelets due to the reduced viscosity (reduction in shear forces).[38] In addition, with $scCO_2$ extrusion, low viscosity from a high content of dissolved $scCO_2$ creates a tough environment for sustaining a seal (via melted polymer layer) in the screw zone containing dissolved $scCO_2$.[38]

Melt-mixing with extruders employing $scCO_2$ has enabled improvements in the intercalation, exfoliation, and dispersion of clay in nanocomposites. Twin-screw extrusion with $scCO_2$ has been used to create nanocomposites with PP, polypropylene grafted with MA (PP-g-MA) and organoclay Cloisite 20A (C20A). The $scCO_2$ extrusion of PP/C20A nanocomposites resulted in nanocomposites with larger distances between clay platelets and smaller tactoid sizes than those extruded without $scCO_2$. To further improve dispersion, PP-g-MA was used as a compatibilizer and more exfoliated and dispersed clay structures were obtained as concentrations of PP-g-MA increased. The most pronounced difference between $scCO_2$ and non-$scCO_2$ extrusion was seen at a PP-g-MA concentration of 10 wt.% in which the $scCO_2$-treated nanocomposite revealed almost no XRD peak as compared to non-$scCO_2$, which showed a distinct peak. This indicated that the $scCO_2$ nanocomposite was almost completely exfoliated whereas the non-$scCO_2$ still contained mostly intercalated clays. This near complete exfoliation of $scCO_2$ nanocomposites was confirmed by TEM images. In addition to enhancing the clay structure, $scCO_2$ was also determined to have increased the thermal stability of samples.[38]

Nguyen et al. employed a single-screw extruder attached to a modified pressure chamber to create PP/C20A nanocomposites. The pressure chamber was used to pretreat C20A by soaking the C20A in $scCO_2$ for a set duration and then rapidly depressurizing the chamber. Once the clay pretreatment was complete, a solution of pretreated C20A and $scCO_2$ was directly injected from the pressure chamber into the single-screw extruder containing molten PP. This resulted in PP/C20A nanocomposites (6.6 wt.% C20A) containing well-dispersed clay with a high degree of exfoliation, increasing the modulus by 54%. In comparison, nanocomposites created by the conventional extrusion (without $scCO_2$ pretreatment of C20A) of PP and C20A, with and without $scCO_2$, produced intercalated nanocomposites (6.7 wt.% C20A) exhibiting an increase in modulus of 32% and 28%, respectively.[37]

Melt-Mixing with ScCO₂: Foaming

Extrusion with $scCO_2$ is an effective method for creating foams. The foaming process occurs as the polymer containing dissolved $scCO_2$ is extruded through the die and experiences a pressure drop (from supercritical to subcritical) due to expanding CO_2 molecules. Both the magnitude and rate of the pressure drop influence the nucleation rate, a determinant of foam cell size and cell density. Since clay is a good nucleating agent, the addition of clay increases the nucleating efficiency of pure polymers to increase cell density and decrease cell size when foaming. The effect of clay particle size due to its configuration also plays a role in foam morphology, with smaller particles of exfoliated clay exhibiting higher cell densities and smaller cell sizes than larger particles of intercalated clay.[30] Compared to intercalated clay, a higher pressure drop rate manifests from exfoliated clay since the exfoliated nanocomposite melt has a greater viscosity.[30] $ScCO_2$-assisted single-screw extrusion foaming experiments have been reported to increase the distance between clay platelets for modified and natural clay in high-density polyethylene (HDPE), with natural clay expected to exhibit no intercalation without $scCO_2$.[29]

Solvent Casting

Solvent casting permits the integration of clay into liquefied polymer by dissolving the polymer with solvents. Using this technique, the solid polymer is dissolved in an excess of organic liquid solvent, and clay is dispersed in the resultant solution. The concentration of solids in the liquid solvent is typically from 5% to 40%. If needed, the solution can be stirred and heated to temperatures between ambient and solvent boiling point to aid in the dissolution of the polymer. For harder-to-dissolve polymers, pressurization may be required in addition to heating for dissolution, as used for PVC in tetrahydrofuran (THF). Dissolution of the polymer usually occurs within a few hours. Various elements can be incorporated into the solution, e.g., plasticizers, additives, cosolvents, and dyes. Other measures, such as degassing, filtration, over-pressurization, or under-pressurization, can be employed to improve the final product by preventing bubbles, enhancing clarity, and preventing solvent loss.[49] Lastly, the solution is put into a mold or cast as a film and the solvent is evaporated, leaving the polymer–clay composite behind. Kim et al.[50] used solvent casting to prepare intercalated (with some exfoliation present) nanocomposites with organoclay, poly(ethylene oxide) (PEO) and PMMA. The PEO/PMMA/organoclay nanocomposites exhibited good clay dispersion.

Vessel (Batch) Infusion/Dispersion in $ScCO_2$

Similar to solvent casting, $scCO_2$ can replace liquid solvents as the swelling and dissolving agent (for certain polymers) to create improved PCNs.[16,36,39] The batch process can be conducted as an impregnation of clay into the polymer, intercalation of the polymer between clay platelets, or as a method to improve dispersion of clay already within a polymer. In any case, this method involves a period of soaking in $scCO_2$ to enable the CO_2 molecules to dissolve in the polymer, plasticize the polymer, and diffuse between polymer chains and clay platelets. The dissolved CO_2 in the polymer induces swelling, increasing free volume, and plasticization to improve mobility of the polymer chains and clay. Diffusion of CO_2 aids in the impregnation of clay into the polymer, the dispersion of clay throughout the matrix, and the diffusion of polymer chains between clay platelets. The overall effect of these processes increases the mobility of clay and polymer chains to enhance the impregnation of the clay into the polymer, clay dispersion throughout the polymer matrix, and clay intercalation by polymer chains.

Using the $scCO_2$ batch process, organoclays impacted by CO_2 treatment, C20A and C93A, respectively, underwent intercalation by powdered PP-g-MA[36] and impregnation into pelletized linear LLDPE grafted with MA (LLDPE-g-MA).[39] Although natural clay MMT is not CO_2-philic, this $scCO_2$ processing method was shown to be effective at improving MMT dispersion into PDMS, a CO_2-philic amorphous polymer, and enhancing its properties as compared to non-$scCO_2$ processed PDMS/MMT nanocomposites.[16]

Vessel (Batch) Infusion/Dispersion in ScCO₂: Foaming

Using certain polymers, foams can be created by rapid depressurization after batch soaking in scCO₂. As mentioned previously, foaming occurs due to the rapid expansion of CO₂ molecules that are dissolved within the polymer. Employing this technique, Zeng et al. foamed polystyrene (PS)/organoclay and PMMA/organoclay nanocomposites, observing that clay particle size is directly proportional to foam cell size and inversely proportional to cell density. The cell densities for PS/organoclay and PMMA/organoclay nanocomposites containing exfoliated organoclay (5 wt.%) were respectively 5× and 93× higher than the cell densities for their pure counterparts. The higher foam cell density for PMMA is attributed to its more favorable interaction with CO₂, due to the strong affinity between CO₂ and the PMMA carbonyl groups. The enhanced nucleating efficiency supplied by this process is expected to lead to a reduction in processing conditions, and thus reduce production costs.[31]

References

1. Kojima, Y., Usuki, A., Kawasumi, M., Okada, A., Fukushima, Y., Kurauchi, T., Kamigaito, O. Mechanical properties of nylon 6-clay hybrid. J. Mater. Res. **1993**, 8, 1185–1189.
2. Kawasumi, M. The discovery of polymer–clay hybrids. J. Polym. Sci. Part A: Polym. Chem. **2004**, 42, 819–824.
3 Utracki, L.A. *Clay-Containing Polymeric Nanocomposites.* Rapra Technology Limited: Shrewsbury, UK, Vol. 1, 2004.
4 Fried, J.R. *Polymer Science and Technology.* Prentice Hall Professional Technical Reference: Upper Saddle River, NJ, 2003.
5. Ngo, T.D., Wood-Adams, P.M., Hoa, S.V., Ton-That, M.T. Modeling the delamination process during shear premixing of nanoclay/thermoset polymer nanocomposites. J. Appl. Polym. Sci. **2011**, 122, 561–572.
6. Peprnicek, T., Kalendova, A., Pavlova, E., Simonik, J., Duchet, J., Gerard, J.F. Poly(vinyl chloride)-paste/clay nanocomposites: Investigation of thermal and morphological characteristics. Polym. Degrad. Stab. **2006**, 91, 3322–3329.
7 Lyatskaya, Y., Balazs, A.C. Modeling the phase behavior of polymer–clay composites. Macromolecules. **1998**, 31, 6676–6680.
8. Park, C.I., Park, O.O., Lim, J.G., Kim, H.J. The fabrication of syndiotactic polystyrene/organophilic clay nanocomposites and their properties. Polymer. **2001**, 42, 7465–7475.
9. Lee, S., Lanterman, H.B., Sardesai, A., Wenzel, J., Marshall, B., Yen, J., Amin-Sanayei, R., Moucharik, M. Polymerization of poly(vinylidene fluoride) in a supercritical fluid medium. U.S. Patent 7,091,288, August 15, 2006.
10. Boyle, T.B., Carroll, J.J. Study determines best methods for calculating acid–gas density. Oil Gas J. **2002**, 100, 45–53.
11. Brantley, N.H., Kazarian, S.G., Eckert, C.A. In situ FTIR measurement of carbon dioxide sorption into poly(ethylene terephthalate) at elevated pressures. J. Appl. Polym. Sci. **2000**, 77, 764–775.
12. Shieh, Y.T., Su, J.H., Manivannan, G., Lee, P.H.C., Sawan, S.P., Spall, W.D. Interaction of supercritical carbon dioxide with polymers: I. Crystalline polymers. J. Appl. Polym. Sci., 59, 695–705.
13. Shieh, Y.T., Su, J.H., Manivannan, G., Lee, P.H.C., Sawan, S.P., Spall, W.D. Interaction of supercritical carbon dioxide with polymers: II. Amorphous polymers. J. Appl. Polym. Sci. **1996**, 59, 707–717.
14. Michalik, S.J., Design, synthesis, and optimization of non-fluorous, co₂-philic polymers: A systematic approach. M.S. thesis, University of Pittsburgh, Pittsburgh, PA, 2003.

15. Ganapathy, H.S., Yuvaraj, H., Hwang, H.S., Kim, J.S., Choi, B.C., Gal, Y.S., Lim K.T. CO$_2$-soluble semiconducting polymers synthesized in supercritical carbon dioxide. Synth. Met. **2006**, *156*, 576–581.

16. Horsch, S., Serhatkulu, G., Gulari, E., Kannan, R.M. Supercritical CO$_2$ Dispersion of nano-clays and clay/polymer nanocomposites. Polymer. **2006**, *47*, 7485–7496.

17. McHugh, M.A., Park, I.H., Reisinger, J.J., Ren, Y., Lodge, T.P., Hillmyer, M.A. Solubility of CF$_2$-modified polybutadiene and polyisoprene in supercritical carbon dioxide. Macromolecules. **2002**, *35*, 4653–4657.

18. McHugh, M.A., Garach-Domech, A., Park, I.H., Li, D., Barbu, E., Graham, P., Tsibouklis, J. Impact of fluorination and side-chain length on poly(methylpropenoxyalkylsiloxane) and poly(alkyl methacrylate) solubility in supercritical carbon dioxide. Macromolecules **2002**, *35*, 6479–6482.

19. DeSimone, J.M., Zihibin, G., Elsbernd, C.S. Synthesis of fluoropolymers in supercritical carbon dioxide. Science. **1992**, *257*, 945–947.

20. Harrison, K., Goveas, J., Johnston, K.P. Water-in-carbon dioxide microemulsions with a fluoro-carbon–hydrocarbon hybrid surfactant. Langmuir. **1994**, *10*, 3536–3541.

21. Nalawade, S.P., Picchioni, F., Janssen L.P.B.M. Supercritical carbon dioxide as a green solvent for processing polymer melts: Processing aspects and applications. Prog. Polym. Sci. **2006**, *31*, 19–43.

22. Areerat, S., Funami, E., Hayata, Y., Nakagawa, D., Ohshima, M. Measurement and prediction of diffusion coefficients of supercritical CO$_2$ in molten polymers. Polym. Eng. Sci. **2004**, *44*, 1915–1924.

23. Duarte, A.R.C., Martins, C., Coimbra, P., Gil, M.H.M. Sorption and diffusion of dense carbon dioxide in a biocompatible polymer. J. Supercritical Fluids **2006**, *38*, 392–398.

24. Berens, A.R., Huvard, G.S., Korsmeyer, R.W., Kunig, F.W. Application of compressed carbon dioxide in the incorporation of additives into polymers. J. Appl. Polym. Sci. **1992**, *46*, 231–242.

25. Muth, O., Hirth, T., Vogel, H. Investigation of sorption and diffusion of supercritical carbon dioxide into poly(vinyl chloride). J. Supercritical Fluids **2001**, *19*, 299–306.

26. Alessi, P., Cortesi, A., Kikic, I., Vecchione, F. Plasticization of polymers with supercritical carbon dioxide: Experimental determination of glass-transition temperatures. J. Appl. Polym. Sci. **2003**, *88*, 2189–2193.

27. Margon, V., Agarwal, U.S., Bailly, C., de Wit, G., van Kasteren, J.M.N., Lemstra, P.J. Supercritical carbon dioxide assisted extraction from the polycarbonate depolymerization mixture. J. Supercritical Fluids **2006**, *38*, 44–50.

28. Lopez-Periago, A.M., Vega, A., Subra, P., Argemi, A., Saurina, J., Garcia-Gonzalez, C.A., Domingo, C. Supercritical CO$_2$ processing of polymers for the production of materials with applications in tissue engineering and drug delivery. J. Mater. Sci. **2008**, *43*, 1939–1947.

29. Garcia-Leiner, M., Lesser, A.J. Polymer–clay nanocomposites prepared in supercritical carbon dioxide. Polym. Mater. Sci. Eng. **2003**, *88*, 92–93.

30. Han, S., Zeng, C., Lee, L.J., Koelling, K.W., Tomasko, D.L. Extrusion of polystyrene nanocomposite foams with supercritical CO$_2$. Polym. Eng. Sci. **2003**, *43*, 1261–1275.

31. Zeng, C., Han, X., Lee, L.J., Koelling, K.W., Tomasko, D.L. Polymer–clay nanocomposite foams prepared using carbon dioxide. Adv. Mater. **2003**, *15*, 1743–1747.

32. Zhao, Q., Samulski, E.T. In situ polymerization of poly(methyl methacrylate)/clay nanocomposites in supercritical carbon dioxide. Macromolecules **2005**, *38*, 7967–7971.

33. Manitiu, M., Bellair, R.J., Horsch, S., Gulari, E., Kannan, R.M. Supercritical carbon dioxide-processed dispersed polystyrene–clay nanocomposites. Macromolecules **2008**, *41*, 8038–8046.

34. Manke, C.W., Gulari, E., Mielewski, D.F., and Lee, E.C. System and method of delaminating a layered silicate material by supercritical fluid treatment, U.S. Patent 6,469,073 B1, October 22, 2002.

35. Thompson, M.R., Liu, J., Krump, H., Kostanski, L.K., Fasulo, P.D., Rodgers, W.R. Interaction of supercritical CO$_2$ with alkyl-ammonium organoclays: Changes in morphology. J. Colloid Interface Sci. **2008**, *324*, 177–184.

36. Liu, J., Thompson, M.R., Balogh, M.P., Speer, R.L., Jr., Fasulo, P.D., Rodgers, W.R. Influence of supercritical CO_2 on the interactions between maleated polypropylene and alkyl-ammonium organoclay. J. Appl. Polym. Sci. **2011**, *119*, 2223–2234.
37. Nguyen, Q.T., Baird, D.G. An improved technique for exfoliating and dispersing nanoclay particles into polymer matrices using supercritical carbon dioxide. Polymer **2007**, *48*, 6923–6933.
38. Hwang, T.Y., Lee, S.M., Ahn, Y., Lee, J.W. Development of polypropylene–clay nanocomposite with supercritical CO_2 assisted twin screw extrusion. Korea–Aust. Rheol. J. **2008**, *20*, 235–243.
39. Factor, M., Lee, S., Confirmation and quantification of linear low-density polyethylene (LLDPE) and maleated linear low density polyethylene (LLDPE-g-MA) Infused with nanoclay in supercritical carbon dioxide. Paper SPE 0458 presented at Society of Plastic Engineers Annual Technical Conference, Boston, MA, May 2, 2011.
40. Cervantes-Uc, J.M., Cauich-Rodriguez, J.V., Vazquez-Torres, H., Garfias-Mesias, L.F., Paul, D.R. Thermal degradation of commercially available organoclays studied by TGA-FTIR. Thermochim. Acta **2007**, *457*, 92–102.
41. Lee, M., Park, C.B., Tzoganakis, C. Measurements and modeling of PS/supercritical CO_2 solution viscosities. Polym. Eng. Sci. **1999**, *39*, 99–109.
42. Clark, K., Lee, S. Removal of ungrafted monomer from polypropylene–graft–maleic anhydride via supercritical carbon dioxide extraction. Polym. Eng. Sci. **2004**, *44*, 1636–1642.
43. DeSimone, J.M., Maury, E.E., Menceloglu, Y.Z., McClain, J.B., Romack, T.J., Combes, J.R. Dispersion polymerizations in supercritical carbon dioxide. Science **1994**, 265, 356–359.
44. McHugh, M.A., Krukonis, V.J. *Supercritical Fluid Extraction Principles and Practice.* Butterworth: Boston, MA, 1986.
45. Zhao, Q., Samulski, E.T. A comparative study of poly(methyl methacrylate) and polystyrene/clay nanocomposites prepared in supercritical carbon dioxide. Polymer **2006**, *47*, 663–671.
46. Wang, K.H., Choi, M.H., Koo, C.M., Choi, Y.S., Chung, I.J. Synthesis and characterization of maleated polyethylene/clay nanocomposites. Polymer **2001**, *42*, 9819–9826.
47. Hasegawa, N., Okamoto, H., Kato, M., Usuki, A. Preparation and mechanical properties of polypropylene–clay hybrids based on modified polypropylene and organophilic clay. J. Appl. Polym. Sci. **2000**, *78*, 1918–1922.
48. Ottaviani, R.A., Rodgers, W.R., Fasulo, P.D., Okonski, D.A. Methods for Making Nacomposite Materials, General Motors, USA, U.S. Patent 7,462,666 B2, December 9, 2008.
49. Siemann, U. Solvent cast technology — A versatile tool for thin film production. Prog. Colloid Polym. Sci. **2005**, *130*, 1–14.
50. Kim, H.B., Choi, J.S., Lee, C.H., Lim, S.T., Jhon, M.S., Choi, H.J. Polymer blend/organoclay nanocomposite with poly(ethylene oxide) and poly(methyl methacrylate). Eur. Polym. J. **2005**, *41*, 679–685.

16

Infusion of Volatile Corrosion Inhibitor into Thermoplastic Resins and Films in a Supercritical Fluid Medium

Leah A. Taylor and Sunggyu Lee

CONTENTS

Introduction 252
Background 253
    Corrosion 253
        Corrosion Control 254
        Corrosion Inhibitors 255
    Supercritical Fluid 259
        Supercritical Carbon Dioxide 260
        Supercritical Infusion 260
    Biodegradable Polymers 262
Experiment 263
    Materials Selection 263
    Experimental System 263
        Supercritical Infusion Reactor System 263
        Extrusion Process 264
        Scanning Electron Microscopy 265
    Experimental Procedures 265
        Film Preparation 265
        Supercritical Infusion 265
        Corrosion Studies 266
Results and Discussion 267
    Effects of Fluid Density on Infusion 267
        Results 268
    Corrosion Prevention 268
    Supercritical Infusion 271
        Polymer Infusion 271
        Scanning Electron Microscopy 273
Conclusion 274
Acknowledgment 275
References 275

Introduction

The development of functional materials for specialty markets is a growing industry throughout the global economy. An increased need for materials to fulfill consumer and manufacturing demands continues to evolve and new products must be produced to meet these requirements. Plastic products can be modified for use in a number of different applications. The most widespread technique utilized to modify plastics is the incorporation of additives through polymer blending. Common additives, in both particulate and fluid forms, include ultraviolet (UV) stabilizers, antioxidants, lubricants, pigments, anticorrosives, and fillers.[1,2] For most applications, the additives are incorporated into the polymeric matrix through melt compounding. Some examples of melt compounding include extrusion and melt mixing.[1,3] During melt compounding, however, certain drawbacks may inevitably occur. This is because the processing is based solely on the substrate polymeric resin's melt properties and not on the additive's properties. Typical problems include structural incompatibility that may cause eventual phase separation of molded objects, undesirable cross-linking, alteration of additive's functionalities, morphological irregularities, and chemical and thermal degradation of functional additives. In order to alleviate these problems, new techniques may be desired to integrate the additives into the polymer substrate. One such technique is supercritical infusion or supercritical fluid (SCF) infusion.

Supercritical infusion is a process in which a SCF is used as a medium to disperse a functional additive into a polymer matrix. The word "infusion" is more applicable to a situation when this functional additive is in a particulate form. SCF is an attractive medium for infusion for a number of reasons. The solvent strength is significantly higher for SCFs compared to traditional subcritical fluids or ambient air.[4] Most thermoplastic polymers become softened in appropriate SCF media at lower temperatures than their respective (or reported) glass transition temperatures (T_g).[5,6] Further, SCF infusion prevents or helps prevent degradation of both the substrate and additive by operating at lower temperatures than traditional methods such as melt compounding. This allows for processing of temperature-sensitive materials such as volatile additives or biologically active ingredients. Finally, the properties of the SCF, such as fluid density, can be controlled relatively easily by regulating the temperature and pressure. For these reasons, SCF infusion was developed as a method of choice for infusing a volatile corrosion inhibitor (VCI) into a polymer matrix.

Corrosion prevention is an application that receives a great deal of industrial attention. Loss of product or shortened service life due to corrosion is a seriously detrimental economic constraint in metal processing and parts manufacturing industries. This is particularly true for international trade of metallic parts such as automotive, electronic, and mechanical parts and goods. Economically speaking, estimates indicate that approximately 7–8% of the United States' income is spent to either replace corroded parts or on corrosion prevention.[7] The development of a simple, effective, anticorrosion packaging can help prevent large losses of manufactured metallic parts. Current packaging options consist mainly of plastic materials. The environmental impact of plastics waste became an increasingly important issue for many consumers and manufacturers during the past decades. Estimates show that approximately 29.9 million tons of plastic waste was generated in the United States in 2009,[8] of which only very little was recycled. Additionally, the annual growth rate of the plastic packaging industry is estimated at 25%.[9] Thus, the environmental concerns arising from plastic waste are considerable. With today's focus on environmental issues, biodegradable packaging has become more appealing.

Biodegradable plastics are made from polymers that are susceptible to microbial attack and can degrade in the environment. Degradation occurs relatively quickly (i.e. ~3–4 mo) under most practical circumstances, and the byproducts are environmentally benign. The biodegradable plastics that are currently available in the market can be processed at conditions similar to their non-biodegradable counterparts. This also makes biodegradable polymers a reasonable candidate as polymer substrates for supercritical infusion.

The primary focus of this work is to synergistically combine the fields of corrosion prevention/inhibition with biodegradable polymer film packaging by using SCF technology, thus resulting in novel packaging materials with dual functionalities, viz., corrosion inhibition and biodegradability. The secondary objective is to develop a more robust process that can be used to infuse temperature-sensitive functional additives into a thermoplastic polymeric matrix by using a suitable SCF medium. In order to achieve these objectives, a packaging film was prepared using polymeric resin and then infused with a VCI. For this work, both non-biodegradable and biodegradable polymers were supercritically infused with the VCI sodium nitrite ($NaNO_2$). Additionally, the feasibility of supercritical carbon dioxide ($scCO_2$) infusion as a commercially viable process was evaluated. The effect of SCF density upon the infusion penetration depth of VCI into a polymeric substrate was investigated. The effects of process parameters on the extent and infusion depth of VCI were also examined. The phenomenological results of supercritical infusion were characterized and elucidated. The ability of the VCI-infused films to protect cut iron piece specimens from corrosion was established.

Background

Corrosion

Corrosion is the result of a reduction-oxidation reaction in which a metal is unintentionally attacked by another chemical substance and converted into an unwanted compound. Metal corrosion is a thermodynamically favored process that results in a decrease in the Gibbs free energy of the system.[11] For most metals, oxidation is a spontaneous process in ambient air. The generalized reaction for corrosion may be represented by the following reaction:

$$M \rightarrow M^{n+} + ne-$$
(1)

Different metals are varyingly susceptible to diverse oxidation processes depending upon the specific environment to which the metal substrate is subjected. One of the most common corrosion mechanisms is found in the corrosion of iron (or steel) due to the interaction with moisture and oxygen, which is more commonly referred to as rusting.[12]

Rusting is an electrochemical reaction in which the iron substrate serves as an anode on which oxidation occurs.[12] During an electrochemical reaction, electrons are transferred from the metal to the oxidant by conduction through the metal itself. Initially, the onset of oxidation results from the difference in electrode potential between two locations on the metal, the anode and the cathode.[11] The reduction potential for iron in the presence of water [$Fe^{2+}_{(aq)}$] is less than that of O_2. Therefore, Fe(s) can be oxidized by O_2(g). This relationship is shown in the following equations:

$$\text{Cathode: } O_2 \text{ (g)} + 4H^+\text{(aq)} + 4e^- \rightarrow 2H_2O \text{ (l)} \quad E_{red} = 1.23 \text{ V} \qquad (2)$$

$$\text{Anode: Fe (s)} \rightarrow Fe^{2+}\text{(aq)} + 2e^- \quad E_{red} = -0.44 \text{ V} \qquad (3)$$

The oxidation of Fe^0 (s) to Fe^{2+} (aq) results in the production of electrons. These electrons migrate to another area of the metal which acts as a cathode. This cathodic portion of the metal is the surface that has the most O_2 available. The oxidized iron (Fe^{2+}) continues to undergo further oxidation to become Fe^{3+}. Ultimately, Fe^{3+} forms hydrated iron(III) oxide, otherwise known as rust. The overall process is represented by the following:

$$4Fe^{2+} \text{ (aq)} + O_2 \text{ (g)} + 4H_2O \text{ (l)} + 2H_2O \text{ (l)} \rightarrow 2Fe_2O_3 \cdot H_2O \text{ (s)} + 8H^+ \text{ (aq)} \qquad (4)$$

A schematic detailing the oxidation of iron to produce rust is shown in Figure 16.1. In this specific illustration, water vapor condenses on the substrate surface to serve as the anode. Oxygen at the substrate/air interface acts like a cathode for the electrochemical reaction.

The rate and type of corrosion process which a metal undergoes depends upon a number of factors. The physical and chemical characteristics of the metal itself play a role in the corrosion process. These characteristics may include the chemical composition of the metal, the metal structure, and any stresses imparted on the metal sample. Environmental factors, such as temperature, fluid velocity, and humidity, influence metal corrosion. For iron placed in the open atmosphere, air pollution may also accelerate the rate of corrosion. For example, the presence of acid rain catalyzes the corrosive process.[13] In addition to the open atmosphere, other corrosive environments include aqueous solutions, soils, and inorganic solvents. Depending upon these different environmental and metal characteristics, different types of corrosion may be present. Even though not elaborated in this chapter, metallic corrosion[14] can be classified into eight different forms, viz., uniform, galvanic, crevice, pitting, intergranular, selective leaching, erosion–corrosion, and stress corrosion.

Corrosion Control

Corrosion control is an important aspect of materials engineering, and prevention (including inhibition and deterrence) may be achieved in a number of ways. Some general

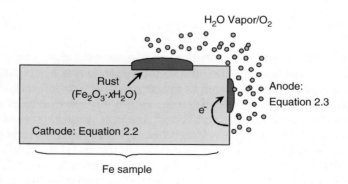

FIGURE 16.1
A Schematic detailing the corrosion of iron in the presence of water and oxygen to form hydrated iron(III) oxide.
Source: Adapted from Brown et al.[12]

techniques include cathodic protection, coatings, passivation, and environmental alteration. Cathodic protection functions by providing an external source which supplies electrons to the protected metal. The continuous supply of electrons forces the metal to act
as a cathode. The electron supply forces the generalized oxidation reaction in the reverse
direction and reduction is favored over oxidation. Cathodic protection is often employed
in buried objects such as pipelines and tanks. The pipe or tank is electrically connected
to another metal that is more reactive. The metal undergoes oxidation, and the pipe or
tank is protected. Design considerations also help protect objects from corrosion. In addition to cathodic protection and design aspects, coatings are often used to prevent metal
corrosion. Coating the metal surface provides a physical barrier to species that promote
corrosion. A number of materials including metals, ceramics, and polymers can be used
as coatings. Important considerations for suitable coatings include a high degree of adhesion to the metal surface, strong resistance to salt penetration, and nonreactivity with the
metal. Another way to retard corrosion is through passivation. Passivation is an interaction between the metal and its environment in which a normally active metal loses its
chemical reactivity.[14] Naturally occurring passivity is believed to be caused by the formation of thin oxide layer that is strongly adhered to the surface.[14] This thin layer on the
surface, referred to as the passivating layer, prevents further oxidation of the substrate
metal.[11] Altering the corrosive environment also helps prevent oxidation. Controlling
humidity, temperature, and fluid movement around the metal object are common techniques to control oxidation. The addition of species that prevent oxidation or alter the
corrosive environment in a positive way may also be useful.[14] These are referred to as
corrosion inhibitors (CIs).

Corrosion Inhibitors

CIs are substances whose presence results in the decreased oxidation of a metal, thus slow-
ing down the process. These inhibitors are chemical species that can have either an inorganic or organic chemical structure. There are two main types of CIs: contact inhibitors or
volatile inhibitors. Contact inhibitors must be applied directly to the surface of the metal
object. Alternatively, VCIs can easily transfer to the vapor phase and adsorb onto the surface of the metal object, thereby protecting it from corrosion. Both contact CIs (CCIs) and
VCIs can function in three different ways: the CI can inhibit anodic reactions, cathodic
reactions, or both. Anodic inhibitors contain oxidizing anions that help enrich the metal
electrode potential, thereby controlling corrosion. Cathodic inhibitors prevent corrosion
by diminishing the oxygen concentration in the medium surrounding the metal object.
Mixed inhibitors contain species that function as both anodic and cathodic inhibitors.[11]
Table 16.1 lists some common inhibitors and briefly outlines the method of protection used
for each.

The nature of the CI itself influences the method of application to the metal substrate.
The metal can be exposed to the CI during manufacture, shipping, or storage. Corrosion-
inhibited storage/packaging materials are made of paper or polymer infused with inhibitors. Some common types of corrosion-inhibited plastics are listed in Table 16.2. VCIs
are the inhibitor of choice for packaging materials. Corrosion-inhibited packages are
essentially composite materials consisting of two components: the polymer matrix and
the VCI. The polymer matrix serves as the carrier material for the VCI, and the VCI provides protection for the metal parts.[11] The VCI contained within the matrix diffuses out
of the polymer and into the environment surrounding the metallic object. The rate of diffusion depends upon the permeability of the polymer with respect to the VCI; the vapor

TABLE 16.1

List of Inhibitor Mechanism and Chemical Structure for Some Common CIs

Chemical Type	CI Molecules	Inhibition Mechanism	Common Protected Metals
Inorganic inhibitors	Sodium nitrite	Anodic	Iron, steel, aluminum
	Ammonium carbonate	Hybrid	Steel
	Calcium bicarbonate	Cathodic	Steel
	Strontium chromate	Anodic	Iron, steel, aluminum, magnesium, titanium
	Sodium metasilicate	Hybrid	Steel, aluminum, lead, brass, copper
Organic inhibitors	Cyclohexylamine (a.k.a., hexahydroaniline)	Hybrid	Steel
	Hexamethylenimine (HMI)	Hybrid	Steel, nickel, copper, tin, lead, silver
	Morpholine	Hybrid	Steel, aluminum, nickel, silver, brass
	Imidazole	Hybrid	Steel

Source: Adapted from Goldade et al.[11]

pressure of the VCI at the application also influences the rate of diffusion. A number of models have been developed to clarify the mechanism of VCI desorption from a polymer. Fujita described desorption as a cooperative movement of the VCI permeant and the polymer matrix. In this model, the VCI molecules "jump" from one void space within the matrix to another.[7] Dorfman et al.[15] studied the desorption of VCI from polyethylene (PE) and substantiated the belief that the VCI diffusion from the polymer carrier is controlled by solid-phase diffusion. The relationship governing the concentration gradient with time is described as:

$$m/m_0 = (4/l)(Dt/\pi)^{0.5} \tag{5}$$

where m and m_0 is the amount of VCI desorbed at $t = $ t and $t = \infty$, respectively. The term l denotes the film thickness, and D is the VCI diffusivity. This study also showed that the diffusion rate of VCI from the polymer substrate increases with increasing temperature and VCI concentration.[15] Following desorption of the species into the metal environment, the VCI can then adsorb onto the surface or interact with other present species to prevent corrosion. A sealed volume with relatively constant pressure is necessary for the VCI-infused package to function correctly. A diagram representing a corrosion system in which a metal object is packaged in VCI-infused polymer film is shown in Figure 16.2.

TABLE 16.2

Different Varieties of Corrosion-Inhibited Plastics that may be Used during Transport or Operation

Inhibited Plastics	Description
Inhibited polymer coating	CI–polymer composite bonded to substrate surface
Inhibited polymer film	Corrosion-inhibited packaging material in a form of polymeric film
CI carriers	Pellets, fibers, and foams that are used as packaging material inside traditional shipping containers
Sealants, glues, greases	Corrosion-inhibited compounds that can be used in specific areas or applications where corrosion potential may be high

Source: Adapted from Goldade et al.[11]

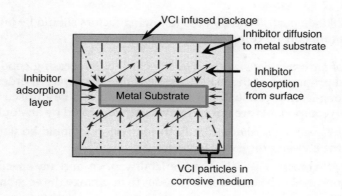

FIGURE 16.2
A schematic of anticorrosion system for a metal sample protected by a VCI-infused polymer package. **Source:** Adapted from Goldade et al.[11]

As shown, a metal specimen is enclosed in a VCI-infused package. The inhibitor slowly diffuses out of the polymer matrix and into the airspace within the package. Next, the inhibitor diffuses to the metal substrate and adsorbs onto the surface to form a protective layer.

Desorption of the VCI species from the surface of the metal may occur with time. After the system reaches equilibrium, the adsorption of new inhibitor molecules from the gas phase will readily fill any free adsorption sites vacated by desorbed VCI. The metal item in the package will be protected from corrosion if the surface is adequately covered with VCI molecules. This relationship is shown:

$$t_1(m_1 - m_3) + t_T(m_2 - m_3) \geq Q_{min} \tag{6}$$

where t_1 is the time immediately following packaging when atmospheric corrosion on the metal substrate ceases, and t_T is the amount of time the item has spent inside the package. The release rate of the VCI within the package immediately following packaging and following time t_1 are given as m_1 and m_2, respectively. The value m_3 is the release rate of the VCI during packaging.[11] The minimum concentration of VCI in the package that is required to prevent corrosion is designated by Q_{min}. The release rate is greatest immediately following packaging, as readily expected. This deters any corrosion that may have already started and increases the VCI concentration of vapors in the environment surrounding the metal substrate. Following the initial release of the VCI, the release rate abates somewhat, but the VCI continues to diffuse into the vapor phase. The release rate of the VCI depends on the association between the polymer matrix and the VCI as well as the total infused amount of VCI.

The production of an effective VCI package depends on a number of factors. One of the most important factors is the degree of binding between the VCI and the polymer. The adhesion should be strong enough that the VCI does not dissipate before the package is done being used. However, if the adhesion is too strong, the concentration of the VCI around the metal substrate may be too low and corrosion will still occur. The polymer and VCI are bound together during the manufacturing process. The degree of binding depends upon the compatibility between the polymer and the VCI. Manufacturing conditions can also affect the adhesion between the two components as well as alter the properties of the final product. Goldade et al.[11] cite that the following factors should be taken into consideration when manufacturing composite VCI materials.

1. *Durability of the composite material.* The effect of the processing conditions on the mechanical properties of the material should be considered in order to make an effective, long-lasting product. Further, the VCI should be infused well enough that the two components are not spontaneously separated by any outside effects.

2. *Processing behavior of the components.* Both components should be stable and manageable at the chosen processing temperature.

3. *Feasibility of the chosen materials.* The availability, price, and any special processing requirements need to be considered in order to maximize the economic aspects of the product.

4. *Other factors.* Environmental constraints or other potential operational problems may determine specific design elements of the process.[11]

The use of corrosion-inhibited packaging began with the introduction of VCI-infused paper products in the late 1970s. Heidemann[16,17] described the use of VCI papers for the corrosion protection of various metals including copper, aluminum, and silver. Since then, VCI packaging products grew to include plastic products. Today, a number of VCI packaging materials are available on the market.

Polymer films containing VCI are regularly used with good results to retard corrosion in metal parts. These films are advantageous over other anticorrosion approaches for a number of reasons. For example, films are clean, easy to use, and simple to manufacture. Additionally, the polymer resins commonly used as the carrier matrix are readily available. Corrosion-inhibited packages mainly consist of polyolefins such as PE and polypropylene (PP). The majority of films are produced from linear low-density polyethylene (LLDPE) because LLDPE has a high degree of chemical resistance and excellent mechanical properties such as a high value of elongation at break and resistance to puncture or tearing.[11] A number of different kinds of corrosion-inhibited films are manufactured using LLDPE or LDPE.

As previously discussed, the processing method for manufacturing VCI-infused films is important, and properly mixing the VCI with the polymer is the most important step.[11] Currently, four different methods exist for incorporating the VCI into the polymer matrix. They are as follows:

1. *Melt Blending.* This method consists of processing the VCI with the melt phase of the polymer. After the VCI and polymer melt are thoroughly mixed, the blend is cast into films.

2. *Diffusive Saturation.* During diffusive saturation, a precast polymer film is impregnated with a CI in either the liquid or gas phase.

3. *Adhesive Fixation.* Adhesive fixation involves entrapping CI particles to premade polymer films or objects. The CI particles can be fixed to the surface or encased within the film or object.

4. *Formation at polymerization.* In some cases, the monomers can be mixed with VCI species at the time of polymerization to form corrosion-inhibited materials.

For each of these methods, the most efficient process should be chosen based on the nature of the VCI and polymer components as well as desired efficiency. Overall, the process that is most capable of preserving the properties of both ingredients is ideal.[11]

The most common method for producing VCI-inhibited films is melt blending via extrusion. Extrusion is favorable for a number of reasons. First, it is relatively easy to perform and allows for a large quantity of films to be produced. Also, extrusion is currently used for a number of other applications; therefore, the infrastructure is already in place.[2] Despite the advantages of extrusion for infusing VCI films, a number of drawbacks also exist that make this process less favorable. First, the high temperatures required to melt and blend the polymer can cause premature degradation of VCI and polymer or cross-linking. Cross-linking can lock up VCI molecules in the polymer matrix, thus making them unavailable for the desired function of corrosion inhibition, while the degradation of polymer generates lighter molecular weight molecules and leads to an inferior product. In addition to high temperatures, shear forces are also present and the combined effect of these results in polymer breakdown. This can lead to lower molecular weight and the formation of adverse byproducts. Another difficulty arising from melt blending is the tendency of the filler particles, including crystalline VCI, to agglomerate. Disparities in viscosity of the melt blend often result in increased compounding difficulties. One common cause for viscosity variations is a large variation in additive particle sizes. Different particle sizes can create micro-sized areas of varying viscosity within the compounding section during melt blending. The sections of lower viscosity tend to migrate to the outer walls where the shear forces are higher. This creates an inconsistent dispersion of filler particles within the melt and may ultimately result in variable mechanical properties of the final product. Another source of viscosity variations during melt blending is due to the moisture content of the additives and polymer resin.[18] The degree of viscosity variation within the melt is directly affected by the moisture content of the raw materials.[19] The polymer and filler should both be free of water, or dried prior to melt blending, in order to ensure optimal mixing. Insufficient moisture removal also results in poor polymer-filler adhesion.[20,21] This can lead to surface defects and inferior mechanical properties. For these reasons, alternative methods for VCI infusion are being investigated. One of these techniques is SCF infusion.

Supercritical Fluid

Most of the earlier applications of SCFs were more focused on selective extraction of particular ingredients from a substrate that contains a heterogeneous array of diverse constituents. Today, SCF applications include food processing, pharmaceutical manufacturing, polymerization,[22,23] post-polymerization grafting,[24–26] hydrocarbon cracking,[27,1] devulcanization,[28] waste treatment, disinfection,[29] dry cleaning,[30] and powder formation.[31] SCF technology remains a practical but revolutionary technique to perform a number of engineering operations that would be otherwise difficult or inconceivable.

An SCF is any fluid or fluid mixture which exists above its critical point in both temperature and pressure. Once a fluid exceeds its critical temperature, the resulting gaseous fluid cannot be converted back into a liquid regardless of the pressure exerted on the system.[32] Figure 16.3 shows the supercritical region in a phase diagram as it relates to temperature and pressure. At the critical point, the density of the liquid phase equals that of the vapor phase and a single, equal density phase results. The SCF exhibits both gas-like and liquid-like properties, and the physicochemical, transport, and electrochemical properties typically associated with that particular molecule begin to change. Some properties commonly attributed to the supercritical region are increased solvent strength, liquid-like density, gas-like diffusivity, gas-like viscosity, and substantially reduced surface tension.[6] These properties can be easily exploited and advantageously controlled by manipulating the pressure and temperature of the system. A number of different substances have been investigated for

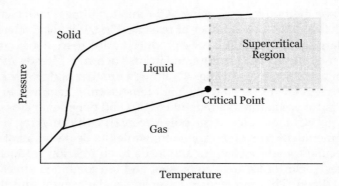

FIGURE 16.3
A generalized *T*–*P* phase diagram detailing the location of each phase including the supercritical fluid region.

potential SCF use. These include but are not limited to methanol, ethanol, acetone, propane, benzene, toluene, water, nitrous oxide, and carbon dioxide. An SCF system can be comprised of a single species or a combination of two or three miscible components. An example of frequently used binary or ternary systems is carbon dioxide/acetone, carbon dioxide/methanol, and carbon dioxide/ethanol/water.[32] The type and number of components used in a supercritical system depends on the intended application and nature of the solute. The most common component of SCF systems is carbon dioxide, mainly due to its low critical temperature (31.1°C) and low cost.

Supercritical Carbon Dioxide

Supercritical carbon dioxide ($scCO_2$) is usually the solvent of choice for many applications because it is inexpensive and nontoxic. Supercritical carbon dioxide is characterized by a low critical temperature (31.1°C), a moderate critical pressure (73.1 atm), and a higher density that is more readily achievable than most SCFs.[5] In addition to the previously described characteristics of SCFs, $scCO_2$ also possesses unique effects on polymeric materials. Some polymers, such as fluoropolymers and silicones, are soluble in $scCO_2$. However, other polymeric materials, such as polyolefins, rarely dissolve into $scCO_2$. Instead, these polymers undergo softening and swelling at a temperature far below the normal glass transition temperature (T_g). The T_g is the temperature at which the long-range segmental motion of a polymer chain ceases.[33] The T_g of many polymers is substantially reduced in the environment of $scCO_2$. The decreased T_g results in softening, which means that the polymer can be more easily mixed with other constituents at a relatively low temperature. This results in a more mild processing condition and a procedure which can prevent polymer degradation. The ability to easily change a material's thermomechanical properties involving boundary rigidity and fluidity while preserving the properties of the substrate makes SCF infusion ideal.

Supercritical Infusion

Supercritical infusion as a process is a technology still in its infancy. The first infusion applications were derivatives of other SCF applications. For example, in 1995, Chen and Levasalmi[9] described using SCF to introduce monomers and initiator into a polymer substrate to produce a polymer blend. This work resulted in the successful preparation of a polystyrene–poly(chlorotrifluoroethylene) blend. Chen and Levasalmi determined that the final product retained the basic geometry and crystallinity of the original substrate. Furthermore, this study illustrated the ability of an SCF to provide "fine tuning" of reaction parameters such as reactant concentration and reaction time. In the same year, Watkins and McCarthy[4] produced a polymer/metal nanocomposite using $scCO_2$. Here, platinum was infused into polytetrafluoroethylene (PTFE) to produce a conductive nanocomposite. The distribution and size of the metal clusters within the composite could be controlled by adjusting the experimental parameters. Also, the successful infusion of particles into PTFE demonstrated the ability of $scCO_2$ to penetrate resilient polymers. The research by Boggess et al.[34] focused on infusing a silver additive into a polyimide (PI). PI is designed to perform in extreme temperature conditions, and there are no known organic solvents for PI. These properties not only make PI ideal for specific applications but also create obstacles for processing. Boggess et al. demonstrated that $scCO_2$ can be used for processes that may otherwise be unfeasible. Other infusion projects include the preparation of starch–silicon–poly(styrene-*co*-allyl alcohol) composites[35] and the infusion of flavor compounds into a polymer substrate.[36] Research interests also focused on the infusion of insect control substances into building materials or pharmaceuticals into medicines.[6] Despite the advances in SCF technology, the infusion of VCIs into a polymer substrate has not been studied sufficiently. A recent patent by Gencer et al.[1] details a method for infusing VCIs into a polymer using a number of different SCFs. Otherwise, VCI packaging materials consist of melt-blended plastics or other VCI-infused paper products.

The supercritical infusion process maintains desirable material characteristics better than traditional melt blending. The VCI particles suspended in the $scCO_2$ are much finer than those mixed into a melt blend. The discharge property of VCI infused by supercritical infusion is believed to be superior. Supercritical infusion allows for a uniform distribution of small particles into the polymer matrix. Further, VCI can also be infused into a polymeric object or article that has been molded into the final form. An additional benefit of the $scCO_2$ infusion process is that any moisture present within the polymer or inhibitor is mixed into the SCF and removed with the effluent stream. This alleviates the problems associated with moisture control during melt compounding. Finally, by reducing or eliminating a portion of the melt compounding and extrusion during preparation, the mechanical properties of the film can be preserved and better retained. Structural degradation due to cross-linking or thermal decomposition can be avoided.[32] This results in preservation of the original molecular weight, and the breathability of the polymer is not as greatly reduced. Furthermore, premature breakdown, agglomeration, or reaction of the VCI during compounding can be minimized, since the supercritical infusion occurs at a much lower temperature than conventional thermal compounding; the processing temperature may be easily reduced by 100°C.

In addition to the previously mentioned advantages, the fluid properties of $scCO_2$ make it an ideal medium for SCF infusion. First, the diffusion coefficients of the VCI molecules are at least one order of magnitude greater in $scCO_2$ than in liquid.[4,36] Additionally, the surface tension of the interface between the polymer and $scCO_2$ is much lower than that with liquid solvents or air. These factors permit the VCI-rich $scCO_2$ to easily penetrate or smear into the polymer matrix.[36] As previously mentioned, the polymer matrix swells and softens while in a $scCO_2$ environment. The swelling creates additional void spaces within the polymer matrix, which in turn allows for easier passage of the VCI molecules. During system depressurization, the $scCO_2$ rapidly returns to CO_2 gas and quickly diffuses out of the polymer. The VCI particles are no longer soluble in gaseous CO_2 and thus are deposited within the matrix. The swollen polymer slowly returns to the original shape and the VCI remains trapped, i.e., infused. Desorption of the VCI particles from the polymer matrix is determined by the diffusion coefficients of the specific VCI within that particular polymer matrix.

Biodegradable Polymers

The packaging industry in particular was significantly impacted by the introduction of petroleum-based polymers. These polymers included PP, PE, poly(ethylene terephthalate) (PET), and poly(vinyl chloride) (PVC). These plastic products are used for their versatility, excellent mechanical properties, and durability. Despite these advantages, two difficulties still exist for the future growth of the polymer industry. First, exhaustion of the nonrenewable resource petroleum for polymer production is a concern, even though new efforts of tapping shale gas for PE production appear to be quite promising. Due to the lack of natural degradation, plastic disposal is also a problem. Ultimately, about 70.7% of all plastics are disposed to a landfill. Polyolefins, which comprises the majority of packaging polymers, are not considered biodegradable.[8] In light of these issues, polymer material research began to focus more on the development of biodegradable polymers. Currently, biodegradable polymers only fill a niche in the overall market, thus making them more expensive to produce.[37]

Biodegradable polymers are designed to degrade by hydrolysis or by the enzymatic action of living organisms such as bacteria, fungi, and algae.[38] These polymers can come from two sources: either biodegradable natural materials or synthetic biodegradable polymers. Naturally occurring biodegradable polymers include starch-based polymers or soy protein-based polymers. Synthetic biodegradable polymers are made using chemical building blocks to produce a polymeric carbon backbone, similar to naturally-occurring polymers. A list of biodegradable polymers is shown in Table 16.3. Initial degradation is caused by either chemical damage, biological attack, heat, or UV damage. The carbon chain is enzymatically broken down during a biological attack. The high molecular weight polymer slowly degrades into low molecular weight molecules. The ultimate products of biodegradation are carbon dioxide and water. Biomass, methane, or other residues may also be produced depending upon the type of polymer and microorganisms present.[39,40] A greater degree of degradation will occur if the polymer is functionalized with groups prone to hydrolysis. This may include groups such as amides, esters, urea, or urethanes.[39]

TABLE 16.3

Common Biodegradable Plastics and their Potential Applications

Biodegradable Polymer	Applications	Manufacturer	Trade Name
Cellulose acetate	Adhesive tape, tubing	Mazzucchelli	BIOCETA®
Poly(ester amide)	Plastic containers, bags	Bayer	BAK 1095®
Aromatic–aliphatic copolyester	Film, food packaging	BASF	Ecoflex®
Poly(lactic acid) (PLA)	Coatings, films, packaging	Cargill-Dow	EcoPLA®
Polycaprolactone	Films, packaging, coatings	Union Carbide	TONE®
Starch-based blends	Bags, packaging	Novon	Poly-NOVON®

Source: Adapted from U.S. EPA[8] and Mohanty et al.[37]

Experiment

The principal objective of the experimental portion of this study is twofold. First, the effects of the SCF conditions, primarily $scCO_2$ density, on the infusion depth were determined. Second, a VCI-infused film is produced, and the capability of this film to prevent oxidation of iron specimen was established. The corrosion prevention ability of the VCI-infused, biodegradable film was compared to a non-biodegradable thermoplastic polymer film infused with VCI. The research results demonstrated the importance of temperature and, to a lesser extent, pressure on the infusion depth of the VCI into the polymer. For this reason, the effect of the $scCO_2$ fluid density, which is dependent on both temperature and pressure, was investigated. Polymer films were supercritically infused with VCI, and the corrosion prevention capability of each film was determined.

Materials Selection

The chief materials required in this study were relatively straightforward: a VCI, a biodegradable polymer resin, a commodity polymer resin, and CO_2 supply. The VCI chosen for infusion was $NaNO_2$, with a trace amount of yellow colorant. $NaNO_2$ is an inorganic anodic inhibitor with a decomposition temperature of 271°C. $NaNO_2$ is most commonly used to protect iron, steel, and aluminum. The $NaNO_2$ was provided by Northern Technologies International Corporation (NTIC, Circle Pines, MN). The biodegradable polymer substrate used was Ecoflex® F BX 7011 and was provided by BASF Plastics. It is a semi-crystalline transparent polymer with mechanical and processing properties similar to that of linear low-density PE.[41] Ecoflex® is a synthetic copolyester resin formed from adipic acid, butanediol, and terephthalic acid. The structure of the Ecoflex® polymer backbone is shown in Figure 16.4. The commodity polymer used as a basis of comparison for this study was Exact 4049 (ρ 0.872 g/cc, MI 4.5 g/10 min). Exact 4049 is a film and packaging-grade resin manufactured by ExxonMobil. The CO_2 used to generate the SCF was Instrument Grade 4.0, 99.9% pure, and was supplied by Airgas (Tulsa, OK).

Experimental System

Supercritical Infusion Reactor System

The setup for the $scCO_2$ experiments consists primarily of a high-pressure reactor with a CO_2 delivery system. A schematic of the process system is shown in Figure 16.5. The lines in

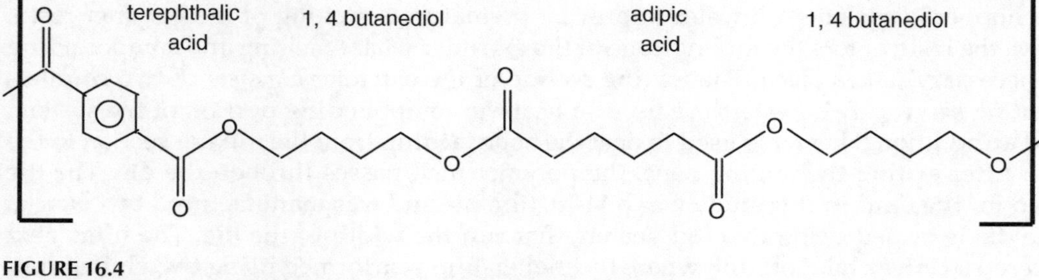

FIGURE 16.4
Molecular structure of Ecoflex® resin used to prepare VCI-infused packaging films.

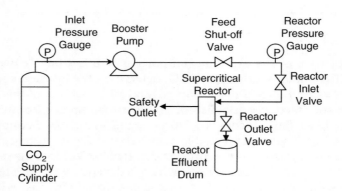

FIGURE 16.5
A schematic of the scCO₂ infusion process system.

the system are made of 316 stainless steel and rated at 7500 psig at 200°F. The reactor is a 300 ml, 316 stainless steel autoclave vessel manufactured by Autoclave Engineers. A specially-made metal thimble is fitted into the reactor and can be used to hold the reactants. The reactor is equipped with a Magnedrive-II agitation system also manufactured by Autoclave Engineers. The reactor is equipped with a series of clamp heaters and type-J thermocouples for temperature control. The inlet line to the reactor is also wrapped with heating tape, but these heating tapes are controlled separately from the reactor heaters and can be used at the operator's discretion. The reactor heaters and the inlet line heating tapes are manually controlled using an Omega temperature controller. The internal portion of the reactor also has a set of optional-use cooling coils. The CO_2 supply pressure is monitored using an Ashcroft pressure gauge. A booster pump is used to achieve pressures in excess of that supplied by the CO_2 cylinder. The booster pump pressure is also observed by using an Ashcroft pressure gauge. The reactor pressure is monitored with a Heise pressure gauge. This experimental setup was used for both the scCO₂ infusion of VCI into the polymer resins as well as for scCO₂ infusion of VCI into the preformed polymer films.

Extrusion Process

The films produced for this study were prepared using a twin-screw extrusion system. A schematic of this system is shown in Figure 16.6. The drive system for the unit is a Haake Rheocord torque rheometer, and the heat control unit is a Eurotherm 808. The polymer resin enters the system through a gravity fed hopper. The feed throat of the hopper is cooled with water to prevent premature softening of the polymer resin. Next, the resin enters the mixing zone of the extruder where melting and compounding (if necessary) takes place. The mixing section of the extruder consists of two counter-rotating screws. Two heaters are used to heat the compounding portion of the system, and an additional heater is used to heat the zone leading from the mixing section to the die. After exiting the mixing zone, the polymer melt passes through the die. The die used for the films in this study was a 14-in. film die and was manufactured by Leistriz. The die is heated using five rod heaters that run the width of the die. The films next entered a Univex take-off unit where the molten film was formed using two chilled roll-

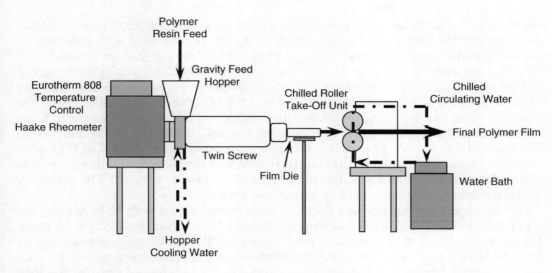

FIGURE 16.6
A schematic of twin-screw extrusion system and take-off unit.

ers. The counter-rotating rollers were cooled using a chilled water bath. The final film is gathered on a spool and stored for later use.

Scanning Electron Microscopy

Scanning electron microscopy (SEM) is an imaging technique in which a beam of electrons is passed over the surface of a sample in order to form a three-dimensional image of the sample. The microscope used for this analysis was a Hitachi S 4700 FESEM. This microscope uses a tungsten electron gun as the electron source. Film samples were imaged at 1000× to determine if any damages or changes occurred on the surface of the film. The primary use for SEM was to examine the surface morphology to verify that the integrity of the films had not been compromised by SCF infusion.

Experimental Procedures

Film Preparation

The films were prepared using the system described in "Materials Selection" section above. For the Ecoflex® resin, the melt temperature was 190°C in the melt zone and 185°C in the die zone. The Engage® resin was melt-processed at a temperature of 183°C in the melt zone and 180°C in the die zone. Both films were processed using the Univex take-off unit with rollers chilled to 15°C. The average thickness of the films was 0.100 mm.

Supercritical Infusion

Fluid density studies. The fluid density studies were performed using the experimental apparatus described in the Experimental Introduction section above. The fluid densities examined were 0.15, 0.20, 0.25, 0.30, 0.35, and 0.40 g/cm^3. The reactor and thimble were cleaned prior to each use. Ecoflex® raw resin (10 g) and $NaNO_2$ (5 g) were loaded into the thimble then placed inside the reactor. The reactor head was mounted on the reactor and bolted down to 55 ft lb_f of torque. Next, the reactor was charged with CO_2 to a pressure of 500 psig. The pressure was monitored for any pressure drops indicative of a leak within the system. After a stable pressure was established, the reactor was heated to the desired temperature, and the agitation was provided. After the intended temperature was achieved, additional CO_2 was added to the reactor to bring the reactor to the final pressure. At this point, the fluid inside the reactor lies within the supercritical region. The reactor was run for a prespecified period of time after which the heaters were turned off and the reactor was slowly vented. The reactor was allowed to cool to 80°F and was then unbolted and the sample removed. The necessary measurements were taken and observations were noted.

Film infusion. The Ecoflex® and Engage® films were infused with $NaNO_2$ using the $scCO_2$ infusion system described in the Experimental Introduction section. The temperature and pressure of the supercritical reactor were 130°F and 1300 psig, respectively, for all runs. $NaNO_2$ (5 g) were loaded into the thimble then placed in the reactor. Next, a 3.3 × 4 $in.^2$ film patch was weighed to determine the mass prior to infusion. The film was then suspended from the unused reactor cooling lines on the reactor closure. Care was taken to ensure that the film sample was not damaged and that both sides of the film could be exposed to the $scCO_2$ fluid. The film was also not in direct contact with the solid $NaNO_2$ powder. The reactor was then sealed and operated in the same manner described in the "Supercritical Infusion Reactor System" section. The film sample was weighed again after being removed from the reactor. The change in mass due to infusion was recorded.

Corrosion Studies

Sample preparation. Iron samples were used as the substrate for the corrosion studies. The iron used was cast grey iron purchased from McMaster–Carr. The original sample was an iron bar with dimensions of 24 × 2 × 0.25 $in.^3$. This portion was cut down into smaller sections with dimensions of 0.5 × 2 × 0.25 $in.^3$. These smaller sections were then cleaned prior to corrosion testing. In order to clean the samples, the iron specimens were polished with two sets of abrasive paper. The first was 600 mesh abrasive paper, and the second was 1000 mesh. Afterward, the samples were placed in an ultrasonic acetone bath for 1 hr to remove any remaining loose fines/particles. The iron sections were removed from the ultrasonic bath, rinsed with deionized water, and dried in a low-humidity oven for 20 min. The iron specimens were weighed and the mass of each was recorded.

Corrosion testing. The samples were removed from the drying oven and placed in a section of the infused film. The film was large enough to be used to craft a small package. The edges of the film were folded together and sealed using a heated press. Each sample was placed in a closed environment for 30 days. Control samples, which were not packaged, were also placed in the oven for the same amount of time. After the required time passed, the samples were removed from the oven. Each sample was polished to remove any corrosion that formed and then placed in the acetone ultrasonic bath. The samples were dried and reweighed to determine the loss of mass due to corrosion. The procedure used to evaluate the corrosion on the samples was based on ASTM Standard G1-03 — Standard Practice for Preparing, Cleaning, and Evaluating Corrosion Test Specimens.[42]

Results and Discussion

Effects of Fluid Density on Infusion

SCF density is an important consideration in SCF applications. The ability to easily modify the fluid density by changing the temperature and pressure is one of the most attractive characteristics of SCF. The solvent strength is directly related to fluid density; thus, by changing the density, the solvent strength is also altered. Table 16.4 shows the SCF density calculated using a modified Peng–Robinson equation for selected supercritical conditions. The influence of fluid density on different SCF applications has been reported by numerous sources. Generally speaking, the solvent strength of the fluid increases as the fluid density increases.[32] For example, Shah et al.[43] determined that increasing the temperature and maintaining a constant pressure results in decreased dispersibility of nanocrystals in ethane. The effects of increasing the temperature of supercritical water result in a decrease in local density, according to Aizawa et al.[44]

The SCF within the reactor was modeled using the Peng–Robinson equation of state, which is given as follows:

$$P = \frac{RT}{v-b} - \frac{a(T)}{v(v+b)+b(v-b)} \tag{7}$$

where:

$$b = 0.07780\,(RT_c/P_c) \tag{8}$$

$$a(T) = \alpha(T)\,0.45724\left[\left(R^2 T_c^2\right)/P_c\right] \tag{9}$$

$$\alpha(T) = \left[1 + m\left(1 - T_r^{1/2}\right)\right]^2 \tag{10}$$

$$m = 0.37464 + 1.54226\omega - 0.26992\omega^2 \tag{11}$$

TABLE 16.4

Density of scCO$_2$ at Selected Supercritical Conditions Calculated using Peng–Robinson Equation

	40°C	60°C	80°C	100°C	120°C
100 bar	0.565	0.293	0.225	0.191	0.169
200 bar	0.830	0.695	0.563	0.461	0.390
300 bar	0.929	0.832	0.735	0.645	0.568

The density values are in the units of g/cm^3. The CO$_2$ density at its critical point is 0.292 g/cm^3.

Here, P is pressure, R is the universal gas constant, V is molar volume, a is the attractive factor, b is the covolume, and ω is the acentric factor. The subscript c is used to denote critical parameters. The critical pressure (P_c) and critical temperature (T_c) for carbon dioxide are 73.1 atm and 31°C, respectively.[44]

Results

The effect of scCO$_2$ fluid density on the infusion depth of the VCI into the polymer was investigated. The infusion penetration depth was determined by using a dissecting microscope and a micrometer. The operating conditions used in this study and the respective infusion depths are shown in Table 16.5.

The results of the study are shown in Figure 16.7. As can be seen, the fluid density does have a slight effect on the infusion depth of the VCI; however, the operating temperature seems to play a more significant role. The effect of temperature on infusion depth is shown in Figure 16.8. Analysis of variance (ANOVA) was used to verify this observation. The effect of SCF density on infusion depth was not statistically significant, whereas the temperature was statistically significant at 95%.[46] Changes in fluid density mainly affect the solubility of the VCI in the scCO$_2$, whereas changes in temperature affect the polymer softening. Generally speaking, both fluid density and temperature are important for SCF operations; however, the specific nature of the process determines the effect each property will impart on the final product. For example, SCF extraction is greatly affected by the ability of the SCF to solubilize the solute of choice. On the other hand, SCF infusion is more dependent on the polymer softening and swelling which is mainly related to temperature. For this reason, temperature has exhibited a stronger effect on the infusion depth than the fluid density.

Corrosion Prevention

The other objective for this study was to compare the corrosion prevention ability of the infused biodegradable films to other films manufactured with non-biodegradable

TABLE 16.5

Experimental Conditions Used for Fluid Density Studies

Fluid Density (g/cm³)	Temperature (°c)	Infusion Depth (cm)
0.15	102	0.0787
0.20	65	0.0624
0.24	60	0.0675
0.25	55	0.0602
0.30	49	0.0570
0.30	71	0.0713
0.36	43	0.0521
0.40	41	0.0515
0.40	60	0.0744
0.4	71	0.0668
0.5	71	0.0681
0.5	41	0.0539
0.62	43	0.0603

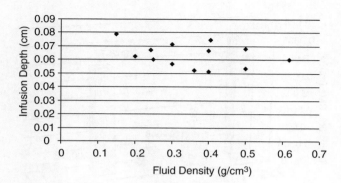

FIGURE 16.7
The infusion depth as a function of fluid density for VCI-infused Ecoflex® resin.

polymers. To date, no studies compared the performance of VCI-infused biodegradable packaging to similar non-biodegradable products. Establishing biodegradable products as good candidates for VCI infusion and corrosion prevention applications will make these products more competitive in the packaging market.

The damage due to corrosion was assessed using three measurements: corrosion rate (CR), protective factor (γ), and protective value (z). The first measurement, the corrosion rate, was evaluated using the following relationship:

$$CR = (K \times W)/(A \times T \times D) \tag{12}$$

where K is the constant 8.76×10^4 (mm/yr), W is the mass loss (g), A is the area (cm^2), T is the time of exposure (hr), and D is the density of the substrate (g/cm^3).[10] The corrosion rate for each type of sample is shown in Figure 16.9. Two additional factors are also used to gauge the effectiveness of the protective package. The protective factor indicates how much lower the corrosion rate is in the presence of the packaging. The protective factor is calculated according to the following equation:[47]

$$\gamma = CR_C / CR_S \tag{13}$$

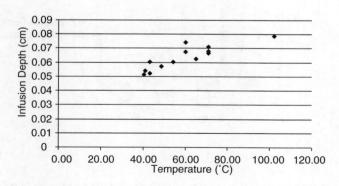

FIGURE 16.8
The effect of temperature on infusion depth of VCI-infused Ecoflex® resin.

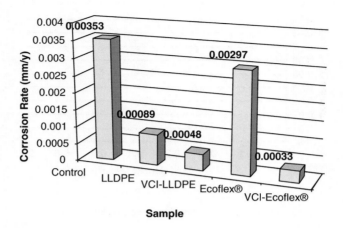

FIGURE 16.9
Calculated corrosion rates for iron samples resulting from different packaging techniques.

where CR_C and CR_S are the corrosion rates of the control and experimental samples, respectively. The protective value shows the percent by which the corrosion is inhibited in the presence of the VCI-infused package. This value is calculated according to the following relationship:

$$z = [1 - (1/\gamma)] \times 100\% \qquad (14)$$

The calculated values of γ and z are shown in Figure 16.10. As expected, the corrosion rate for the control sample that was exposed to the air had the greatest rate of corrosion (0.00353 mm/yr). The samples contained in the uninfused polymer bags had the next greatest corrosion rate. The LLDPE and Ecoflex® bags had corrosion rates of 0.00089 and 0.00297 mm/yr, respectively. The sample in the biodegradable Ecoflex® bag exhibited a corrosion rate of about 10 times greater than that of the non-biodegradable LLDPE bag.

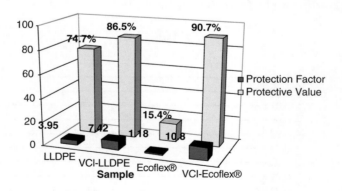

FIGURE 16.10
Calculated protective factor (γ) and protective value (z) for corrosion of iron samples in polyethylene (LLDPE), VCI-infused polyethylene (VCI-LLDPE), biodegradable polymer (Ecoflex®), and VCI-infused biodegradable polymer (VCI-Ecoflex®) packaging.

This is most likely due to a higher amount of water molecules present in the Ecoflex® sample. First, Ecoflex® has a greater tendency to absorb moisture from the environment. Also, biodegradable polymers permit moisture to move in and out more than other hydrophobic films. Thus, the metal corrosion is promoted by the presence of additional moisture on the metal surface. The VCI-infused storage bags proved to provide the greatest amount of corrosion protection. The VCI-Ecoflex® package performed marginally better than the VCI-LLDPE, and the respective corrosion rates for samples inside the VCI-LLDPE and VCI-Ecoflex® samples were 0.00048 and 0.00033 mm/yr, respectively. This illustrates that biodegradable Ecoflex® packaging can be at least competitive with non-biodegradable LLDPE packaging. However, the added function as a biodegradable packaging makes it superior to conventional non-biodegradable VCI packaging film.

Supercritical Infusion

Polymer Infusion

The films and resins were both successfully infused with $NaNO_2$ using $scCO_2$. The infusion was evident based on the color change that was observed. Figure 16.11 shows the films before and after the $scCO_2$ infusion process. As shown, the films changed from a transparent film with a slight whitish color to a transparent yellow film. The degree of infusion was also assessed by evaluating the amount of VCI infused into the films. The total amount of $NaNO_2$ infused into the polymer films was determined by calculating the change in mass before and after infusion. These changes are shown in Table 16.6. The average weight change for each type of film is similar, and statistical analysis confirms that the differences due to type of film are insignificant. The overall mass increases were fairly substantial, thus indicating good infusion of VCI into the polymer matrices. The LLDPE films exhibit a slightly higher increase in weight compared to the Ecoflex® films. LLDPE has a lower material density than Ecoflex®, and this difference may result in a higher average wt.% increase for LLDPE. The Ecoflex® resin showed a greater increase in weight when compared to the Ecoflex® films. This was most likely due to the direct contact of the resin pellets with the $NaNO_2$. The films were suspended within the VCI-rich SCF, whereas the resin was in direct contact with the $NaNO_2$ powder. Some residual VCI may be present on

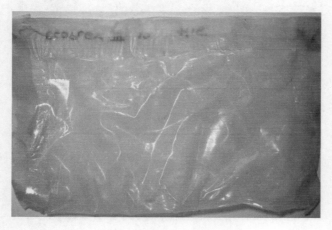

FIGURE 16.11
Films infused with VCI (upper photo) and uninfused (lower photo) films.

TABLE 16.6

Average Change in Weight and wt.% Due to $NaNO_2$ Infusion in Supercritically Infused LLDPE Films, Ecoflex® Films, and Ecoflex® Resins

Polymer	Average wt. Change (g)	Average wt.% Increase
LLDPE film	0.09153 ± 0.018	5.92
Ecoflex® film	0.10325 ± 0.011	3.97
Ecoflex® resin	0.8545 ± 0.143	7.88

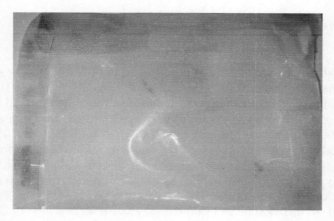

FIGURE 16.12
Electron micrograph of Ecoflex® film prior to $scCO_2$ infusion.

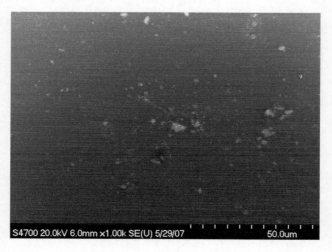

FIGURE 16.13
Electron micrograph of Ecoflex® film following $scCO_2$ infusion.

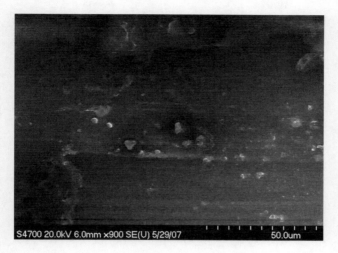

FIGURE 16.14
Electron micrograph of LLDPE film prior to scCO$_2$ infusion.

the surface of the resin and cause an increase in the average weight. Any surface residual VCI present was not separately measured or accounted for.

Scanning Electron Microscopy

SEM was performed to determine if any changes occurred on the surface of the film samples. Ideally, the SCF infusion should not result in any surface changes because the infusion temperature is below the melting and softening temperatures of the polymer resin. The results of SEM analysis are shown in Figures 16.12–16.15. The changes in Ecoflex® film before and after infusion are shown in Figures 16.12 and 16.13, respectively. The film structure of the Ecoflex® samples exhibits minimal degradation. Some surface inconsistencies exist; however, these are most likely due to mechanical stresses, such as bending, that

FIGURE 16.15
Electron micrograph of LLDPE film prior to scCO$_2$ infusion.

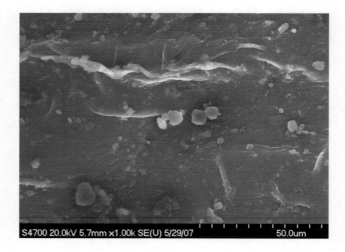

S4700 20.0kV 5.7mm ×1.00k SE(U) 5/29/07   50.0um

FIGURE 16.16

were necessary in order to fit the film in the reactor. Mechanical stress is not anticipated for products infused in a process designed to accommodate films. The SEM results for the LLDPE films are shown in Figures 16.14 and 16.15. The LLDPE films showed slightly more degradation than the Ecoflex® films following infusion. Visual examination of the infused films shows that the LLDPE films lost some clarity and appeared to look "softened" following infusion. Additionally, some of the defects are due to mechanical stresses rather than the supercritical infusion itself. However, none of the imperfections compromised the barrier properties of the LLDPE films. Overall, the films were successfully infused with VCI and the integrity of the films was not compromised.

The absence of surface defects following supercritical infusion is a significant result. A VCI infusion process that can be performed after the polymer is molded (or extruded) means that SCF infusion can be performed on a number of substrates without concern that infusion will damage the substrate. Plastic containers, foams, and other preformed objects can be used as substrates. Additionally, the infused substrates can be recharged with VCI after a certain period of use without undergoing another cycle of extrusion or molding.

Conclusion

The focus of this work was twofold. First, SCF was used to infuse polymer films and resins with VCI particles in order to produce packaging materials that can prevent metal oxidation. The success of this research enables low-temperature compounding of VCI-infused polymeric resins as well as low-temperature infusion of VCI into preformed polymeric films and articles, thus preventing undesirable thermal degradation of macromolecules and VCI during processing as well as avoiding moisture-sensitive melt processing. The benefits of this study were further amplified by using a biodegradable film as the packaging material of choice as opposed to a traditional commodity resin. Thus, the final product attains dual functionalities, viz., corrosion prevention and biodegradation. The comparative results of this study showed that the Ecoflex®-VCI packaging performed slightly better than the non-biodegradable LLDPE-VCI packaging in terms of corrosion prevention capability. This outcome illustrates the potential of Ecoflex® as an excellent alternative for manufacturing CI packaging, and validates the development of processes designed to supercritically infuse biodegradable resin.

The second objective of this work was to devise a number of processes that can be used to perform SCF infusion of polymer resins or films with inorganic crystalline materials such as $NaNO_2$. The supercritical infusion process developed here can be applied in two distinct processing modes, viz., preforming (i.e. pre-extrusion or premolding) and postforming. The former provides an excellent alternative to melt compounding of specialty resin additives, whereas the latter renders the unique capability of charging and recharging VCI without morphologically deforming the premolded polymeric articles.

Overall, this study was also successful in demonstrating that $scCO_2$ infusion is a realistic approach for manufacturing functional products that incorporate materials with specific limitations, i.e., temperature and moisture sensitivity.

Acknowledgment

The authors are grateful for the support of Northern Technologies International Corporation (Circle Pines, MN).

References

1. Gencer, M.A., Lee, Sunggyu, L., Sardesai, A., Kubik, D.A. Process for incorporating one or more materials into a polymer composition and products produced thereby, U.S. patent 7,217,750, May 2007.
2. Selke, S., et. al. *Plastics Packaging: Properties, Processing, Applications, and Regulations.* Hanser Publishers: Munich, 2004.
3. Pinchuk, L.S., Neverov, A.S. *Polymer Films Containing Corrosion Inhibitors.* Khimiya, Moscow, 1993.
4. Watkins, J.J., McCarthy, T.J. Polymer/metal nanocomposite synthesis in supercritical CO_2. *Chemistry of Materials* **1995**, 7(11), 1991–1194.
5. Sun, Y.-P. *Supercritical Fluid Technology in Materials Science and Engineering: Synthesis, Properties, and Applications.* Marcel-Dekkar: New York, 2002.
6. McHugh, M., Krukonis, V. *Supercritical Fluid Extraction*, 2nd ed., Butterworth-Heinemann, Boston, 1994.
7. Sangaj, N.S., Malshe, V.C. Review: Permeability of polymers in protective organic coatings. *Progress in Organic Coatings* **2004**, 50, 28–39.
8. U.S. EPA, "Wastes–Nonhazardous Waste–Municipal Solid Waste," http://www.epa.gov/osw/nonhaz/municipal/, accessed June 2011.
9. Chen, W., Levasalmi, J.M. New approaches to polymer modification, Society of Professional Engineers—Annual Technical Conference, 1995.

10. Koch, G.H., Brongers, M.P.H., Thompson, N.G., Virmani, Y.P., Payer, J.H. Direct costs of corrosion in the United States. *Corrosion: Fundamentals, Testing, and Protection*, Vol. 13. ASM Handbook, ASM International, 2003, 959–967.

11. Goldade, V.A., Pinchuk, L.S., Makarevich, A.V., Kestelman, V.N. *Plastics for Corrosion Inhibition*, Springer-Verlag, Berlin, Germany, 2005.

12. Brown, T., LeMay, H.E., Bursten, B. *Chemistry: The Central Science*, 7th ed., Prentice-Hall, Upper Saddle River, NJ, 1997.

13. Kaesche, H. *Corrosion of Metals: Physicochemical Principles and Current Problems*. Springer-Verlag, Berlin, 2003.

14. Callister, W.D. *Materials Science and Engineering: An Introduction*. Wiley, New York, 2000.

15. Dorfman, A.M., Zolotovitskii, Y.M., Lyakhovich, A.M., Chalykh, A.E., Reshetnikov, S.M. The desorption of a volatile corrosion inhibitor from polyethylene. *Protection of Metals* **2001**, *37*(1), 9–12.

16. Heidemann, G. Corrosion protection through VCI paper. *Anti-Corrosion Methods and Materials* **1975**, *26*(12), 12–13.

17. Heidemann, G. Corrosion Protection of copper, aluminum, and silver with Protective Paper. *Anti-Corrosion Methods and Materials* **1981**, *28*(12), 5–6.

18. Walia, P.S., Lawton, J.W., Shogren, R.L., Felker, F.C. Effect of moisture level on the morphology and melt flow behavior of thermoplastic starch/poly(hydroxyl ester ether) blends. *Polymer* **2000**, *41*(22), 8083–8093.

19. Utracki, L.A. *Commercial Polymer Blends*, Chapman & Hall, London, 1998.

20. Marcovich, N.E., Reboredo, M.M., Aranguren, M.I. Composites from sawdust and unsaturated polyester. *Journal of Applied Polymer Science* **1996**, 61, 119–124.

21. Cutillo, R., Jackson, S. Compounding wood–polymer composites with in-line drying technology. *Journal of Vinyl and Additive Technology* **2005**, *11*(14), 166–169.

22. Lee, S., Lanterman, H.B., Fullerton, K.L., Pettit, P. Polymerization, compatibilized blending, and particle size control of powder coatings in a supercritical fluid, U.S. patent 6,340,722, January 22, 2002.

23. Lee, S., Lanterman, H.B., Sardesai, A., Wenzel, J., Marshall, B., Yen, J., Amin-Sanayei, R., Moucharik, M. Polymerization of poly(vinylidene fluoride) in a supercritical fluid medium, U.S. patent 7,091,288, August 15, 2006.

24. Lee, S., Kwak, S., Azzam, F.O. Graft copolymerization in supercritical fluid media, U.S. patent 5,663,237, September 2, 1997.

25. Trivedi, A.H., Kwak, S., S. Lee, S. Grafting of poly(vinyl chloride) and polypropylene with styrene in a supercritical co₂ solvent medium: synthesis and characterization. *Polymer Engineering & Science* **2001**, *41*(11), 1923–1937.

26. Clark, K., Lee, S. Supercritical fluid process for synthesis of maleated poly(vinylidene fluoride). *Polymer Engineering & Science* **2005**, *45*(5), 631–639.

27. Azzam, F.O., Lee, S. Selective oxidation of hydrocarbons via supercritical wet oxidation. *ACS Fuel Preprint* **1992**, *37*(4), 1280–1283.

28. Benko, D., Beer, R., Lee, S., Clark, K. Devulcanization of cured rubber, U.S. patent 6,992,116, January 31, 2006.

29. Lee, S., Garcia, A., Wootton, J. Systems for water purification through supercritical oxidation, U.S. patent 7,186,345, March 6, 2007.

30. Racette, T.L., Damaso, G.R., Schulte, J.E. Cleaning system utilizing an organic cleaning solvent and pressurized fluid solvent, U.S. patent 7,097,715, August 29, 2006.

31. Lee, S. Lanterman, H.B., Fullerton, K.L., Pettit, P. Polymerization, compatibilized blending, and particle size control of powder coatings in a supercritical fluid, U.S. patent 6,849,678, February 1, 2005.

32. Lee, S., Gencer, M.A., Azzam, F.O., Fullerton, K.L. Depolymerization process, U.S. patent 5,386,055, January 31, 1995.

33. Fried, J.R. *Polymer Science and Technology*, 2nd ed., Prentice-Hall Professional Technical Reference, New Jersey, 2003.

34. Boggess, R.K., Taylor, L.T., Stoakley, D.M., St.Clair, A.K. Highly reflective polyimide films created by supercritical fluid infusion of a silver additive. *Journal of Applied Polymer Science* **1997**, *64*(7), 1309–1317.

35. Green, J.W., Rubal, M.J., Osman, B.M., Welsch, R.L., Cassidy, P.E., Fitch, J.W., Blanda, M.T. Silicon-organic hybrid polymers and composites prepared in supercritical carbon dioxide. *Polymers for Advanced Technologies* **2000**, *11*(8–12), 820–825.

36. Avison, S.J., Gray, D.A., Davidson, G.M., Taylor, A.J. Infusion of volatile flavor compounds into low-density polyethylene. *Journal of Agriculture and Food Chemistry* **2001**, *49*(1), 270–275.

37. Mohanty, A.K., Misra, M., Hinrichsen, G. Biofibres, biodegradable polymers and biocomposites: An overview. *Macromolecular Materials and Engineering* **2000**, *276–277*(1), 1–65.

38. Gross, R.A., Bhanu, K. Biodegradable polymers for the environment. *Science* **2002**, *297*(5582), 803–807.

39. Jayasekara, R., Harding, I., Bowater, I., Lonergan, G. Biodegradability of a selected range of polymers and polymer blends and standard methods for assessment of biodegradation. *Journal of Polymers and the Environment* **2005**, *13*(3), 231–251.

40. Kleeberg, I., et. al. Biodegradation of aliphatic–aromatic copolyesters by *Thermomonospora fusca* and other thermophilic compost isolates. *Applied and Environmental Microbiology* **1998**, *64*(5), 1731–1735.

41. Ecoflex® F BX 7011, Product data sheet, September 2001, http://iwww.plasticsportal.com/products/ecoflex.html.

42. Standard practice for preparing, cleaning, and evaluating corrosion test specimens, ASTM Standard G1-03, ASTM International, 2003, 1–9.

43. Shah, P.S., Hanrath, T., Johnston, K.P., Korgel, B.A. Nanocrystal and nanowire synthesis and dispersibility in supercritical fluids. *Journal of Physical Chemistry B* **2004**, *108*(28), 9574–9587.

44. Aizawa, T., Kanakubo, M. Hijima, Y., Ikushima, Y., Smith, R.L. Jr. Temperature dependence of local density augmentation for acetophenone *N,N,N',N'*-tetramethylbenzidine exciplex in supercritical water. *Journal of Physical Chemistry A*, **2005**, *109*(33), 7353–7358.

45. Smith, J.M., Van Ness, H.C., Abbott, M.M. *Introduction to Chemical Engineering Thermodynamics*, 6th ed., McGraw-Hill, 2001.

46. Box, G.E.P., Hunter, J.S., Hunter, W.G. *Statistics for Experimenters: Design, Innovation and Discovery*, 2nd ed., Wiley Series in Probability and Statistics, Wiley-Interscience, Wiley, 2005.

47. Tyr, S.G., Moshura, O.V., Tyr, E.V. The prospects of using volatile corrosion inhibitors. *Protection of Metals* **2001**, *37*(6), 534–538.

48. Lambert, J.M., Hukvari, S.C. High-pressure reactor design, in Encyclopedia of Chemical Processing, Vol 2, Sunngyu Lee, Ed., Taylor and Francis, New York, 2006.

49. Perry, R.H., Green, D.W., Maloney, J.O. *Perry's Chemical Engineers' Handbook*, 6th ed., McGraw-Hill, 1984.

50. Geankoplis, C. *Transport Processes and Unit Operations*, 3rd ed., Prentice-Hall, New Jersey, 1993.

Section V

Environment and Safety

17

Electrostatic Precipitation

Kenneth R. Parker

CONTENTS

Introduction 281
Background 282
Basic Operating Principles 282
The Physics of Precipitation 285
    Ion Production 285
    Particle Charging 287
    Particle Migration 288
    Particle Deposition and Removal from the Collector Electrodes 291
Factors Affecting the Design and Performance of Precipitators 292
    Gas Composition 292
    Gas Temperature 293
    Gas Pressure 293
    Gas Flow Rate 293
    Viscosity and Density 294
    Particle Concentration 294
    Particle Composition and Electrical Resistivity 294
    Particle Sizing 295
    Particle Shape 295
    Particulate Surface Properties 295
Sizing of Electrostatic Precipitators 296
Practical Approach to Industrial Precipitator Sizing 297
Versatility of the Electrostatic Precipitator 298
Conclusions 299
References 299

Introduction

Although electrostatic precipitators have been employed almost for a century to remove particulates, fumes, and mists from a wide range of industrial processes, the main approach for assessing the size and design of a new plant has been based on measurements from units operating under similar conditions. Recent developments using computerized fluid dynamics approaches have enabled the physics to be theoretically studied in greater detail, which leads to an improved understanding of precipitation. In this chapter, the precipitator operation is reviewed from theoretical and practical aspects and some of the gas and particulate characteristics, which have to be considered when designing a precipitator for a specific duty, have been indicated. While it is mainly used currently for pollution control purposes, where efficiencies of 99.8%+ are required to comply with emission legislation, the precipitator can be used for the recovery of valuable materials, cleaning process gases for subsequent reuse, air cleaning duties, and the cleaning of process oil streams. The recent issue by the EPA of the ESPVI Series 4W precipitator program enables power station precipitators to be proactively studied to speedily evaluate low-cost enhancement scenarios for precipitators that require upgrading to meet more stringent emission regulations.

Background

Electrostatic precipitators have been used commercially for almost a century for the collection of particulates present in the gas streams on many industrial processes. The early plants were installed to recover valuable material that, otherwise, would have been lost to atmosphere, rather than in preventing atmospheric pollution. These plants, mainly in the sulfuric acid, bullion, and nonferrous smelting industries, were designed and priced to produce a commercial "pay back" in the order of 3–5 years. In recent years, with the increasing awareness and worldwide recognition of the problems associated with atmospheric pollution, most industrialized nations have enacted legislation to limit emissions from all sources. This legislation is continuously reviewed and is becoming steadily more stringent, such that, in general, particulate emissions are presently controlled to a maximum of 50 mg/Nm^3 for inert materials and a maximum of 10 mg/Nm^3 for substances that are considered hazardous to health.

The features of the electrostatic precipitation process that make it an ideal vehicle for the removal of particulates are

1. Versatility and effective performance on a wide range of industrial processes. Can be designed to satisfy any required efficiency and gas flow rate.
2. Designs can be produced to cover a temperature range from ambient up to 850°C.
3. Can collect particles over the complete size range spectrum.
4. Material is usually recovered in its original state. (Plants can be designed to operate as a wet phase device if required.)
5. Low pressure loss; typically less that 1 mbar.
6. Acceptable electrical power consumption for required efficiency level.
7. Robust and reliable construction and a life expectancy of more than 20 years.
8. Low maintenance requirements.

Basic Operating Principles

The basic principle of operation is that the particulates are passed through an electric field where they initially receive an electric charge to become charged particles and are deflected across the field to be retained on the counter electrode. Most industrial precipitators are

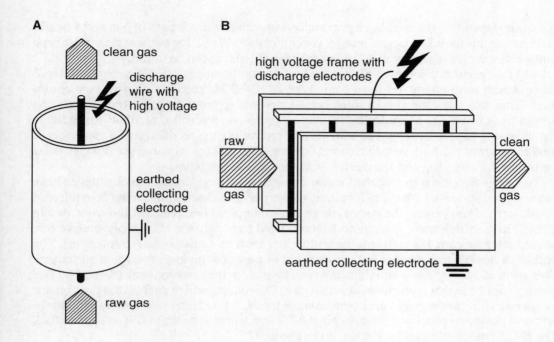

FIGURE 17.1
Basic single-stage ESP arrangements: (A) a wire tube arrangement and (B) and parallel plate arrangement.

based on a single-stage approach where both charging and migration (precipitation) take place within the same set of electrodes, as illustrated in Figure 17.1. Another type of unit, particularly used for air cleaning duties, is two-stage precipitation, where there is a separate charging field followed by a collection field, as illustrated in Figure 17.2.

In general, the basic components of any precipitator consist of some form of small diameter discharging element and, for a parallel flow unit, a flat plate counter electrode, or, in the case of vertical flow mist type of installation, a large diameter tube.

In practice, the form of discharge element ranges from a simple weighted wire, typical of tubular units, to specially designed controlled emission electrodes, used in a number of industries, not only to mitigate corona suppression from high concentrations of fine particulates but also to provide a long and trouble-free operating life.[1] The receiving, or collecting, electrodes range from stiffened flat plates to some form of channel construction for multifield high efficiency parallel flow applications. The size and the form of receiving

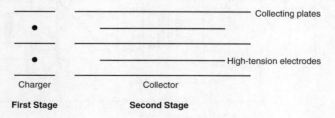

FIGURE 17.2
Two-stage ESP arrangement.

electrode depend on the supplier's preference, but plates with a length of 5 m and a height of 15 m can be found in power station precipitators.[2] While for vertical flow mist type units, tubes with a diameter of 300 mm and a length of 6 m can be found.

Until the recent development of switched mode power supplies (SMPS), most industrial precipitators were energized from some type of 50–60 Hz rectified high voltage equipment. This basically comprised a suitably designed and insulated step-up transformer, the output from which can be controlled by employing silicon controlled rectifiers to modulate the incoming supply voltage. The transformer output is then rectified using silicon rectifiers to produce the high negative direct current (DC) voltage required for the necessary ionization for charging and the electric field for particle migration.

Recent developments in switched mode or high frequency direct current supplies have been applied to new and some older plants where performance enhancement is required.[3] In this form of equipment, the normal three-phase supply is first rectified and converted to a 25–50 kHz square wave, which can be modulated to supply the HT transformer, which, having a ferrite core, is much smaller and lighter than its conventional counter part. The output of the HT transformer is then rectified to produce the high voltage requirement. This form of supply has a number of advantages over the conventional (50–60 Hz) frequency supply, in that it produces almost a pure DC voltage, which enables the precipitator to operate at higher voltages and performance levels, while being much more efficient in terms of power conversion (95% as against 85%) and a power factor of 0.94 as against 0.63. The SMPS basic building block is shown in Figure 17.3.

For air cleaning duties, although energization is normally derived from conventional supplies, the voltages are much lower, 6–12 kV, and typically positive, because of a lower ozone production, resulting from the reaction between the ions and the oxygen in the air.

In operation, the high electrical field adjacent to the energized electrode elements ionizes the gas molecules, forming both positive and negative ions. With negative energization, the positive ions are immediately captured, while the negative ions and any electrons, generally referred to as a corona discharge, migrate under the influence of the electric field into the interelectrode space. As the gas borne particles pass through the interelectrode space, the larger particles receive an electric charge either by collision with the ions/electrons or by diffusion charging for the smallest particles. The charged particles then move under the influence of the electric field and migrate to the collecting electrodes, where the charge subsequently leaks away to earth. In the interim, before becoming neutral, the charged particles are retained on the collector surface by a combination of van der Waals and electric forces.

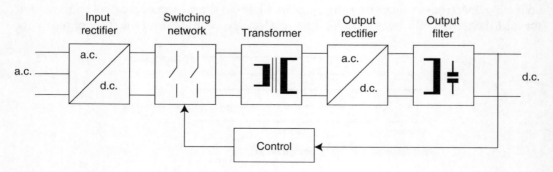

FIGURE 17.3
Basic building block for a switch mode power supply.

In a dry precipitator to ensure that the collection process is continuous, after a period of time the collector electrodes are generally mechanically rapped to remove the deposited material. This time aspect, together with the frequency and intensity of the rapping, is important to minimize rapping re-entrainment and hence maximize collection efficiency. In wet precipitators, the particles are deposited on a film of running water or removed by flushing, while for mist precipitators, the droplets are usually selfdraining.

The Physics of Precipitation

Ion Production

Although there are various methods of particle charging, for example, triboelectric, ultraviolet, and radiation effects, for industrial precipitator applications corona charging is universally used, as it is the most efficient and cost-effective approach. The physics of corona or ion production therefore occupies an essential position in the practice of electrostatic precipitation. Investigations into the physics of ionization date back to the middle of the last century and are still ongoing, particularly since the development and application of computational fluid dynamics (CFD) using fast computers capable of deriving satisfactory solutions to the Poisson and La Place equations.

Early investigations were concerned with developing voltage-current relationships of corona discharge and the effects of positive and negative energization systems using wire and tube arrangements. Gaugain[4] found that the breakdown potential for a given outer diameter R depends on the diameter of the central discharge element r and derived the following relationship:

$$E = A + (C/r)^{1/3} \qquad (1)$$

where E is the electrical breakdown field, r, the radius of inner electrode, and A and C are experimental constants (for $r \ll R$).

Röentgen,[5] working with point/plane electrodes, found that a certain voltage had to be applied to initiate a corona current flow. This corona onset voltage was dependent on the sharpness of the point, gas pressure, and polarity of the point electrode. Further investigations produced a parabolic relationship for negative corona current flow as:

$$I = AV(V - M) \qquad (2)$$

where I is the corona current flow, V the applied voltage, and A and M are experimental constants.

Townsend[6] noted that, with negative ionization of the discharge electrode, the appearance of the corona discharge was very different to when it was positively energized, when the corona appeared as a diffuse glow surrounding the emitter. With negative energization, the corona appeared as bright flares, which tend to move across the surface of the electrode, and had a distinctive hissing sound, a lower initiation voltage, and a higher breakdown potential. The visual appearance of negative corona takes the form of "tufts"—bright glow points—or Trichel pulses.[7] The number of tufts and their luminescence increase as the energizing voltage is raised, producing a higher corona current discharge.

The conduction of electricity through gases is fundamentally different to that in solids and liquids, which contain "charge carriers" that move under the influence of the electric field to produce the current flow. With gases, however, ions need to be provided from some outside force or agency to induce a current flow. For precipitators, this outside agency is high voltage applied across the electrode system. In Figure 17.4, a typical electric field distribution between a small diameter emitter and a much larger passive electrode is shown (i.e., r ≪ R).

This shows the electric field adjacent to the emitter is extremely high and it is this stress that excites any free electrons in the immediate vicinity. These fast moving electrons acquire sufficient energy from the applied electric field so as to collide with other gas molecules to produce further free electrons and positive ions. Townsend,[6] working in this area, proposed the concept of a chain reaction or electron avalanche, in which each new electron produced generates new electrons by ionization in ever increasing numbers.

The number of ions at a distance x from the active zone can be represented by:

$$n = n_0 e^{ax} \tag{3}$$

where n_0 represents the number of ions at a distance $x = 0$ and a is the Townsend ionization coefficient, which varies with the gas temperature, pressure, and electric field strength.

In general terms, the field strength varies with distance across the field, x, as in a precipitator. Hence the equation takes the integral form:

$$n = n_{oe} \int_0^x a\,dx \tag{4}$$

The term $1/a$ is the mean free electron path between collisions.

In Figure 17.5, the relationship between the Townsend ionization coefficients (ion pair production) for air at atmospheric pressure at different temperatures and field strengths is given.

In the figure, it can be seen that, at 20°C, doubling of the field strength results in the number of ion pairs, or Townsend coefficient, increasing by a factor of 20. The impact of a rise in gas temperature, which increases the mean free electron path and produces significantly increased ionization, is also indicated.

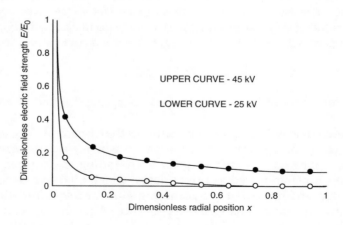

FIGURE 17.4
Relative field strength between the electrodes of a tubular precipitator arrangement. (*View this art in color at www.dekker.com.*)

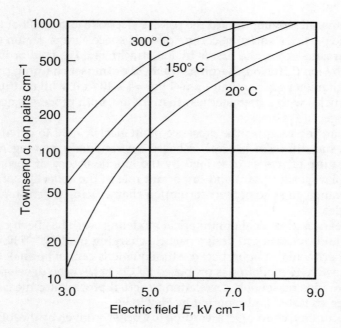

FIGURE 17.5
Effect of electric field strength and temperature on ion pair production.

In practice, when the system is negatively energized, the electrons collide with and attach themselves to gaseous molecules to produce negative ions as they rapidly move across the field area, although there are a large number of ion pairs immediately adjacent to the discharge element. Concurrently, the positive ions are attracted toward the discharge element and, although during transit they produce further ion pairs on reaching the element take no further part in the process. As the distance from the element increases, because of attachment, the number of electrons decreases, and there is a corresponding increase in the number of negative ions. The net number of electrons at a distance x from the electrode is represented by the following equation:

$$n = n_{oe} \int_{o}^{x} (a-\eta)dx$$

(5)

where η is the coefficient of ion attachment or the Townsend second coefficient.

In the case of positive energization, normally found in air cleaning applications, the primary electrons produced at the boundary of the visible glow are attracted toward the emitter. In moving through the field, they collide with and produce new ion pairs by impact ionization, with the positive ions migrating toward the passive earthed electrode. Except for the collection of electrons, the emitter by itself plays little part in the ionization phenomenon, which is essentially a gas process with the primary electrons being released from the gas molecules through photoelectric effects in the plasma region.

Particle Charging

Particle charging occurs in the area between the active plasma region and the passive electrode surface. This area comprises a high space charge having neutral ions, negative ions, and some free electrons, all moving toward the passive electrode as a result of the electric field. As the gas borne particles enter the corona derived space charge region of the field, two charging mechanisms occur: the first is by ion attachment, i.e., field or impact charging, and the second by ion diffusion charging. The field or impact charging predominates for particles with a diameter greater than 1 μm (1 μm = 1×10^{-6} m), while diffusion charging is essential for particles with a diameter less than 0.2 μm; both processes occur in the intermediate size range.

While field charging requires the presence of an electric field to drive the free mobile charge carriers, the diffusion process is based on randomly moving gas ions arising through temperature effects as described by the kinetic theory of gases, i.e., Brownian motion, which plays a significant and important role in the collection of particles in the submicron size range, in spite of their saturation charge being much smaller than that of the larger particles.

Over recent years, a great deal of numerical modeling work has been carried out using computational fluid dynamics to derive particle charging models;[8–10] however, the basic models need experimental support because the equations cannot be analytically solved. A reasonable alternative to modeling is proposed by Cochet,[11] who developed an equation, which appears to give reasonable correlation to actual precipitator measurements in the critical size range, as tested and reported by Hewitt.[12]

The particle saturation charge Q_p^∞, according to Cochet, is given by the following formula:

$$Q_p^\infty = \left\{ \left(1 - \frac{2\lambda}{d_p} \right)^2 + \frac{2}{1 + 2\lambda/d_p} \times \left(\frac{\varepsilon_r - 1}{\varepsilon_r + 2} \right) \right\} \times \pi \varepsilon_0 d_p^2 E \tag{6}$$

where ε_r is the electrical permittivity of the particle, ε_o the electrical permittivity of the gas, d_p the particle diameter, E the electric field strength, and λ the mean free path electron path.

The equation can be rewritten in simplified terms:

$$Q_p^\infty = p E d_p^2 \tag{7}$$

where p is a constant, which varies between 1 and 3 fornonconductive particles and is 2 for conductive particles.

In practice, both field and diffusion charging occur simultaneously and are basically inseparable. It can be shown that all particles reach around 90% of their full saturation charge in less than 0.1 sec. Hence, for all practical conditions, it can be assumed that on entering the precipitation field, having a typical exposure/treatment time of 3–5 sec, the particles rapidly achieve their saturation charge.

Particle Migration

Within the precipitation field, a particle experiences the following forces acting upon it: a momentum force, $F_m = ma$; an electrical force, $F_e = Q_p E$; and a drag force, $F_d = ReAC_o$ (Re is the Reynolds number and C_o the Cunningham coefficient).

Under a steady state condition

$$F_m + F_e + F_d = 0 \tag{8}$$

Prior to solving this equation, the drag force F_d has to be calculated. In the case of low Reynolds number, the drag coefficient is given by

$$C_o = 24/\text{Re} \tag{9}$$

As the particle size d_p reduces and approaches the region where the fluid loses its continuum (mean free path of molecules = λ), Stokes law needs to corrected by the Cunningham correction factor, C_o.

$$C_o = 1 + 1.246 \times 2\lambda \, / \, dp + 0.42 \times 2\lambda/dp \times e^{(0.87dp/2\lambda)} \tag{10}$$

This relationship is plotted in Figure 17.6, in which the significant correction factor that arises for submicron sized particles is shown.

The drag force (or Stokes law) can be written as

$$F_d = 3\pi\eta d_p \omega_{th} \times \frac{1}{C_0} \tag{11}$$

where η is the gas viscosity, d_p the particle diameter, C_0 the Cunningham correction, and ω_{th} the theoretical particle migration velocity.

Assuming the fluid has no component acting toward the passive electrode and all particles achieve their saturation charge, the equation of motion of a charged spherical particle in an electric field is characterized by

$$\frac{d\omega_{th}}{dt} + \frac{3\pi\eta d_p \omega_{th}}{mC_0} = \frac{Q^{\infty}{}_p E}{m} \tag{12}$$

Taking $\omega_{th} = 0$ at $t = 0$, the solution of the above equation can be readily found, i.e.,

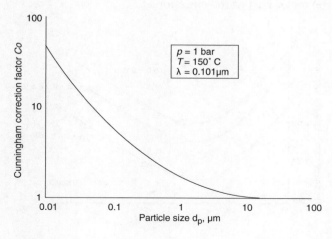

FIGURE 17.6

Relationship between particle size and the Cunningham correction factor.

$$\omega_{th} = \frac{Q^{\infty}_{p} E}{3\pi\eta d_{p}} \times C_0 \tag{13}$$

As $Q^{\infty}_{p} = pEd^2_{p}$,

$$\omega_{th} = E^2 \times d_{p} \times C_0 \tag{14}$$

The significance of this relationship is

    a. As the limiting charge on the particle is proportional to the radius squared, the migration velocity of the particle will increase with particle size.

    b. As the electric field is proportional to the applied voltage, the migration velocity is proportional to the voltage squared.

It will be appreciated from this fundamental approach that the operation of an electrostatic precipitator requires a voltage high enough to produce an electric field to precipitate the particles and deliver sufficient corona current to satisfy ion production for initially charging the particles.

In Figure 17.7, the relationship between the theoretical migration velocity ω_{th} against particle size for three different field strengths in a typical flue gas at 150°C, for particles having an electrical permittivity of 10, is given. Because of the logarithmic scale, the effect of the field strength squared factor is not apparent, unless one examines the actual migration velocity figures.

Although the foregoing portrays the theoretical approach to the migration of charged particles through an electric field, the theoretical migration velocity should not be confused with the Deutsch "effective migration velocity," which is derived from plant efficiency measurements and the specific surface area of the precipitator. The effective migration velocity derived from measured efficiencies and the specific collection area for the precipitator should more realistically be considered as a measure of a "performance factor" because it applies to the material that has been collected and not to the finer material that invariably forms the major part of the emission.

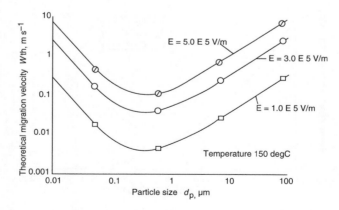

FIGURE 17.7
Relationship between the theoretical particle migration velocity and particle size.

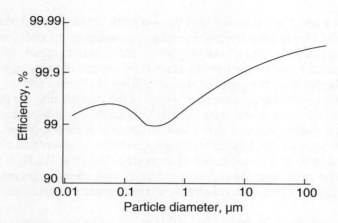

FIGURE 17.8
Precipitator fractional efficiency curve.

Regardless of which method is considered, the general trend of a much reduced migration velocity/efficiency at around 0.5 μm region is apparent. The subsequent increase in efficiency of smaller particles is the result of Brownian motion, which aids their charging and migration. To derive a fractional efficiency curve in practice, it is necessary to determine the particle sizings at both the inlet and the outlet of the precipitator and then evaluate separate grade efficiencies from the overall mass precipitator efficiency. The separate grade efficiencies can then be transposed into effective migration velocities from the Deutsch relationship.

In Figure 17.8, a typical particle size penetration (1 – efficiency) relationship is illustrated through penetration curves based on both particle mass and number relationships; both demonstrate the high penetration window in the 0.5 μm region.

Particle Deposition and Removal from the Collector Electrodes

When considering the motion of a charged particle in an electric field, until the particle has been removed from the collector plate and transferred to the receiving hopper, it should not be considered as being collected. This is one of the many reasons why the theoretical values of migration velocity are usually far higher than those based on actual measured efficiency/plate area calculations. The theoretical approach also assumes that the particle on arriving at the collector losses its charge and plays no further part in the process. In practice, charged particulates arriving at the collector can cause a serious impact on the electrical operating conditions and hence the performance.

As the particle arrives at the collector, the first reaction is that it is held by a combination of van der Waals and electrical forces. Depending on the thickness of the deposited layer and the electric resistance of the layer, the charge on the particle leaks away to earth, leaving the particle to be retained only by van der Waals forces. Unless the deposit is periodically removed, the precipitation process slowly degrades, as the deposit has a deleterious impact on the electrical operating conditions. In the case of a dry precipitator, this is achieved by some form of mechanical rapping, and in the case of a wet precipitator, depending on the design, this is achieved by periodic water washing (flushing), a continuous spray irrigation system, or having a film of water continuously flowing over the collectors (film flow).

For a dry application, the frequency and the intensity of the rapping must be such as to shear the deposit as a layer from the plates, rather than attempting to disintegrate the layer completely, as this would lead to severe rapping re-entrainment. Ideally for the particles to reach the hopper and to overcome the horizontal component of velocity, the effective size of agglomerate should be in the order of at least 500 μm in diameter.[13]

As the rate of deposition varies along the length of the precipitator, the frequency of rapping reduces from the inlet to the outlet fields. This enables the deposited layer between blows to achieve a thickness, such that it can be sheared from the plates as an optimum sized agglomerate to reach the hoppers with the minimum of re-entrainment. Because the mean particle size of the deposited material decreases along the length of the precipitator, ideally, and because of the increased packing density and cohesion between particles, the intensity of the rapping blow should increase as the particle size becomes smaller toward the outlet.

Although the majority of the dust is deposited on the collectors, some is deposited on the discharge electrodes, and, to optimize corona production, the discharge electrodes should be kept deposit free. On a dry plant, this is achieved by rapping systems similar to those employed for the collectors, while on wet systems, water washing of the internals is usually effective in maintaining corona emission.

The relationship, indicated in Eq. (14), is important in practice, because to optimize performance it is important that the voltage applied across the electrode system is maintained at the maximum possible level. The actual operating voltage across the electrode system is largely dependent on the properties of the gas, the material suspended in the gas, the layer of deposited material, and the design of the electrode system. Though it would be ideal to consistently operate just below the electrode breakdown potential, as the complete system is dynamic and the conditions are constantly changing, an automatic voltage control system is used to maintain the voltage and, hence efficiency, as high as possible.

Factors Affecting the Design and Performance of Precipitators

To obtain a better understanding of the factors that affect precipitator design and operation, one must examine the characteristics of both the gas and particulates as they are presented to the precipitator.

Gas Composition

The main requirement of the gas carrying the particulates is that it must be capable of maintaining as high an electric field as possible and permit the flow of corona current. The composition of some gases can, however, affect the electric operating conditions of the precipitator; generally the corona characteristics are modified by the presence of electropositive or electronegative gases, i.e., gases that readily absorb or reject negative ions.

Although one normally thinks of electrostatic precipitators only being applied for the removal of particulates from gas streams, the process has been satisfactorily applied for the removal of particulates from oil streams; again the main criterion is that the oil must be sufficiently insulating to maintain an electric field. The arrangement of the plant is somewhat different to a conventional gas application. Because the electrodes are normally perforated and the oil passes freely through them, the particles are contact charged before they migrate to the area of maximum field intensity between the electrode edge and casing wall. Here they are allowed to agglomerate to a sufficient size to fall under gravity into the base of the processing vessel from where they can be removed. Some installations use high voltage AC for initial treatment, followed by high voltage DC for final cleaning of the oil stream; others use DC treatment only.

Gas Temperature

The gas temperature has impacts mainly on the materials of construction; ordinary carbon steel is adopted as the cost-effective material for most applications, which limits the operating temperature to approximately 400°C, while for higher temperatures, stainless or high nickel alloys need to be used. In the case of plants operating close to acid dew point temperature, particularly for wet type environments, when choosing the material for fabrication, one must take corrosion into account to ensure durability.[14]

The main advantage of the dry type of precipitator is that the gas in most applications can be delivered directly to the precipitator from the process without the need for additional cooling or pretreatment. This means that the collected material is usefully captured in a dry state for subsequent disposal/reuse and that the cleaned gases are emitted from the chimney buoyant and usually at a temperature high enough to result in a steam free discharge.

Gas Pressure

Because most process plants operate close to ambient pressure conditions, gas pressure is not a major effect, except that one has to ensure the casing is "gas tight" and will withstand the operating conditions, i.e., to prevent either the egress of process gas or the in leakage of ambient air.

Gas Flow Rate

In the initial sizing of a precipitator, while it is important to have accurate knowledge of the total gas flow such that the correct contact time can be assessed to meet the required efficiency, there is also an optimum operating gas velocity to be considered. This optimum velocity is determined to some extent by the particle characteristics; too high a design velocity for a dry application can result in particle scouring and potential rapping re-entrainment, while too low a velocity will detract from the overall collection efficiency, as the deposition along the collector plate, most being collected close to the inlet, adversely distorts the electric precipitation field.

Generally with wet or mist applications, as the deposited material is retained on a wetted surface, operational gas velocities can be appreciably higher than that of a dry precipitator, as the risk of particle scouring is significantly reduced.

It is important that, to optimize the collection efficiency, the gas distribution across the frontal area of the precipitator must be as uniform as possible, although some recent work has indicated that for certain conditions a "skewed" velocity profile can be advantageous.[15] It is not easy to produce a typical operational gas velocity of around 1.5 m/s, which has been decelerated from approximately 15 m/s in the inlet approach ductwork, and an acceptable standard of distribution is an RMS deviation of 15%.[16] This standard can be achieved through field corrective testing, large-scale model tests, or the more recent CFD approaches.[17]

Viscosity and Density

These parameters are determined by the compositionand temperature of the gases and affect the precipitation process as follows:

a. As the charged particles migrate through the interelectrode space, they are retarded by the effects of gas viscosity and density.
b. As the temperature rises, although the gas density decreases assisting ion movement, the viscosity rises and hinders the particle transportation mechanism.

Particle Concentration

For many applications, the main effect of inlet fly ash concentration is minimal and normally has impact only on the overall removal efficiency requirement. Generally, increases in the inlet loading over that specified tend to arise from an "upset condition" on the process plant and are normally associated with the carryover of larger sized particles, which are more readily precipitated than the finer ones. Nevertheless, it is possible that the upset condition can modify other characteristics of the gas and result in an unacceptable emission transgression.

Particle Composition and Electrical Resistivity

For most dry process applications, the major impact of particle composition and precipitator performance is related to the electrical resistivity of the deposited material. The value of the electrical resistivity of the deposited particles can result in the required exposure/contact time within the precipitation fields varying by a factor of 4 or more.

For resistivities greater than approximately 10^{12} Ω cm, a phenomenon known as back or reverse ionization arises, which severely detracts from the overall efficiency. Because of the increasing resistance of the deposited material as the precipitation process proceeds, particles subsequently arriving only slowly lose their charge. Consequently, a voltage begins to build up on the deposit and in the worst case reaches a point where positive ions begin to be emitted from the surface of the layer. These positive ions not only neutralize any negative charge on the arriving particles but also considerably modify the electric field such that the overall efficiency is compromised. The onset of reverse ionization can be readily recognized by a significant increase in current flow and an apparent fall in average operating voltage.

With fly ash having a slightly lower dust resistivity, an alternate operational condition arises, where instead of the voltage continuing to build up on the surface of the deposited layer to initiate positive ion flow, it reaches a value where electrical break down occurs within the deposited material, producing a "leader," which results in flashover across the interelectrode area. This condition produces an electrical condition typified by a slightly reduced operating voltage with a significant fall in corona current; any attempt to increase the current into the plant only results in further sparking and a decreasing performance level.

With conductive materials having electrical resistivities below 10^8 Ω cm, such as unburnt carbon or metallic particles, the particles are readily charged by the corona, and, as they reach the collector, they lose their charge so rapidly as to be repelled back into the gas stream. Although the charging and repelling process can occur several times during their transit through the precipitator, some particles can leave the plant without being captured.

An important thing as regards chemical composition and format is to know if the material is likely to produce a sticky deposit, or is in a liquid phase, when it reaches the collector electrode, such that the correct type of precipitator can be supplied. For wet or mist type applications, the materials of construction need to be corrosion resistant to protect it from either the gas or the particulate components.

Another difficulty, usually associated with particulate composition, is the cohesive strength of the deposited layer. If, on a dry plant, the material lacks cohesion and is only lightly bonded together, it is likely to be readily re-entrained during rapping and thus detract from the overall performance.

Particle Sizing

The electrostatic precipitator can effectively collect particulates having diameters from 0.01 μm to approximately 100 μm. The fractional efficiency, however, is not constant, as there is a higher penetration window in the 0.5 μm diameter range, which, coincides with the change from collision to induction charging of the particles (Figure 17.8). This in itself, provided the particle sizing is available for a specific application/duty, means that the precipitator size needs to be increased to cater for particles falling in this range. This is of prime importance in enhancing existing precipitator performance levels since the emitted material, because the penetration window contains a major proportion of the finer particles.

Particle Shape

For most applications, the particle shape can be either granular resulting from the comminution of the feed material by grinding, milling, etc., or, in the form of a fume. These usually result from the material being initially volatilized in the high temperature zone of the process to subsequently condense into a spherical fume upon cooling. Neither of these forms, provided they are known or assessed beforehand and the size aspect has been considered in the design parameters, has a serious impact on the overall performance. For those processes producing a large mass of fine fume, space charge effects need to be addressed in the form of discharge element used to overcome potential corona suppression effects.

Operational problems can arise with platelet type materials, which have a fairly large surface area but virtually no thickness. These are very light in terms of mass and tend to attach themselves on to the collectors and each other, thereby reducing electrical clearance, which leads to flashover.

The presence or carryover of siliceous cenospheres, which sometimes arise during combustion, is understood to result from burning carbon particles, which are being trapped within a larger fused siliceous particle, producing sufficient gas to inflate the molten mass into a hollow sphere. This problem is not because of their shape, which is usually spherical, but because being hollow; they have little mass and poor cohesive properties, so can be readily re-entrained by rapping after being deposited.

Particulate Surface Properties

As far as precipitator performance is concerned, the surface condition/properties of the particle are generally more important than the chemical composition of the actual matrix. In many combustion applications using sulfur bearing carbonaceous fuels, the waste gases on cooling can pass through an acid dew point temperature, and any resultant sulfuric acid mist produced subsequently uses the particles as condensation nuclei depositing a thin layer of highly conductive material on their surface. At higher temperatures of operation, e.g., above 300°C, there is little possibility of reaching any dew point temperature and it is the chemical matrix of the particle that primarily governs the electrical resistivity.

Where high resistivity particles are met in practice, the electrical operating characteristics and the difficulties of the reverse ionization phenomenon can be mitigated by the injection of chemical reagents into the flue gases ahead of the precipitator to modify the electrical resistivity.[18] (Other reagents can be used to increase the cohesive properties of easily re-entrained materials if necessary to minimize re-entrainment.)

An alternative approach to flue gas conditioning is to modify the method of electrical operation, either by intermittent energization or by pulse charging. In these, instead of the continuous production of corona, the precipitator is pulsed, such that ions are only produced during the application of the voltage, thereby enabling the charge on the particles reaching the collector to discharge prior to the next set of ions arriving.[19] The capacitive component of the precipitator maintains the field voltage high enough to precipitate particles during the nonpulsing phase.

Sizing of Electrostatic Precipitators

The development of computerized programs for resolving the Poisson and La Place equations has greatly assisted in gaining a better understanding of the physics of precipitation. To date, however, no computer program is commercially available for determining precipitator design/sizing parameters from first principles. In practice, most precipitators are sized by the suppliers from interpretation of test data from plant operating under similar duties.

Deutsch,[20] working in the mid-1920s, proposed that the performance or collection efficiency of a precipitator took the form of an exponential equation:

$$\text{Efficiency} = 1 - e^{-x} \tag{15}$$

where x is dependent on factors related to the precipitator design and physical properties of the gas and dust.

In deriving the above relationship Deutsch assumed, with infinite turbulence producing a homogeneous distribution of the particles within the gas stream, that the particles were fully charged and the gas velocity was uniform, as was the corona current distribution on the collector plates; none of which are strictly true in practice. During his investigations, Deutsch, by changing the gas velocity through the precipitator and keeping all other factors constant, derived the following efficiency relationship:

$$\text{Efficiency} = 1 - e^{-2l/v} \tag{16}$$

where l is the field length and v the gas velocity. It should be recognized that l/v is the reciprocal of the precipitation contact time.

The above relationship was later transposed by the precipitation industry and used for many years to relate the size of the precipitator, collection efficiency, and gas flow rate for plants operating on similar duties and inlet conditions. Although the formula is often assumed to be theoretically based, it is no more than a useful method of comparing precipitator performance levels.

The relationship is probably more recognizable in the following form:

$$\text{Efficiency} = 1 - e^{-(A\omega/V)} \tag{17}$$

where ω is the effective migration velocity (m/s), A the area of collecting plate (m²), and V the gas flow rate, (m³/s).

Rearranging

$$\omega = \log_e\left(\frac{1}{1-\text{efficiency}}\right) \times V/A \tag{18}$$

It is important to note that the value ω, the effective migration velocity, derived from Deutsch, is not equal to the theoretical value ω_{th}, derived from Eq. (14).

Practical Approach to Industrial Precipitator Sizing

Although the Deutsch relationship was used for many years as a design tool for comparing and sizing precipitators to cater for different gas flow rates and efficiencies, it was not until the 1960s, when difficulties arose in meeting more stringent emission requirements, that modifications to the traditional Deutsch relationship were considered. The lower emission requirements demanding more of the finer, more difficult, particulates were collected.

One of the better known modifications was that derived by Matts and Onhfeldt,[21] who produced the following relationship, termed the modified Deutsch formula, for deriving an improved working figure:

$$\omega_k = \log_e\left(\frac{1}{1-\text{efficiency}}\right)^2 \times V/A \tag{19}$$

where ω_k is termed the modified effective migration velocity.

An alternative approach is that proposed by Petersen,[22] who developed the following relationship:

$$\log_e\left(\frac{1}{1-\text{efficiency}}\right) = \left(1 + b\omega A/V\right)^{1/b} \tag{20}$$

where b is an empirically derived constant (=0.22), which holds true for both the power and cement industry applications.

The practical implication of these modified approaches is that, for a change in efficiency, based on an installed plant whose operational data is known, the required plate area increase is significantly higher than that derived from the straight Deutsch formula, because the need to remove the finer particulates in the penetration window.

The recently updated and released Precipitator Computer Programme ESPVI Series 4W[23] for power plant applications can accurately track the performance of existing precipitators, because its algorithms use the established mathematical physics reviewed earlier. In its operation, the program divides the particle sizing into 27 discrete fractions and derives the charging and migration velocity for each, before summating them to derive the overall efficiency. By using theplant's design and operating data as input, the program can be used to quickly investigate any upgrade scenario and hence derive a low-cost approach to enhance the performance of any existing installation.[24]

Versatility of the Electrostatic Precipitator

The use and versatility of the electrostatic precipitator can be appreciated from the following generalized applications.

1. Steam raising: utility and industrial boilers; firing (e.g., anthracite, bituminous, sub-bituminous, lignite and brown coals, heavy and light fuel oils, orimulsion, petroleum coke, etc.); biomass (wood, straw, chicken litter, grain husks, etc.).

2. Iron and steel manufacturing: blast furnace gas cleaning for combustion (basic iron and ferromanganese units); steel converters, ladles, and electric arc furnaces; scarfing and deseaming machines; sinter plants and iron ore pelletizers; foundry and cupola applications.

3. Metallurgical process plant: dryers, smelters, roasters, and refining on nonferrous plants (e.g., copper, lead, nickel, and zinc); gold and silver bullion refining operations.

4. Coal and gas operations: coal drying, carbonization and treatment, milling, and grinding; detarring, coal gasification, and distillation processes.

5. Cement and lime manufacture: vertical and rotary kilns—wet, semidry, and dry processing; clinker coolers, limestone crushing, etc.; raw meal mills, grinders, and general feed preparation units.

6. Waste incineration: municipal, chemical, clinical, and hazardous waste disposal units; sewage sludge fired installations.

7. Pulp and paper manufacture: black liquor and Kraft recovery furnaces.

8. Chemical processing and production: dust and fume from roasters, crushers, and dryers; fumes arising during the production of fine chemicals and dyes; mist collection from sulfuric and phosphoric acid plants; gas cleaning ahead of acid production units.

9. Miscellaneous applications: carbon black collection; catalyst recovery on refinery "cat crackers;" penultimate gas cleaning stage on uranium reprocessing plant; capture and recovery of oil mists; "clean room" applications.

This list does not cover all possible applications but relates to many common industrial processes. As each application produces different waste gas flows, temperatures, particulate inlet loading, etc., the precipitator size and its design tend to be site specific for a given duty.

As indicated, the performance of the electrostatic precipitator, whether it is in material recovery or for prevention of pollution, can deliver emissions down to the mg/Nm³ region and, for air cleaning, emissions in the low µg/Nm³ region.

Conclusions

Electrostatic precipitation is perhaps the most versatile and cost effective of all particulate collecting devices and can be applied to any process where there is a need to remove solid particulate and mist of fume sized particles from the gas stream, whether it be for recovery or pollution control duties. It can be designed to deliver any efficiency for any gas flow rate and temperature and has a low pressure drop and a life span of more than 20 years.

The process of collection can be considered in the following stages:

- The production of a corona field to create ions.
- The charging of the particles by the ions.
- The migration of the charged particles through the field.
- The arrival of the charged particle at the receiving electrode.
- The removal of the deposited particles from the receiving electrode.

Each of these has been addressed from both theoretical and practical aspects, as they relate to the design and performance of any installation, so as to give even a nonspecialist an idea of the role played by electrostatic precipitation, particularly for controlling emissions from coal fired electricity generation plant where collection efficiencies in the order of 99.8%+ are now required to satisfy environmental legislation.

References

1. Grieco, G.J. Electrostatic precipitator electrode geometry and collector plate spacing—selection considerations. In Proceedings PowerGen 94 Conference, Orlando, FL, U.S.A., December 1994; Paper TP 94-58.
2. Riehle, C. Basic and theoretical operation of ESPs (Chapter 3). *In Applied Electrostatic Precipitation;* Parker, K.R., Ed.; Blakie Academic and Professional: London, **1997**;.ISBN 0-7514-0266-4.
3. Parker, K.R. High frequency energisation systems (Chapter 8). In *Electrical Operation of Electrostatic Precipitators;* Parker, K.R., Ed.; IEE (UK) Power and Energy Series 41U.K., **2003**;. ISBN 0-85296-137-5.
4. Gaugain, J.M. On the disruptive discharge. Annales de Chimie et de Physique **1862**, *64,* 175
5. Röentgen, V. Göttinger Nach. **1878,** 390.
6. Townsend, J.S. *Electricity in Gases;* Oxford University Press: Oxford, U.K., 1915.
7. Trichel, G.W. The mechanism of the negative point-to-plane corona near onset. Phys. Rev. **1938,** *54,* 1078
8. Murphy, A.T.; Adler, F.T.; Penny, G.W. A theoretical analysis of the effects of an electric field on the charging of fine particles. Trans. AIEE **1959,** *78,* 318–326.
9. Liu, B.Y.H.; Yeh, H.C. On the theory of charging aerosol particles in an electric field. J. Appl. Phys. **1968,** *39,* 1396–1402.
10. Liu, B.Y.H.; Kapadia, A. Combined field and diffusion charging of aerosols in the continuum regime. J. Aerosol Sci. **1978,** *9,* 227–242.
11. Cochet, R. Lois Charge des Fines Particules (submicroniques) Etudes Theoriques—Controles Recents Spectre de Particules. Coll. Int. la Physique des Forces Electrostatiques et Leurs Application (Centre National de la Recherche Scientifique, Paris) **1961,** *102,* 331–338.

12. Hewitt, G.W. The charging of small particles for electrostatic precipitation. Trans. AIEE **1957**, *76*, 300.
13. Baylis, A.P.; Russell-Jones, A. Collecting electrode rapping designed for high efficiency electric utility boiler electrostatic precipitators. In Proceedings of the Fourth EPA/EPRI Symposium on the Transfer and Utilization of Particulate Control Technology Houston, TX, U.S.A., October 1982; EPRI: Palo Alto, CA.
14. Parker, K.R. WESPS and fine particle collection. Proceedings of the Eighth ICESP Conference, Birmingham, AL, U.S.A., May 2001; Paper B5-1.
15. Hein, A.G.; Gibson, D. Skewed gas flow technology improves precipitator performance— Eskom operating experience in South Africa. In;. Proceedings of the Sixth ICESP Conference, Budapest, Hungary, 1996; 238–243.
16. ICAC Publication No. EP 7. *Gas Flow Model Studies*; Institute of Clean Air Companies, 1993.
17. Schwab, M.J.; Johnson, R.W. Numerical design method for improving gas distribution within electrostatic precipitators. In;. Proceedings of the American Power Conference 56th Annual Meeting, Chicago, U.S.A., April 25–27, 1994.
18. Chandan, R.; Parker, K.R.; Sanyal, A. A review of flue gas conditioning technology in meeting particulate emission standards. In;. Proceedings of the Electric Power 1999 Conference, Baltimore, U.S.A., 1999.
19. Parker, K.R. Alternative mains frequency energisation systems (Chapter 7). In *Electrical Operation of Electrostatic Precipitators*; Parker K.R. IEE (UK) Power and Energy Series 41: U.K., **2003**;.ISBN 0-85296-137-5.
20. Deutsch, W. Bewegung und Ladung der Elektrizitatstrager im Zylinderkondensator. Ann. Der Phys. **1922**, *68*, 335–344.
21. Matts, S.; Onhfeldt, P.O. Efficient gas cleaning with SF electrostatic precipitators. Flakten **1963/4**, *1–12*, 93–110.
22. Petersen, H.H. A precipitator sizing formula. In;. Proceedings of the Fourth ICESP Conference, Beijing. China, September 1990; Int. Academic Pub.: Beijing, 1993; 330–338.
23. Parker, K.R.; Plaks, N. *Electrostatic Precipitator Training Manual.*;.US Environmental Protection Agency Report No. EPA-600/R-04/072, July 2004.
24. Parker, K.R.; Plaks, N.; Zykov, A.M.; Kolchin, K.I.; Konovalov, V.K. ESP performance enhancement analysis by advanced modelling techniques. In;. Proceedings Electric Power 2004 Conference, Baltimore, U.S.A., April 2004.

18

Dust Explosion Hazard Assessment and Control

Vahid Ebadat

CONTENTS

Introduction 302
Conditions Required for Dust Explosions to Occur..302
Laboratory Testing to Assess Explosion Characteristics of Dust Clouds 303
    Laboratory Tests to Determine the Likelihood of an Explosion....................304
        Explosion Classification Test...304
        Minimum Explosible Concentration 304
        Minimum Ignition Temperature 304
        Minimum Ignition Energy 304
        Electrostatic Volume Resistivity 305
        Electrostatic Chargeability 305
        Limiting Oxidant Concentration 305
    Laboratory Tests to Determine the Consequences of an Explosion...............306
Basis of Safety from Dust Cloud Explosions...306
    Explosion Prevention Measures..306
        An Explosible Dust Cloud is Never Allowed to Form................306
        The Atmosphere is Sufficiently Depleted of Oxidant (Normally the Oxygen in
        Air) so that it Cannot Support Combustion 307
        All Ignition Sources Capable of Igniting the Dust Cloud are Removed 308
    General Precautions 312
        Bonding and Grounding 312
        Use of Insulating Materials 312
        Charge Reduction by Humidification 312
        Charge Reductions by Ionization..312
        Explosion Protection 313
    Explosion Protection Measures 314
        Explosion Containment 314
        Explosion Suppression 314
        Explosion Relief Venting 315
    Explosion Isolation Measures..315
Conclusions...316
Bibliography 316

Introduction

Statistics clearly show that fire/explosion hazards could exist in any plant/equipment that handles or processes a combustible dust. The consequences of a dust explosion can range from disruption of production to loss of plant and injury or fatality of personnel.

To eliminate/control potential dust cloud explosion hazards the following information is required:

- Understanding of the explosion characteristics of the dust(s).
- Identification of locations where combustible dust cloud atmospheres are or could be present during normal and abnormal operating conditions.
- Identification of potential ignition sources that could be present under normal and abnormal conditions.
- Proper plant design to eliminate and/or minimize the occurrence of dust explosions and protect people and plant against their consequences.

The intent of this document is to provide the readers with basic information on how to identify, assess, and eliminate/control potential dust explosion hazards in their facilities. More detailed information on the topics that are covered in this document may be obtained from Bibliography.

Conditions Required for Dust Explosions to Occur

A number of conditions must exist simultaneously and in one location for a dust explosion to occur:

Dust must be combustible: (As far as dust clouds are concerned, the terms "combustible," "flammable," and "explosible" all have the same meaning and could be used interchangeably.)

The first stage of a dust explosion hazard assessment is to determine whether the dust will explode when dispersed as a cloud. The combustibility of a dust can be assessed by conducting an explosion classification test on a representative sample of the dust. In this laboratory test the observation of flame propagation determines whether or not a suspended dust is capable of initiating and sustaining an explosion. It should be noted that some powders that are noncombustible at ambient temperature conditions could become combustible at elevated temperatures. Therefore, if a noncombustible dust cloud is to be subjected to above ambient temperatures as a part of the process the explosion classification test should be conducted at the operating (elevated) temperature.

Dust must be airborne: If a layer of dust on a surface is ignited, the burning process will be relatively slow, releasing heat over a long period of time. However, if a sufficient concentration of finely divided dust particles are suspended in air and ignited, because of the availability of oxygen around each particle, the combustion process will be very rapid giving rise to a dust explosion.

Dust concentrations must be within the explosible range: If the concentration of a dust cloud is below the minimum explosible concentration (MEC) an explosion cannot propagate.

The ease of ignition of a dust cloud and also the resulting explosion violence increase as the dust cloud concentration is increased above the MEC until an optimum concentration is reached causing the highest explosion violence. At higher dust cloud concentrations the explosion violence decreases. The MEC of dust clouds is typically in the range of 10–500 gm^{-3}.

The maximum explosible concentration—the concentration above which an explosion cannot propagate because of lack of sufficient oxidant—is not normally well defined for dust clouds.

The MEC of dust clouds can be determined by laboratory testing.

Dust must have particle size distribution capable of propagating flame: The sensitivity of a dust cloud to ignition and the resulting explosion violence (severity) increase with a decrease in particle size. This is because the combustion process involves chemical reaction at the solid–oxidant interface. Therefore, as the dust particles get finer the total surface area that is available for oxidation will increase. It should be noted that, in practice, very often dust clouds are made up of particles with sizes ranging from fine to coarse. As the fines become airborne more readily, they play a more prominent role in the initial ignition and explosion propagation.

The atmosphere in which the dust cloud is present must be capable of supporting combustion: A dust cloud explosion will only occur if there is a sufficient amount of oxidant available. In practice, the oxygen in air is normally the most common oxidant. Other oxidants include chlorine, nitrous oxide, nitric oxide, and nitrogen tetraoxide.

Explosion prevention can be accomplished by the depletion of oxidant. The concentration of oxidant below which combustion cannot occur in a specific mixture is referred to as the limiting oxidant concentration (LOC).

Generally, combustible organic compounds are unlikely to propagate flame if the oxygen content of the atmosphere in which the dust cloud is present is below 8% v/v, using nitrogen or carbon dioxide as the inert gas. Other inert gases include argon, helium, steam, and flue gas (waste gas from on-site processes).

The LOC should be determined by laboratory testing.

An ignition source with sufficient energy to initiate flame propagation must be present: The ignition sources that have been found to be the cause of the majority of explosions in dust handling/processing plants include welding and cutting, heating and sparks generated by mechanical failure, sparks generated by mechanical impacts, flames and burning materials, self heating, electrostatic discharges, electrical sparks, etc.

The sensitivity of a dust cloud to ignition by different ignition sources could be determined through appropriate laboratory tests.

Laboratory Testing to Assess Explosion Characteristics of Dust Clouds

To assess the possibility of an explosion in a facility and to select the most appropriate basis of safety, explosion characteristics of the dust(s) that are being handled/processed in the facility should be determined.

The explosion characteristics of powders normally fall within one of two groups, "likelihood of an explosion" and "consequences of an explosion." These two groups of tests are discussed below.

Laboratory Tests to Determine the Likelihood of an Explosion

The following tests provide information on the likelihood of a dust explosion:

Explosion Classification Test

The explosion classification test determines whether a dust cloud will explode when exposed to an ignition source, the test results in a material being classified as either combustible or noncombustible.

The explosion classification test is usually conducted in a modified Hartmann tube apparatus. The apparatus consists of a 1.2 L vertical tube mounted onto a dust dispersion system. Dust samples of various quantities are dispersed in the tube and attempts are made to ignite the resulting dust cloud by a 10 J electrical arc ignition source. If the material fails to ignite in the modified Hartmann tube apparatus, the testing is continued in the 20 L sphere apparatus. Dust samples of various quantities are dispersed inside the sphere and are exposed to a 10,000 J ignition source.

Minimum Explosible Concentration

The MEC test determines the lowest concentration of dust cloud in air that can give rise to flame propagation on ignition. The test involves dispersing a sample of the dust in a 20 L sphere apparatus and attempting to ignite the resulting dust cloud with an energetic ignition source. Trials are repeated for decreasing sample concentrations until the MEC is determined. The MEC of a given dust cloud is influenced by the size of the ignition source. An increase in the size of the ignition source will result in a lower MEC value.

The MEC test is performed in accordance with International Standards Organization (ISO) method 618411 or American Society for Testing and Materials (ASTM) E1515.

Minimum Ignition Temperature

The minimum ignition temperature (MIT) test determines the lowest temperature capable of igniting a dust dispersed in the form of a cloud. The MIT is an important factor in evaluating the ignition sensitivity of dusts to such ignition sources as heated environments, hot surfaces, electrical apparatus, and friction sparks.

The MIT test is performed in accordance with ASTM E-2021 and the European Standard 61241-2-1. Dust samples of various sizes are dispersed into the furnace and the minimum furnace temperature capable of igniting the dust cloud at its optimum concentration for ignition is determined.

The MIT value is influenced by particle size and moisture content of the dust. A decrease in particle size and moisture content of the dust particles results in a lower MIT value.

Minimum Ignition Energy

The minimum ignition energy (MIE) test determines the lowest electrostatic spark energy that is capable of igniting a dust cloud at its optimum concentration for ignition. The test is used primarily to assess the potential susceptibility of dust clouds to ignition by electrostatic discharges.

Minimum ignition energy test is performed in accordance with ASTM E2019, British Standard 5958, and International Standard: IEC 1241-2-3 using the modified Hartmann tube apparatus. Dust samples of various sizes are dispersed in a 1.2 L vertical tube and attempts are made to ignite the resultant dust cloud with discrete capacitive sparks of known energy.

The MIE value is influenced by particle size and moisture content of the dust and by process conditions, such as temperature and oxidant content. A decrease in particle size and moisture content of the dust particles results in a lower MIE value. An increase in the temperature of the atmosphere in which the dust cloud is suspended will result in a decrease in MIE.

Electrostatic Volume Resistivity

Volume resistivity is a measure of the electrical resistance for a unit volume of material and is the primary criterion for classifying powders as low, moderate, or high insulating. Insulating powders have a propensity to retain electrostatic charge and can produce hazardous electrostatic discharges when exposed to grounded plant, equipment, or personnel.

Volume resistivity can be measured in accordance with British Standard 5958. The method involves placing a powder sample into a standardized electrode cell. A voltage is applied to the cell and the current through the powder is measured. Volume resistivity is calculated using the known voltage, the measured current, and the geometrical relationship between the electrodes.

Because of the effect of atmospheric and absorbed moisture on volume resistivity, it is usually suggested that this test be performed at ambient and low relative humidity conditions.

Electrostatic Chargeability

The concept of electrostatic chargeability refers to the propensity of powder particles to become charged when flowing through conveyances or when handled in containers. Electrostatic chargeability is measured by having samples flow through a pipe and measuring the resultant electrostatic charge. The test provides data that can be used to develop appropriate material handling guidelines.

Because of the effect of atmospheric and absorbed moisture on powder chargeability, this test is usually performed at ambient and low relative humidity conditions.

Limiting Oxidant Concentration

The LOC test determines the minimum concentration of oxygen (displaced by an inert gas such as nitrogen) capable of supporting combustion. An atmosphere having an oxygen concentration below the LOC is not capable of supporting combustion and thus cannot support a dust explosion. The LOC test is used to study explosion prevention or severity reduction involving the use of inert gases and to set oxygen concentration alarms or interlocks in inerted plants and vessels.

Limiting oxidant concentration testing can be performed using the 20 L sphere apparatus. Dust samples of various sizes are dispersed in the vessel and attempts are made to ignite the resulting dust cloud with an energetic ignition source. Trials are repeated for decreasing oxygen concentrations until the LOC is determined.

Limiting oxidant concentration of a given dust cloud is dependent on the type of the inert gas that is used to replace the oxidant of the atmosphere as well as some process conditions such as temperature. Therefore, LOC testing should simulate the process conditions and be performed by using an inert gas that is representative of the inert gas used in practice.

Laboratory Tests to Determine the Consequences of an Explosion

There are two laboratory tests (measurements) to determine the consequences of an explosion: maximum explosion pressure and maximum rate of pressure rise.

The maximum explosion pressure and maximum rate of pressure rise values are determined by using the 20 L sphere apparatus. The dust sample is dispersed within the sphere, ignited by chemical igniters, and the pressure of the resulting explosion is measured. The sample size is varied to determine the optimal dust cloud concentration. The maximum pressure and rate of pressure rise are measured and used to calculate the explosion severity (K_{st}) value of the dust cloud. These data can be used for the purpose of designing dust explosion protection measures.

Explosion severity testing is performed in accordance with the current ASTM Method E 1226, National Fire Protection Association (NFPA) Standard 68, German Society of Engineers (VDI) Method 3673, and ISO Method 6184/1.

Basis of Safety from Dust Cloud Explosions

Safety from potential dust cloud explosions could include taking measures to avoid an explosion (explosion prevention) or designing plant and equipment so that in the event of an explosion people and plant are protected (explosion protection). Selection of explosion prevention and/or protection measures is usually based on:

- How much information is available on the sensitivity of the powder(s) to ignition and the resulting explosion severity.
- Nature of the processes and operations.
- Level of personnel's knowledge and appreciation regarding the consequences of a potential dust explosion and adherence to preventive measures.
- Environmental effects of a potential dust explosion.
- Business interruptions resulting from a potential dust explosion.

Explosion Prevention Measures

The risk of an explosion is removed when one of the following measures is taken.

An Explosible Dust Cloud is Never Allowed to Form

There are two main methods of ventilation for eliminating or controlling the spread of explosible atmospheres (fuel):

Dilution Ventilation

Dilution ventilation provides a flow of fresh air into and out of the building. This normally results in a reduction of the background concentration of the flammable atmosphere in the working area but there is no control of the flammable atmosphere at the source of release. This method is not practical for controlling the concentration of dust cloud atmospheres but is often used to control vapor concentrations.

Local Exhaust Ventilation

Local exhaust ventilation is designed to intercept the flammable atmosphere at the source of release and directs it into a system where air is safely separated from the fuel. Correctly designed local exhaust ventilation systems could be very effective in limiting the spread of dust cloud atmospheres beyond the source of release. Local exhaust ventilation is generally less expensive to run than dilution ventilation because less air is used.

A local exhaust ventilation system generally includes hood, ductwork, filter, and fan.

To ensure that the local exhaust system itself does not become an explosion hazard; the following suggestions should be considered:

- Exhaust air velocity in the ductwork should be high enough to prevent the powder from depositing in the ductwork.
- While extracting from several hoods or sources into a common duct, each branch duct should deliver the intended air volume at the required conveying velocity. It is preferable that the air balancing is achieved without using dampers. Air balancing without dampers safeguards the system against tampering with air-flow rates, and also eliminates the possibility of dust accumulation behind the dampers.
- The degree of enclosure provided by the hood and the air capture velocity should be such that dust particles are prevented from entering the workplace.
- The air cleaning devices (air filters, dust collectors, cyclones, etc.) should be protected against the consequences of dust cloud explosions. This includes measures to prevent propagation of the explosion back through the exhaust ducting to other areas of plant and equipment.
- All the components of the local exhaust system should be made from conductive and/or static dissipative materials and electrically grounded. This includes the hoods, ductwork, filter housing, and metal support cages for filter socks (if any).

The Atmosphere is Sufficiently Depleted of Oxidant (Normally the Oxygen in Air) so that it Cannot Support Combustion

Safety may be based on reducing the oxidant concentration below a level that will no longer support combustion (LOC), by adding an inert gas. Nitrogen gas is perhaps the most commonly used inert gas. Other inert gases include carbon dioxide, argon, helium, steam, and flue gas (waste gas from on-site processes). Oxidant can also be removed by working under vacuum (safe vacuum pressure should be obtained by testing).

Limiting oxidant concentration for combustion is dependent on the type of dust and type of inert gas used. Once the LOC of the dust has been determined for the inert gas that will be used, the inert gas needs to be introduced into the vessel. Successful inert gas blanketing will only be possible if the entire volume of the vessel is inerted and the inert atmosphere is maintained throughout even when the vessel is opened during the addition of solids and/or liquids to the vessel.

The following techniques are commonly used to achieve an inert atmosphere in a vessel:

Pressure Purging

The vessel is pressurized with an inert gas and then relieved outside. This procedure is repeated until the desired oxygen concentration is reached. The number of pressure purges, n, required to achieve the desired LOC can be calculated by using the following equation:

$$n = \ln(21/LOC)/\ln P \tag{1}$$

where P is purge pressure (bar absolute) and n is the number of pressure purges.
   In the above equation it is assumed that:

- Inert gas contains no oxygen.
- Initial oxygen concentration is 21%.

Vacuum Purging

The vessel is evacuated and then is increased to atmospheric pressure using the chosen inert gas. This procedure is repeated until the desired oxygen concentration is reached. The number of vacuum purges, n, required to achieve the desired LOC can be calculated by using the following equation:

$$n = \ln(LOC/21)/\ln P \tag{2}$$

where P is vacuum pressure (bar absolute) and n is the number of vacuum purges.
   In the above equation it is assumed that:

- Inert gas contains no oxygen.
- Initial oxygen concentration is 21%.

Flow-Through Purging

The vessel is purged with a continuous flow of inert gas. Flow-through purging will only be successful if there is a means of mixing the inert gas within the entire volume of the vessel. The required purge time, t, required to achieve the desired LOC is given by the following equation:

$$t = - V/Q\log(LOC/21) \tag{3}$$

where Q is purge gas flow rate (m^3/sec) and V is vessel volume.

All Ignition Sources Capable of Igniting the Dust Cloud are Removed

The ignition sources that have been found to be the cause of the majority of explosions in dust handling/processing plants include welding and cutting, heating and spark generated by mechanical failure, mechanical impacts, flames and burning materials, self-heating, electrostatic discharges, and electrical sparks. This is not an exhaustive list.
   Elimination of ignition sources involves the following steps.

Control of Heat Sources

Examples of heat sources are the following:

- External surfaces of hot process equipments such as heaters, dryers, steam pipes, and electrical equipment.
- Mechanical failure of equipment such as bearings, blowers, mechanical conveyers, mills, mixers, and unprotected light bulbs.
- Hot work.

A hot surface may directly ignite a dust cloud or first ignite a dust layer that may have settled on it and subsequently ignite a dust cloud.

Measures that may be considered for preventing a dust cloud ignition by heat sources include:

- Maintaining an effective housekeeping program to prevent/remove dust accumulations on potential hot surfaces.
- Maintaining the temperature of the processing equipment below the self-heating temperature of the powder.
- Providing regular inspection and maintenance of the processing plant to prevent overheating due to misalignment, loose objects, belt-slip/rubbing, etc.
- Preventing the overloading of processing equipment, such as grinders and conveyors. Consider installing overload protection devices on drive motors.
- Preventing "foreign objects" from entering the processing equipment by use of suitable separation devices, such as electromagnets or pneumatic separators.
- Isolating/shielding dust layers and clouds from hot surfaces.
- Using approved electrical equipment (correct temperature rating).

Control of Friction/Impact Sparks

The ability of friction/impact sparks to ignite flammable atmospheres is dependent, among other factors, on the composition of the impacting surfaces. In particular, incendive sparks could be expected in any one of the following conditions:

- Items constructed from light alloys strike a rusty steel surface.
- A rusty steel surface that has been coated with a layer of paint containing aluminum is struck by a hard object.
- Striking surfaces containing flint, rock, or grit with a hard object.

In any work where friction/impact sparks could be expected the measures that may be considered for preventing a dust cloud ignition include the following:

- Flammable (gas, vapor, and dust clouds) atmospheres should not be present.
- Hard surfaces, such as concrete, brick, or rock, should be kept wet with water. Alternatively, soft rubber mats may be used to cover the surfaces and act as a cushion for the objects that might fall.

Welding, Cutting, and Similar Hot Work Operations

Flames and sparks that are present during welding/brazing/soldering and cutting and other similar operations could readily ignite dust layers and clouds.

To avoid fires and explosions, measures should be taken to prevent the formation of dust clouds and remove dust deposits from surfaces.

Electrical Equipment and Instruments

Electrical sparks produced during normal working of switches, contact breakers, motors, fuses, etc. can ignite dust layers and clouds.

The requirements for electrical equipment and wiring systems for use in locations where combustible dust clouds or layers may be present during normal and/or abnormal operation of the plant are provided by Article 500 of the National Electrical Code (NEC). The intent of Article 500 is to prevent electrical equipment from providing a means of ignition for an ignitable atmosphere.

Ensuring safety when electrical equipment and wiring systems are present normally involves:

1. Determining the explosion characteristics that are listed below for the dust, considered relative to those for Pittsburgh seam coal:
   a. Minimum explosible concentration, MEC.
   b. Minimum ignition temperature of the dust cloud, MIT.
   c. Minimum ignition energy of the dust cloud, MIE.
   d. Maximum explosion pressure, P_{max}.
   e. Maximum rate of pressure rise, dP/dt.
2. Calculating the "ignition sensitivity" and "explosion severity" as defined below:

$$\text{Ignition sensitivity} = (\text{MIT} \times \text{MIE} \times \text{MEC})_1 / (\text{MIT} \times \text{MIE} \times \text{MEC})_2$$

$$\text{Explosion severity} = (p_{max} \times dP/dt)_2 / (p_{max} \times dP/dt)_1$$

   Subscript 1 refers to the appropriate explosion characteristics for Pittsburgh seam coal (the standard dust used by the U.S. Bureau of Mines) and subscript 2 refers to the appropriate explosion characteristics for the specific dust present in the plant.

   Using the above equations, dusts having ignition sensitivity equal to or greater than 0.2 or explosion severity equal to or greater than 0.5 are considered combustible (for electrical area classification purposes only).
3. Conducting an electrical area classification: An electrical area classification involves identifying the extent (if any) of location(s) in a facility where combustible materials could be present during normal and/or abnormal conditions. According to Article 500 of the NEC, major categories of hazardous locations are:
   a. Class I, in which the combustible material is a gas or a vapor (NFPA 497).
   b. Class II, in which the combustible material is dust (NFPA 499).

Within Class II hazardous locations, Article 500 recognizes two degrees of hazard: Division 1 and Division 2.

Class II, Division 1 locations are considered to be any of the following:

- Combustible dust is in the air under normal operating conditions in quantities sufficient to produce explosible or ignitable mixtures.
- Mechanical failure or abnormal operation of machinery or equipment might cause such explosible or ignitable mixtures to be produced and might also provide a source of ignition through simultaneous failure of electrical equipment, operation of protection devices, or other causes.
- Group E dusts may be present in hazardous quantities. Group E dusts include combustible metal dusts, including aluminum, magnesium, and their commercial alloys, or other combustible dusts whose particle size, abrasiveness, and conductivity present similar hazards in the use of electrical equipment. The NEC does not recognize any Division 2 areas for such dusts.

Note: Dusts having a volume resistivity less than 1 Ωm are considered conductive and therefore Group E.

Class II, Division 2 are locations where combustible dust is not normally in the air in quantities sufficient to produce explosible or ignitable mixtures, and dust accumulations are normally insufficient to interfere with the normal operation of electrical equipment or other apparatus. However, combustible dust may be in suspension in the air as a result of infrequent malfunctioning of handling or processing equipment and where combustible dust accumulations on, in, or in the vicinity of the electrical equipment may be sufficient to interfere with the safe dissipation of the heat from electrical equipment or may be ignitable by abnormal operation or failure of electrical equipment.

Note: The quantity of combustible dust that may be present and the adequacy of dust removal systems are factors that merit consideration in determining the classification and may result in an unclassified area.

The following factors determine the extent of Class II locations:

- Combustible material involved [for example, the dust conductive (Group E)]
- Bulk density of the material
- Particle sizes of material
- Density of the particles
- Process or storage pressure
- Size of the leak opening
- Quantity of the release
- Dust collection system
- Housekeeping
- Presence of any flammable or combustible gas

As discussed in this section, the intent of electrical area classification is to ensure that electrical equipment will not act as an ignition source. However, once the extent of a classified area is determined it is prudent to ensure that all potential ignition sources are eliminated or controlled.

Electrostatic Discharges

In this section it is assumed that the powder does not contain any flammable solvent and it is handled and processed in an atmosphere free from flammable gases and vapors.

Electrostatic charge generation: Although the magnitude and polarity of charge is usually difficult to predict, charge generation should almost always be expected whenever powder particles come into contact with another surface. It occurs, for example, during mixing, grinding, sieving, pouring, and pneumatic transfer. The chemical composition and the condition of the contacting surfaces can often influence the charging characteristics.

Electrostatic charge accumulation: Generally, powders are divided into three groups depending on their ability to retain static charge even if the powder is in contact with an electrically ground conductive object. This ability is known as volume resistivity:

1. Powders with volume resistivities up to about 10^6 Ωm are considered conductive.
2. Powders with volume resistivities in the range 10^6–10^9 Ωm are of medium resistivity.
3. Powders with volume resistivities above 10^9 Ωm are high-resistivity powders.

Charge will accumulate on a powder if the charge generation rate exceeds the rate at which the charge dissipates.

Electrostatic discharges: The accumulation and retention of charge on powder or equipment creates a dust explosion hazard only if the charge is suddenly released in the form of a discharge with sufficient energy to ignite the dust cloud. Potentially incendive discharges resulting from charged powder and equipment include spark discharges, brush discharges, propagating brush discharges, and cone (bulking) discharges.

General Precautions

Bonding and Grounding

Spark discharges can be avoided by electrically grounding conductive items, such as metal plant, fiberboard drums, conductive liners, low-resistivity powders, and people.

Use of Insulating Materials

Where there could be high surface charging processes, nonconductive materials should not be used, unless the breakdown voltage across the material is less than 4 kV. Examples of nonconductive objects include plastic pipes, containers, bags, coatings, and liners.

Charge Reduction by Humidification

High relative humidity can reduce the resistivity of some powders and increase the rate of charge decay from bulk powder in grounded metal containers. However, in most cases this will only be effective if a relative humidity in excess of 65% is maintained.

Charge Reductions by Ionization

Localized ionization (corona discharges) from sharp, grounded, conducting probes or wires can on occasions be used to reduce the level of electrostatic charge from powder particles entering a vessel. Electrostatic ionization devices are not, however, without problems, and should only be used after consulting expert advice.

Explosion Protection

In some powder handling processes it is not possible to avoid the simultaneous presence of an explosible dust cloud and a hazardous buildup of charge. In those situations, measures should be taken to protect against or prevent explosions. These include inerting, use of explosion-resistant equipment, explosion venting, or explosion suppression.

The flow diagram in Figure 18.1 provides a summary of the hazards and laboratory tests required to quantify electrostatic properties and measures that may be taken to ensure safety.

Finally, if the basis of safety from potential dust cloud explosions is the elimination of ignition sources, the answers to the following questions should be "yes":

- Can all ignition sources be identified?
- Is the sensitivity to ignition by these sources known for all process materials?
- Can all ignition sources be eliminated under normal and abnormal conditions?

If the answer to one or more of the above questions is "no," then in addition to taking all reasonable steps to reduce the possibility of formation/spread of dust clouds and exclude potential ignition sources, other measures, such as exclusion of oxidant or explosion protection, should also be considered.

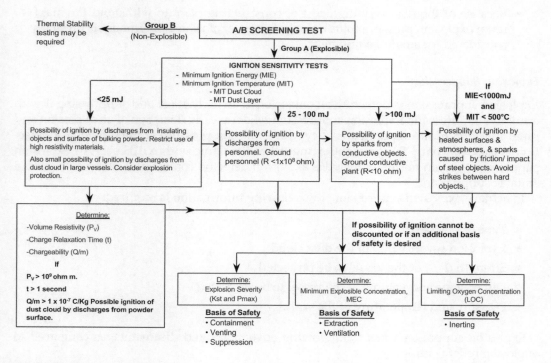

FIGURE 18.1
Electrostatic hazards assessment.

Explosion Protection Measures

If evolution or formation of an explosible atmosphere cannot be prevented and all sources of ignition cannot be reasonably eliminated or excluded, then the possibility of a dust cloud explosion persists. Under such conditions, explosion protection measures should be taken to protect people and minimize damage to the plant. It should be noted that explosion protection measures should be considered in addition to taking all reasonable steps to reduce the possibility of formation/spread of dust clouds and exclude potential ignition sources. Explosion protection measures include the following.

Explosion Containment

Construct the plant to withstand the maximum explosion pressure resulting from the deflagration (propagation of a combustion zone at a velocity that is less than the speed of sound in the unreacted medium) of the dust present in the equipment.

The following guidelines should be applied when considering explosion containment as a basis of safety:

- Equipment should withstand the maximum explosion pressure that is expected for the dust present under the process (pressure, temperature, etc.) conditions.
- All interconnected pipes, flanges, covers, etc. should withstand the maximum explosion pressure of the dust being handled.
- If an explosion-resistant vessel fails, the pressure effects will be more severe than if an extremely weak vessel fails as a result of a dust explosion.
- Because of the relatively high cost of constructing plant to withstand the maximum explosion pressure (plus a safety factor), explosion containment is normally considered for smaller equipment.

Explosion Suppression

Explosion suppression is detecting an explosion at an early stage and suppressing it with a suitable suppressant. Explosion suppression relies on early detection of an explosion and rapid injection of the suppressant. A typical explosion pressure at the moment of detection is 0.035–0.1 barg. Suppressant extinguishes the flame within 0.08 sec. An explosion suppression system normally includes explosion detector, control unit, suppressor, and a suitable suppressant.

To achieve explosion suppression, the following information is required:

- Type of powder
- Explosion severity (K_{st}) of the dust cloud
- Strength (P_{red}) of the vessel to be protected
- Dimensions and volume of the vessel to be protected
- Pressure and temperature of the vessel

Explosion suppression has the following advantages and disadvantages compared to explosion relief venting:

- Extinguishing the flame.

- Reducing the risk of ejecting toxic and/or corrosive materials to the atmosphere.
- Process equipment does not need to be located in an area suitable for explosion relief venting.
- Suppression systems are generally more expensive to install and maintain than explosion relief venting.
- Some suppression systems are not suitable for powders with very high explosion severity (K_{st}) (generally above 300 bar/m/sec, although the limits are being extended all the time).

Explosion Relief Venting

Explosion relief venting is the process of relieving the explosion products (pressure and flame) from the plant to a safe location. The principle of explosion relief venting is that a dust explosion in a vessel causes vent(s) of sufficient area to open rapidly and relieve hot gases and dust to a safe location. In other words, the vessel "fails" in a predictable way such that people and plant are protected from the effects of the dust explosion.

Explosion relief venting has the advantage of being relatively inexpensive compared to other explosion protection options and is simple to install in many cases.

Some of the limitations of explosion relief venting include:

- Not suitable for toxic materials some of which could be released to the atmosphere in the event of explosion.
- Very strongly explosible dusts, such as explosives or industrial powders, with K_{st} values in excess of 600 bar m/sec would be expected to be very difficult to vent because of the lack of available space on the vessel to be protected.
- Venting of explosions to inside a building is not usually acceptable.

Explosion Relief Venting Requirements

Three elements are required for the design and specification of explosion relief vent(s) for a vessel:

- Information on the explosion severity of the airborne dust present is required. This is generally obtained by subjecting representative dust samples to a 20 L sphere explosion test. This measures the explosion severity as expressed by maximum explosion pressure and maximum rate of pressure rise (K_{st} index).
- Shape, volume, pressure rating (P_{red}), and location of the vessel in relation to outside walls, available vent area on the vessel, and vent activation pressures (P_{stat}).
- Vent sizing method based on the dust type and plant to be protected. In the United States, NFPA 68 provides the required information for vent sizing calculations.

Explosion Isolation Measures

Regardless of what type of explosion protection measure is considered, the dust cloud explosion should be prevented from propagating from the location where it originates to other locations in the plant. This is referred to as explosion isolation. Dust explosions can propagate through pipes, chutes, conveyors, etc. The first step in isolating an explosion is to avoid unnecessary connections. If this is not possible, barriers should be created in the path of the explosion.

There are two types of barriers that could be considered for isolating a dust explosion:

Mechanical barriers: Explosion propagation is prevented by some type of physical barrier. Mechanical barriers could include rotary valves that have a sufficient number of blades to form a barrier, screw feeders that are modified to continuously contain a plug of material, and fast acting shutoff valves.

Chemical barriers: Flame front or pressure wave is detected and a suitable suppressant is injected to extinguish the flame. Although chemical barriers extinguish the flame, they cannot prevent the explosion pressure from propagating. Downstream process equipment should therefore be able to withstand the resulting pressure.

Conclusions

The majority of powders that are used in the processing industries are combustible (also referred to as flammable and explosible). An explosion will occur if the concentration of the combustible dust that is suspended in air is sufficient for flame propagation when ignited by a sufficiently energetic ignition source.

A systematic approach to identifying potential dust cloud explosion hazards and taking measures to ensure safety against their consequences generally involves the following:

1. Determining the dust cloud's ignition sensitivity and explosion severity characteristics.
2. Identifying areas of the plant where combustible dust cloud atmospheres could exist under normal and/or abnormal conditions.
3. Identifying potential ignition sources.
4. Taking measures to eliminate/control
   a. Potential ignition sources
   b. Combustible dust cloud concentrations
   c. Oxidant concentrations.
5. Taking measures to protect against the consequences of potential dust cloud explosions. Explosion protection measures include:
   a. Explosion relief venting
   b. Explosion suppression
   c. Explosion containment
   d. Explosion isolation.

Bibliography

1. Abbott, J.A. *British Materials Handling Board Survey of Dust Fires and Explosions in the United Kingdom 1979-1984*; Warren Spring Laboratory: Hertfordshire, U.K., 1986.
2. Bartknecht, W. *Dust Explosions Course, Prevention, Protection*; Springer-Verlag: Germany, 1989.

3. Barton, J. *Dust Explosion Prevention and Protection, a Practical Guide;* Institution of Chemical Engineers: Rugby, Warwickshire, U.K., 2002.

4. Bodurtha, F.T. *Industrial Explosion Prevention and Protection;* McGraw-Hill Book Company: New York.

5. Buschart, R.J. *Electrical and Instrumentation Safety for Chemical Processes;* Van Nostrand Reinhold: New York, 1991.

6. Eckhoff, R.K. *Dust Explosion in the Process Industries;* 2nd Ed.; Butterworth-Heinemann Linacre House: Jordan Hill, Oxford, 1997.

7. NFPA 77. *Recommended Practice on Static Electricity;* 2000.

8. NFPA 68. *Guide for Venting of Deflagrations;* 2002.

9. NFPA 69. *Explosion Prevention Systems;* 1992.

10. NFPA 91. *Standard for Exhaust Systems for Air Conveying of Vapors, Gases, Mists, and Noncombustible Particulate Solids;* 1999.

11. NFPA 497B. *Classification of Class II Hazardous (Classified) Locations for Electrical Installations in Chemical Process Areas;* 1991.

12. NFPA 497M. *Manual for Classification of Gases, Vapors, and Dusts for Electrical Equipment in Hazardous (Classified Locations);* 1991.

13. NFPA 654. *Standard for the Prevention of Fire and Dust Explosions from the Manufacturing, Processing and Handling of Combustible Particulate Solids;* 2000.

14. Van Laar. G.F.M. *Review of Incidents;* Europex Symposium, Dust Explosion Protection, Belgium, Sep 1989, Belgium.

Index

A

Adsorption
    adsorbents, 90–92
    commercial applications, 88–89
    FD model, 99
    heat of adsorption, 96–97
    heat of immersion, 97–98
    hybrid gas separation using, 106
    kinetics, 98–100
    Langmuir model, 94–95
    LDF model, 99–100
    Martinez–Basmadjian model, 95
    and reaction, 107
    SELDF model, 99
    separation process, 88–90, 100–107
        adsorptive drying, 100–101
        air fractionation, 101–102
        bulk liquid mixtures separation, 104
        hybrid gas separation, 106
        hydrogen production, 103–104
        nanoporous carbon membrane, 106–107
        radial and rotary bed adsorbers, 105–106
        rapid PSA cycles, 105
Adsorption equilibria, 92–96
Adsorption isotherms, 92–93
    Brunauer classification, 92
AFM. *See* Atomic force microscopy (AFM)
Agitated mills, for comminution, 136–140
    attritors, 137–139
    vibratory mills, 139–140
Air pollution, adsorption for, 89
Air-swept hammer mills, 130
Algae biodiesel technology, 3
Algae harvesting, 3
Aromatic–aliphatic copolyester, 262
Arrhenius relationship, 99
Atomic force microscopy (AFM), 20–21, 32
Attrition-disk mills
    for comminution, 131–132
Attritors, for comminution, 137–139
Autogenous and semi-autogenous mills
    for comminution, 134–135

B

Back-ionization, of powder, 224–226
Ball mills, for comminution, 132–134

Bed contraction phenomenon, 191–192
Bells, for charging powder, 229
Bingham plastic slurries, 112–113
Biochemical processes, in fluidized beds, 212
Biodegradable plastics, 262
Biomass, properties of, 206
Biomass-based synthesis gas, 3
Bioseparation separation, 88
    by adsorption, 88–89
Blake jaw crushers, 128–129
Bond's theory of comminution, 127
Bowl mills, 135
Breathable films, 4
Brunauer classification, of adsorption
        isotherms, 92
Bubbling fluidization, 183, 185–187
Bubbling fluidized bed (BFB) technology,
        207–209

C

Calcination, in fluidized beds, 206, 210
Carbon Molecular Sieves (CMS), 102
Catalytic reforming, in fluidized bed catalytic
        reactor processes, 205
Cellulose acetate, 262
Centrifugal ball mill, 133–134
Centrifugal pin mills, for comminution,
        130–131
CFD. *See* Computational fluid dynamics (CFD)
Channeling fluidization, 183
Chemical barriers, to explosion, 316
Chemical vapor deposition (CVD), 58
Chlorination of metal oxides, in fluidized bed,
        211
Circulating fluidized bed (CFB) technology,
        207–210
Clay, 4. *See also* Polymer-clay nanocomposites
        (PCNs)
    configurations, 237
    infusion and impregnation in $scCO_2$,
        242–243
    intercalation in $scCO_2$, 242
    polymer–clay compatibility, 237
    $scCO_2$ effect on, 242–243
Coal
    gasification, 121
    properties of, 206

Coal-based synthesis gas, 3
Coal-oil mixtures, 2–3
Coal slurries
  applications, 118–122
    coal gasification, 121
    pipeline transport, 120–121
    slurry fuels, 2–3, 111, 118–120
  particles size distribution, 114–115
  rheology, 112–118
  solid loading, 113–114
  surfactants and dispersants, 115–118
  wastes, 111, 122
Coal slurry fuels, 2–3
Coal water fuel plant, 119
Coal–water mixtures, 2–3
Coaxial mixers. *See also* Solid-liquid mixing
  results and discussion, 170–177
  setup description, 160–162
Cohesion
  continuum theory, 26–27
  DEM simulations, 25–26
  experiments, 20–25
    collision-based, 21
    force measurements, 20–21
    individual particles, 20–21
    many- particles systems, 21–25
    pickup velocity, 21–24
    pneumatic conveying, 24–25
  mathematical models, 25–27
  sources, 18–20
    electrostatics, 19–20
    liquid bridges, 19
    van der Waals forces, 18–19
Cohesive forces, 18
Combustion, in fluidized beds, 206–208
Commercial glass-ceramics, 61
Comminution
  Bond's theory of, 127
  equipment operation, 141–142
  grinding energy requirements, 126–127
  impact mills with internal classification,
    139–141
  intermediate and fine crushers, 130–135
    attrition-disk mills, 131–132
    autogenous and semi-autogenous mills,
      134–135
    ball mills, 132–134
    centrifugal pin mills, 130–131
    critical and operating speed, 132–133
    hammer mills, 130
    impact mills, 130–135
    rod mills, 134
    rolling compression mills, 135

    tube mills, 134
    tumbling mills, 132–135
  machines, 126
  open and closed circuit grinding, 141–142
  primary and secondary crushers, 127–129
    gyratory cone crushers, 127–128
    jaw crushers, 128–129
    roll crushers, 129
  process, 125, 142–143
  rotary cutters, 141
  ultrafine grinders, 135–140
    agitated mills, 136–140
    attritors, 137–139
    fluid-energy mills, 135–136
    vibratory mills, 139–140
Computational fluid dynamics (CFD), 158, 164,
    177, 193–194, 214, 285, 293
Concentration swing adsorption (CSA), 89–90
Conical ball mill, 132–133
Contact corrosion inhibitors (CCIs), 255. *See also*
    Corrosion inhibitors (CIs)
Continuous catalytic reforming (CCR)
    technology, 205
CO-rich synthesis gas, 3
Corona
  charging, 222–224
  discharge, 284–285, 312
Corrosion
  biodegradable polymers and, 262–263,
    270–271
  control, 254–255
  inhibitors, 255–259
  prevention experiment, 263–274
    corrosion studies, 266
    film infusion, 266
    film preparation, 265
    fluid density studies, 265–266
    materials selection, 263
    procedures, 265
    results and discussion, 267–274
    SEM analysis, 265, 273–274
    supercritical infusion, 263–268, 271–273
  process, 253–254
  rate, 269–271
  studies, 266
  supercritical fluid, 252–253, 259–268, 271–275
    scCO$_2$, 253, 260, 263–264, 266–269, 271–274
    supercritical infusion, 260–268, 271–273
  supercritical infusion, 260–268, 271–273
    fluid density effects, 267–268
    polymer infusion, 271–273
    reactor system, 263–264
  testing, 266

Corrosion-inhibited films of LLDPE, 258, 270–275
Corrosion-inhibited plastics, 256–257
Corrosion-inhibited VCI films, 258–259
Corrosion inhibitors (CIs), 255–259

D

Debye–Hückel linearization, 40–42
DEM. *See* Discrete element method (DEM) simulations
Dense-phase fluidization, 183, 189
Depletion flocculation, 55
Derjaguin approximation, 32, 36–38, 40–42, 48
   limits of probed with AFM tips, 46–47
Derjaguin–Landau–Verwey–Overbeek (DLVO) theory, 31–32, 47
Desorption process, 89–90
Deutsch relationship, 291, 297
Diesel engine, coal slurry fuels for, 120
Diffusion–sedimentation phenomenological model, 158
Discrete element method (DEM) simulations, 25–27
Discs, for charging powder, 229
Dispersion–sedimentation phenomenological model, 158
Dodge jaw crushers, 128–129
Downer, 189–191
Dust clouds explosion
   characteristics, 303–306
   conditions to occur, 302–303, 316
   hazards assessment, 302, 313, 316
   isolation measures, 315–316
   laboratory tests to determine
      consequences of explosion, 306
      explosion characteristics, 303–306, 316
      likelihood of explosion, 304–305
   precautions, 312–313
   prevention measures, 306–312
   protection measures, 314–316
   safety, 306–316

E

Ecoflex films, 262–263, 265–266, 269–275
Electronic gas purification, 88
Electrostatic double-layer (EDL) interaction, 39–46
   between two spheres
      approximate models for, 40–42
      exact numerical solutions for, 42–46
Electrostatic fluidized bed technology, for charging powder, 230

Electrostatic forces, 18–20, 27
Electrostatic hazards assessment, 313
Electrostatic precipitation
   ion production, 285–287, 299
   operating principles, 282–285
   particle
      charging, 287–288, 299
      deposition and removal from collector electrodes, 291–292, 299
      migration, 288–291, 299
   physics, 285–292
   process, 282
   single-stage arrangements, 283
   two-stage arrangements, 283–284
Electrostatic precipitator
   applications, 298
   design and performance, 292–296
   effective migration velocity, 290, 297
   impact of gas
      composition, 292–293
      flow rate, 293
      pressure, 293
      temperature, 293
   impact of particle
      composition and electrical resistivity, 294–295
      concentration, 294
      shape, 295
      sizing, 295
      surface properties, 295–296
   industrial, 297–298
   ion production, 285–287, 299
   operating principles, 282–285
   particle
      charging, 287–288, 299
      deposition and removal from collector electrodes, 291–292, 299
      migration, 288–291, 299
   physics, 285–292
   sizing, 296–298
   versatility, 298
   viscosity and density impact, 294
Electrostatic templating mechanisms, 75
Elutriation, 183, 187
Engage films, 265–266
Entrainment, 183, 187
Environmental separation, by adsorption, 88–89
Euler–Lagrange method, 164
Evaporation induced self-assembly (EISA), 77
Explosion. *See also* Dust clouds
   characteristics of dust clouds, 303–306
   classification test, 304
   conditions to occur, 302–303, 316

consequences, 306
containment, 314
electrostatic
    chargeability, 305
    hazards assessment, 313
    volume resistivity, 305
hazards assessment, 302, 313, 316
isolation measures, 315–316
    chemical barriers, 316
    mechanical barriers, 316
likelihood, 304–306
limiting oxidant concentration, 303, 305–308
minimum
    explosible concentration, 302, 304
    ignition energy, 304–305
    ignition temperature, 304
precautions, 312–313
    bonding and grounding, 312
    explosion protection, 313
    humidification, 312
    insulating materials use, 312
    ionization, 312–313
prevention measures, 306–312
    cutting control, 310
    deplete oxidant, 307–308
    dilution ventilation, 306–307
    electrical equipment and instruments
        control, 310–312
    electrostatic discharges control, 312
    elimination of ignition sources, 308–313
    flow-through purging, 308
    friction/impact sparks control, 309
    heat sources control, 309
    hot work operations control, 310
    local exhaust ventilation, 307
    not allow to form dust cloud, 306–307
    pressure purging, 308
    vacuum purging, 308
    welding control, 310
protection measures, 314–316
relief venting, 315
safety measures, 306–316
suppression, 314–315
Extended X-ray absorption fine structure
    spectroscopy, 58

F

Fast fluidization, 183, 188–189
FD. *See* Fickian diffusion (FD) model, of
    adsorption
Fickian diffusion (FD) model, of adsorption, 99
Fickian diffusion kinetics, 240

Fischer-Tropsch synthesis, in fluidized bed
    catalytic reactor processes, 203–204
Fluid-bed combustors
    slurry fuels injection into, 120
Fluid catalytic cracking (FCC), in fluidized bed
    catalytic reactor processes, 203, 205
Fluid-energy mills, for comminution, 135–136
Fluid Energy Processing Micro-Jet grinders,
    135–137
Fluidization
    bed contraction phenomenon, 191–192
    CFD approach, 193–194
    downer, 189–191
    fluidized particles classification, 182
    moving packed bed phenomenon, 192
    regimes, 182–189
        bubbling fluidization, 183, 185–187
        channeling fluidization, 183
        dense-phase fluidization, 183, 189
        dilute transport, 188–189
        entrainment and elutriation, 183, 187
        fast fluidization, 183, 188–189
        gas-liquid-solid system, 190–191, 193
        gas-solid-liquid system, 191
        lean-phase fluidization, 183, 187–189
        liquid fluidization, 184
        minimum fluidization, 183–185
        particulate fluidization, 183–185
        slugging beds, 183, 187–188
        spouted beds, 183, 188
        turbulent fluidization, 183, 187
    Richardson–Zaki equation, 184
    three-phase fluidized bed, 191–192
    two phase theory, 186
Fluidized bed
    catalytic reactor processes, 203–205
        catalytic reforming, 205
        Fischer-Tropsch synthesis, 203–204
        fluid catalytic cracking, 203
        oxychlorination of ethylene, 204
        partial oxidation reactions, 203
        polymerization, 204–205
        propylene ammoxidation, 204
    gas-liquid-solid chemical processes in,
        212–213
        biochemical processes, 212
        hydrocarbon processes, 213
    gas-solid reactions in, 205–211
        calcination, 206, 210
        combustion, 206–208
        gasification, 206, 208–209
        pyrolysis, 206, 209
        roasting, 206, 210

iron ore reduction, 211
liquid-solid reactions in, 211–212
metal oxides chlorination and fluorination, 211
nano- and ultrafine particles, 211
reactor modeling, 213–214
ultrapure silicon, 211
Fluidized bed combustion (FBC) systems, 207
Fluidized bed reactor. *See also* Fluidized bed
catalytic reactor processes, 203–205
gas-liquid-solid reactions, 212–213
gas-solid-fluidized bed reactors, 200–202
gas-solid reactions, 205–211
liquid-solid reactions, 211–213
modeling, 213–214
Fluidized particles classification, 182
Fluorination of metal oxides, in fluidized bed, 211
Fluorocarbon-in-oil emulsion system, 56
Fluorocarbon-in-toluene-in-water multiple emulsions, 56–57
Fluoropolymers, 4
Fractal geometry
aggregates, 12–13
applications, 9–14
diffusion and reaction, 13
dissolution and etching, 13
fractal surfaces, 11–12
nature inspired chemical engineering, 14

G

Gasification, in fluidized beds, 206, 208–209
Gasifier feedstock, 111–112. *See also* Coal slurries
Gas-liquid-solid chemical processes, in fluidized beds, 212–213
Gas-liquid-solid system, 190–191, 193
Gas separation, by adsorption, 88–89
Gas-solid fluidized bed reactors. *See also* Fluidized bed reactor
advantages and disadvantages, 200
configurations, 201
entrainment, 202
flow regime, 202
gas distributor, 202
operating pressure and temperature, 201–202
particle feeding, 202
particle properties, 201
Gas-solid-liquid system, 191
Gas-solid reactions, in fluidized beds, 205–211
calcination, 206, 210
combustion, 206–208

gasification, 206, 208–209
pyrolysis, 206, 209
roasting, 206, 210
Gas-solid system, 185–186
General predictive synthesis approach, 77
Gold nanoparticle catalysts
syntheses and characterization, 58–61
basic studies, 59–60
chemical methods, 58–59
hydrogenation, 60
silane chemistry, 60–61
sputtering methods, 59
Gouy–Chapman theory, 48
Grinding energy
requirements for comminution, 126–127
Gyratory cone crushers, for comminution, 127–128

H

Hamaker–Lifshitz function, 38–39, 48
Hamaker microscopic approach
to van der Waals interactions between colloidal particles, 32–33, 48
Hamaker theory, 18
Hammer mills, for comminution, 130
Henry's law, 92–93, 97, 99
Herschel and Buckley's yield-power law, 112
High-density polyethylene (HDPE), 244, 246
Human history
particle processing in, 2
Hybrid gas separation, using adsorption, 106
Hydraulic fracturing, 3
Hydrocarbon processes, in fluidized beds, 213

I

Impact mills
for comminution, 130–135
centrifugal pin mills, 130–131
hammer mills, 130
with internal classification, for comminution, 139–141
Inorganic framework
strategies for stabilization, 79–80
aging under mild conditions, 79
doping with foreign atoms to reduce size of crystallites, 79
mild template removal, 79
thermal treatments, 79
treatment with vapors of inorganic precursor, 79–80

Integrated gasification combined cycle (IGCC),
        208
    plants, coal slurry for, 121
Intermediate and fine crushers
    for comminution, 130–135
            attrition-disk mills, 131–132
            autogenous and semi-autogenous mills,
                134–135
            ball mills, 132–134
            centrifugal pin mills, 130–131
            critical and operating speed, 132–133
            hammer mills, 130
            impact mills, 130–135
            rod mills, 134
            rolling compression mills, 135
            tube mills, 134
            tumbling mills, 132–135
Internal charging, of powder, 226
Inverse fluidization, 212
Iron ore reduction, in fluidized bed, 211

J

Jaw crushers, for comminution, 128–129

K

Kick's law, 127
Knudsen diffusion, 98
Krieger–Dougherty phenomenological model,
        159, 163

L

Langbein theory, 48
Langmuir model, of adsorption, 94–95
Laplace–Young equation, 19
Larostat®, 24
LDF. *See* Linear driving force (LDF) model, of
        adsorption
Lean-phase fluidization, 183, 187–189
Lifshitz macroscopic approach
    to van der Waals interactions between
            colloidal particles, 33–36, 48
Limiting oxidant concentration (LOC) and
        explosion, 303, 305–308
Linear driving force (LDF) model, of
        adsorption, 99–100
Linear low-density polyethylene (LLDPE),
        204–205, 244, 246
    corrosion-inhibited films, 258, 270–275
Liquid bridges forces, 18–19, 27
Liquid fluidization, 184

Liquid-phase methanol synthesis, 3
Liquid separation, by adsorption, 88–89
Liquid-solid reactions, in fluidized bed,
        211–212
LLDPE. *See* Linear low-density polyethylene
        (LLDPE)

M

Martinez–Basmadjian model, of adsorption, 95
Mechanical barriers, to explosion, 316
Mesoporous carbon synthesis, 80–81
Mesoporous materials syntheses
    templates
            self-assembled arrays of molecules, 71
            self-assembled surfactant, 71–72
Mesostructured materials processing
    environmental variables effect, 78–79
            organic solvent environment, 78–79
            relative humidity (RH) of atmosphere, 78
    morphology control, 80
    synthesis parameters effect, 76–78
            additives effect, 77–78
            complexing agents, 76
            dilute solutions, 77
            inorganic precursor reactivity, 76
            low pH conditions, 76
            mixed precursors, 76
            nanoparticles, 77
            nonaqueous solvents and limited water,
                76–77
            surfactant to inorganic precursor ratio, 77
Metal oxides chlorination and fluorination, in
        fluidized bed, 211
Methanol synthesis, 3
Methyl methacrylate (MMA), 244
Microporous materials syntheses
    single molecules as templates for, 70
Minimum explosible concentration (MEC) and
        explosion, 302, 304
Minimum fluidization, 183–185
Minimum ignition energy (MIE) and explosion,
        304–305
Minimum ignition temperature (MIT) and
        explosion, 304
Molecular dynamics (MD), 25
Molecular imprinting, 70
Molecular packing factor, 71–72
Molecular sieve carbon (MSC) membrane, 106
Molecular sieving, 92
Moving packed bed phenomenon, 192
Municipal Solid Waste (MSW), properties of,
        206, 208

N

Nanobuilding blocks
  use of functional dendrimers for self-assembly of, 72
Nanomaterials. *See also* Nanoparticles
  industrial interest in, 53
Nanoparticles
  dispersion stabilization with, 57
  in fluidized bed, 211
  gold nanoparticle catalysts, 57–61
  nanophase glass-ceramics, 61–63
  in surface and interfacial phenomena, 53–57
Nanophase glass-ceramics, 61–63
Nanostructured materials syntheses
  strategies, 68–70, 72–75
    from bottom-up approach, 68–70
    cooperative self-assembly, 73–75
    ligand-assisted templating, 75
    for stabilization of inorganic framework, 79–80
    templates types, 70
    templating route, 69–70
  templates
    non-rigid templates, 80
    removal, 79–80
    rigid templates, 80–81
    route, 69–70
    self-assembled arrays of molecules, 71
    self-assembled surfactant, 71–72
    single molecules as, 70
    types, 70
Navier–Stokes equations, 168
Newtonian slurries, 112–113
Niobium oxide synthesis, 75
Nozzles for charging powder, 228–229

O

Open and closed circuit grinding
  for comminution, 141–142
Oxychlorination of ethylene, in fluidized bed catalytic reactor processes, 204
Oxyfluoride glass-ceramics, 62

P

Partial oxidation reactions, in fluidized bed catalytic reactor processes, 203
Particle cohesion. *See* Cohesion
Particle–particle interaction
  Derjaguin approximation, 46–47
  electrostatic double-layer interaction, 39–46
  van der Waals interactions, 32–39

Particles
  size distribution in coal slurries, 114–115
Particle science and technology
  applications, 1–5
  definition and scope, 1
  future trends, 5
  historical perspectives, 2
  industrial practices, 2–5
  significance, 5
Particle size, and powder recycling, 230–231
Particulate fluidization, 183–185
Peng–Robinson equation, 267
Pharmaceutical separation, 88
  by adsorption, 88
Pickup velocity, 21–24
Picture frame effect, 223
Pneumatic conveying, 24–25
  dense-phase transport, 149, 152
  experimental observations, 150–152
    particle properties, 150–151
    system behavior, 151–152
  modeling, 152–153
  system configuration, 146–149
    positive *vs.* negative pressure systems, 148–149
    vertical *vs.* horizontal transport, 146–148
  transport regimes, 149
Pneumatic transport. *See* Pneumatic conveying
Poisson–Boltzmann equation, 44–45, 48
Polybed process, 103
Polycaprolactone, 262
Polydimethylsiloxane (PDMS), 239, 244, 246
Poly(ester amide), 262
Polyethylene (PE), 256, 258, 262–263
Poly(ethylene terephthalate) modified with glycol (PETG), 239
Poly(ethylene terephthalate) (PET), 239, 262
Poly(lactic acid) (PLA), 262
Polymer-clay nanocomposites (PCNs)
  clay, 236–237
    infusion and impregnation in $scCO_2$, 242–243
    intercalation in $scCO_2$, 242
  compatibility, 237–238
  creation with $scCO_2$, 243–247
    melt-mixing, 244–246
    in situ polymerization, 243–244
    solvent casting, 246
    vessel (batch) infusion and dispersion, 246–247
  processing with $scCO_2$, 238–243
  $scCO_2$ effect on, 242–243

Polymerization, in fluidized bed catalytic reactor processes, 204–205
Poly(methyl methacrylate) (PMMA), 239, 244, 246–247
Polypropylene (PP), 258, 262
Polytetrafluoroethylene (PTFE), 244, 261
Polyvinylidenefluoride (PVDF), 239, 244
Powder
  charging, 222–230
    back-ionization, 224–226
    bells and discs, 229
    corona charging, 222–224
    electrostatic fluidized bed technology, 230
    Faraday Cage effect, 225–226
    internal charging, 226
    nozzles, 228–229
    tribocharging, 226–228
  coatings, 4
    advantages, 220
    application process, 221–222
    disadvantages, 220–221
    markets, 221
  equipment technology, 232
  recovery, 231–232
  recycling
    particle size, 230–231
    powder recovery, 231–232
  technology, 232–233
Powderizer air-swept impact mill, 139–140
Powdery graft copolymers, 4
Pressure swing adsorption (PSA), 89–90, 100–101, 103, 105
  gas drying process, 101
Primary and secondary crushers
  for comminution, 127–129
    gyratory cone crushers, 127–128
    jaw crushers, 128–129
    roll crushers, 129
Print molecule, 70
Propylene ammoxidation, in fluidized bed catalytic reactor processes, 204
PSA. *See* Pressure swing adsorption (PSA)
Pseudoplastic slurries, 112–113
Pulverized coal combustors
  slurry fuels supplement for, 119
Pulverized coal particles, 2
Pyrolysis, in fluidized beds, 206, 209

Q

β-quartz glass-ceramics, 62

R

Radial and rotary bed adsorbers, 105–106
Rapid PSA cycles, 105
Retsch centrifugal ball mill, 133
Richardson–Zaki equation, 151, 184
Rittinger's law, 127
Roasting, in fluidized beds, 206, 210
Rod mills, for comminution, 134
Roll crushers, for comminution, 129
Roller mills, 135
Rolling compression mills
  for comminution, 135
Rolling-ring pulverizers, 135
Rotary cutters
  for comminution, 141
Roto-Jet fluid-bed jet mill, 135, 137
Rusting, 253. *See also* Corrosion

S

Scanning tunneling microscopy (STM), 59–60
SELDF. *See* Surface excess linear driving force (SELDF) model, of adsorption
Selective surface flow (SSF) membrane, 106–107
Self-assembled arrays of molecules, 71
Self-assembled functional dendrimers, 72
Self-assembled surfactant templates, 71–72
Separation
  adsorption for, 88–90
  adsorptive processes, 100–107
  adsorptive properties for, 88–90
Simulated moving bed (SMB) processes, 104
Single-particle cohesion, 20–21
Single-toggle jaw crushers, 128–129
Size reduction. *See* Comminution
Skarstrom cycle, 101
Slugging beds, 183, 187–188
Slurry, 111. *See also* Coal slurries
Slurry fuels, 2–3, 111, 118–120
  as diesel engine fuel, 120
  and heavy fuel oil, 119
  injection into fluid-bed combustors, 120
  supplement for pulverized coal combustors, 119
Sodium montmorillonite (MMT) clay, 236–237, 243, 246
Solid-liquid mixing
  CFD applications, 158, 164, 177
  coaxial mixers
    results and discussion, 170–177
    setup description, 160–162
  computational model, 163–167

literature on, 157–160
  marine propeller mixer, 162–167
  numerical model validation, 168–169
Solid loading, in coal slurries, 113–114
Sorption enhanced reaction process (SERP), 107
β-spodumene glass-ceramics, 62
Spouted beds, 183, 188
Square-well model, 25–26
Starch-based blends, 262
Starch–silicon–poly(styrene-*co*-allyl alcohol)
    composites, 261
Stokes number, 17, 27
Styrene–ethylene/butylene–styrene block
    copolymer (SEBS), 244, 246
Supercritical carbon dioxide (scCO₂), 238–243
  clay
    effect of, 241–242
    infusion and impregnation in, 242–243
    intercalation in, 242
  diffusion in, 240–241
  effect on
    clay, 241–242
    polymer–clay nanocomposites, 242–243
    polymers, 238–241
  fluid density studies, 267–269
  infusion process and film, 253, 263–264, 266,
    271–274
  melt-mixing with, 244–246
  plasticization by, 240
  polymer–clay nanocomposites
    creation with, 243–247
    processing with, 238–243
  polymer solubility in, 238–240
  in situ polymerization with, 244
  as solvent, 240
  as supercritical fluid (SCF), 253, 260, 263–264,
    266–269, 271–274
  vessel (batch) infusion and dispersion in,
    246–247
Supercritical fluid (SCF), 5
  applications, 259, 267
  fluid density
    effects, 267–269
    studies, 265–266, 268
  infusion, 252–253, 259, 260–268, 271–275
    polymer, 271–273
  reactor system, 263–264
  scCO₂, 253, 260, 263–264, 266–269, 271–274.
    See also Supercritical carbon dioxide
    (scCO₂)
  supercritical infusion, 260–268, 271–273
Surface excess linear driving force (SELDF)
    model, of adsorption, 99

Surface modified nanoparticles (SMNs), 54–55
Surfactants and dispersants, for coal slurries,
    115–118
Sweco vibro-energy mill, 139–140
Switched mode power supplies (SMPS), 284

T

Temperature swing adsorption (TSA), 89,
    100–101, 105
Thermosetting powders, 220
Thixotropic slurry, 112–113
Three-phase fluidized bed, 191–192
Toluene-in-water-in-toluene multiple emulsion,
    56–57
Toothed single roll crushers, 129
Transmission electron microscopy (TEM), 59
Tribocharging, of powder, 226–228
True liquid crystal templating (TLCT), 72–73
Tube mills, for comminution, 134
Tumbling mills
  for comminution, 132–135
    autogenous and semi-autogenous mills,
      134–135
    ball mills, 132–134
    rod mills, 134
    tube mills, 134
Turbulent fluidization, 183, 187

U

Ultrafine grinders
  for comminution, 135–140
    agitated mills, 136–140
    attritors, 137–139
    fluid-energy mills, 135–136
    vibratory mills, 139–140
Ultrafine particles, in fluidized beds, 211
Ultrapure silicon, in fluidized beds, 211

V

Vacuum swing adsorption (VSA), 102
Van der Waals forces, 18–19, 27, 284, 291
Van der Waals interactions
  between colloidal particles, 32–36
    Hamaker microscopic approach, 32–33, 48
    Lifshitz macroscopic approach, 33–36, 48
  between spheres, 36–39
    approximate and simplified equations
      for, 36–39
Vibra-Drum vibratory grinding mill, 139
Vibratory mills, for comminution, 139–140

Virtual finite element method (VFEM), 165
Volatile corrosion inhibitor (VCI), 252–253,
    255–257
  infused
    films, 258–259, 263, 271
    packaging, 257–258, 263, 269–271,
      274–275
  inhibited films, 4, 259
  manufacture, 258
  papers, 258
  in polymer matrix, 258–259, 261–264

W

Waste water treatment, by adsorption, 89
Water-in-oil emulsion system, 56
Water-in-toluene-in-water multiple emulsion, 56
Wrap effect, 223

X

X-ray absorption near edge (XANES)
    spectroscopy, 58